QUIRKS OF HUMAN ANATOMY

An Evo-Devo Look at the Human Body

With the emergence of the new field of evolutionary–developmental biology, we are witnessing a renaissance of Darwin's insights 150 years after his *Origin of Species*. Thus far, the exciting findings from "evo-devo" have only been trickling into college courses and into the domain of nonspecialists. With its focus on the human organism, *Quirks of Human Anatomy* opens the floodgates by stating the arguments of evo-devo in plain English and by offering a cornucopia of interesting case studies and examples. Its didactic value is enhanced by 24 schematic diagrams that integrate a host of disparate observations, by its Socratic question-and-answer format, and by its unprecedented compilation of the literature. By framing the "hows" of development in terms of the "whys" of evolution, it lets readers probe the deepest questions of biology. Readers will find the book not only educational but also enjoyable, as it revels in the fun of scientific exploration.

Lewis I. Held, Jr., earned his B.S. in life sciences from the Massachusetts Institute of Technology in 1973. On completion of his Ph.D. in molecular biology from the University of California, Berkeley, in 1977, he became a Postdoctoral Fellow at the Developmental Biology Center at the University of California, Irvine, from 1977 to 1980. Following a 6-year period as an Assistant Professional Research Biologist at the University of California, Irvine, he joined Texas Tech University as an Associate Professor of Biology in 1987. He was awarded the President's Excellence in Teaching Medal from Texas Tech University in 1995 and is the author of *Models for Embryonic Periodicity* and *Imaginal Discs: The Genetic and Cellular Logic of Pattern Formation.*

QUIRKS OF HUMAN ANATOMY

An Evo-Devo Look at the Human Body

LEWIS I. HELD, Jr.
Texas Tech University

CAMBRIDGE
UNIVERSITY PRESS

32 Avenue of the Americas, New York NY 10013-2473, USA

Cambridge University Press is part of the University of Cambridge.

It furthers the University's mission by disseminating knowledge in the pursuit of education, learning and research at the highest international levels of excellence.

www.cambridge.org
Information on this title: www.cambridge.org/9780521518482

First published 2009

A catalogue record for this publication is available from the British Library

Library of Congress Cataloguing in Publication data
Held, Lewis I., 1951–
Quirks of human anatomy : an evo-devo look at the human body / Lewis I. Held, Jr.
p. ; cm.
Includes bibliographical references and index.
ISBN 978-0-521-51848-2 (hardback) – ISBN 978-0-521-73233-8 (pbk.)
1. Embryology, Human. 2. Human evolution. I. Title.
[DNLM: 1. Evolution. 2. Hominidae – anatomy & histology. 3. Adaptation, Biological – genetics. GN 281 H474q 2009]
QM611.H45 2009
612.6′4 – dc22 200905722

ISBN 978-0-521-51848-2 Hardback
ISBN 978-0-521-73233-8 Paperback

Contents

Preface

Once upon a time, we thought we were created in the image of a perfect deity. Then came Darwin. In the very first chapter of his *Descent of Man* (1871), he took pains to point out the many flaws in our anatomy, including our useless—but dangerous—appendix, our pathetic covering of body hair, and the silly little muscles that let us wiggle our ears. Each of these odd traits is an anachronism inherited from our nonhuman ancestors.

Of course, the very idea that we had nonhuman ancestors was implicit in Darwin's earlier *Origin of Species* (1859). That book shook the world. The present book has a much humbler aim: to honor the 150th anniversary of *Origin* and, coincidentally, the 200th birthday of the man himself (b. 12 February 1809).

In Chapter 13 of *Origin*, Darwin asserted that the evidence from embryology alone was strong enough to convince him of the principle of common descent. Having taught embryology myself for more than 20 years, I know what he meant. Human embryos make many structures we don't need, and we destroy others after we've gone to the trouble of making them. No engineer in his right mind would ever allow such idiocy. Only in the light of evolution do these processes make sense [1265], and only in the language of genetics can we comprehend their tortuous logic [2496]. Unfortunately, genetics blossomed only after Darwin died (19 April 1882). In Chapter 5 of *Origin*, he lamented his failure to trace the causes of heritability.

> Our ignorance of the laws of variation is profound. Not in one case out of a hundred can we pretend to assign any reason why this or that part differs, more or less, from the same part in the parents. But whenever we have the means of instituting a comparison, the same laws appear to have acted in producing the lesser differences between varieties of the same species, and the greater differences between species of the same genus. [559] (p. 167)

Quirks of Human Anatomy examines one of Darwin's favorite topics (oddities) [965] through the lens of his most incisive approach (embryology) to investigate one of the last remaining riddles from his research—namely, how genetic circuitry has facilitated or frustrated anatomical evolution. This puzzle may now be solvable because we've recently learned so much about genomic logic [896,1413,1822]. What better way to celebrate Darwin's life than to revisit his intellectual playground with fresh eyes and new tools to dig for the treasures that eluded him?

Darwin explored everything that caught his fancy, from barnacles to orchids to earthworms. He was often astonished by what he found, and those surprises led him to wonder why nature should so often defy his intuition. To honor his inquisitive spirit, I have approached this subject as if I were an alien seeing humans for the first time. Hence, *Quirks* is full of childlike questions.

This book is mainly intended for college classes in which students are able and willing to do independent scholarly research—for example, capstone courses, honors tutorials, graduate seminars, and journal clubs. For that reason, I have thoroughly documented all statements. The in-text citations (numbers in brackets) provide essential links to published work, including articles too arcane to be located easily through Internet searches. Strings of such citations can serve as ready-made reading lists for discussions or term papers. The References section may seem excessive to some, but not to teachers devising assignments or to researchers compiling their own bibliography databases. Didactically, one of the best papers for a class to study is Pinker and Bloom's "Natural Language and Natural Selection" (plus commentaries) [2038], but any article in the journal *Behavioral and Brain Sciences* will serve to illustrate the fine art of Talmudic debate. Debatable topics are listed under "puzzles" in the Index and posted on Tom Brody's *Interactive Fly* Web site.

Jargon has been purged here wherever possible, and concepts have been simplified wherever feasible. The intent has been to make the material accessible and digestible. Even so, readers may find the text tough going unless they have (1) a familiarity with molecular genetics (*e.g.*, [1167]), (2) an acquaintance with developmental biology (*e.g.*, [908]), and (3) some prior exposure to basic evolutionary concepts (*e.g.*, [815,1336]).

Given its brevity, *Quirks* can only offer a taste of the many discoveries gushing from the new field of "evo-devo" (evolutionary–developmental biology) [1057,1763]. More riddles can be found in (1) Neil Shubin's popular *Your Inner Fish* [2384], (2) the blogs of Olivia Judson (*Wild Side*) and P. Z. Myers (*Pharyngula*), and (3) two timeless classics: *Natural Selection* by George Williams [2824] and *The Human Machine* by R. McNeill Alexander [50] (*cf.* [2599]). Several new books specifically probe the evo-devo of the brain: David Bainbridge's *Beyond the Zonules of Zinn* [120], David Linden's *The Accidental Mind* [1555], Gary Marcus's *Kludge: The Haphazard Construction of the Human Mind* [1643], and Aamodt and Wang's *Welcome to Your Brain* [1].

For a fuller treatment of the "evo" side of human anatomy, see John Langdon's recent *The Human Strategy* [1488] or Wiedersheim's 1895 classic *Structure of Man* [2802]. Regarding the "devo" side, start with Scott Gilbert's splendid *Developmental Biology* [908] or the new edition of *Larsen's Human Embryology* [2311]. Congenital anomalies are usefully annotated in Leslie Arey's old embryology text [91], and exotic curiosities are discussed in Armand Leroi's clever *Mutants* [1524]. For more on genetic gadgetry, try Sean Carroll's *Endless Forms Most Beautiful* [386] or his more advanced *From DNA to Diversity* [392]. Darwin's *Origin* was ably "updated" in *Darwin's Ghost* by Steve Jones [1313]. For definitions, see Hall and Olson's *Keywords and Concepts in Evolutionary Developmental Biology* [1061].

Finally, there is no richer repository of evo-devo narratives than the erudite essays of the late Stephen Jay Gould, who is widely credited with reviving the field via his seminal *Ontogeny and Phylogeny* [2814]. Gould often extolled the merits of using quirks as convenient windows into the evolutionary past (boldface added):

> This common claim for organic optimality cannot be reconciled with a theme that I regard as the primary message of history—the lesson of the panda's thumb and the flamingo's smile: the **quirky** hold of history lies recorded in oddities and imperfections that reveal pathways of descent. [968]

Evo-devo offers more visual appeal than the dry mathematics of population genetics, so it may turn out to be a better way to teach students about evolution in general [906,1160]. With this potential application in mind, I've relied on drawings (*vs.* verbiage) wherever possible. Many of the schematics distill so much data that their legends became overly lengthy. "Reflection" boxes were created to absorb the discursive overflow. They should be consulted whenever readers want more information about the contents or implications of particular figures.

The purpose of this book is not so much to survey what we know as to chart the *boundary* of the known so that we can stroll along that shore, peer into the mists of the unknown, and ponder how we came to be. Contemplation has virtually vanished from the crowded curricula of colleges these days, but that is the dreamy realm where connections are discernable and initiatives are imaginable. It is there where the genies await the right supplicant asking the right question in the right way, and it is there where starry-eyed students may succeed while those of us who teach them the calcified corpus of trite old facts have failed. Reflection boxes are designed to nurture new insights, just as a tour guide might linger by certain paintings to muse about the intentions of the artists. If readers get nothing more than the faint sense of deep secrets beckoning them like buried treasure, then this book will have met its goal.

A handy icon in many of my figures is da Vinci's *Vitruvian Man*, which he drew *ca.* 1492. The choice seems apt, Leonardo was a pioneering anatomist [551]. I have omitted the square and circle that Leonardo used to frame his man (as per the dictates of Vitruvius in *De architectura* 3.1.3 [2696]) because that ideal geometry implies a Platonic perfection that Darwin later disproved [370]. Give Leonardo this much, however: certain aspects of our anatomy *are* astounding from an engineering standpoint. Our two legs, for example, attain equal lengths despite growing independently for decades [106], even though any slight asymmetry at the outset should be amplified greatly by the end [747,2844]. The precision of symmetric growth is as enigmatic mechanistically as it is elegant morphologically [498,1044,1497], although recent evo-devo findings have begun to demystify how we achieve this feat [537,668,2682].

A *symbolic* icon used here is the fishing pole. Indeed, it embodies the book's main theme: evolution is like a fishing expedition in which genes "hook" other genes in abiding causal linkages. The utility of the metaphor is its focus on frivolity [2206]: the genes that have been snared over the eons could not have been predicted in advance and make little sense in hindsight [1606]. The accrual of such arbitrary links, layer upon layer, has culminated in the baroque complexity of our genomic circuitry [1605]. Impressive as the functioning of our genome may be in the aggregate, it is a crazy cobweb in its sundry details, not a coherent tapestry [571].

The challenges that await the next generation of evo-devo researchers are to (1) disentangle this web, (2) decipher its logic, and (3) deduce how it makes the many quirky traits that distinguish our species [390,1823]. (*N.B.*: The hominid lineage that led to us after it diverged from chimps and gorillas is termed the *hominin* subfamily [386,1488], a term used in later chapters.)

Darwin disabused us of the conceit that our outer shell is indicative of our inner workings. What matters for evolution is not beauty but function, and much of our anatomy works adequately but awkwardly—as if it were cobbled together with duct tape and baling wire. (Our clumsy knee comes to mind [10].) With armloads of such evidence, Darwin demoted us from Leonardo's pedestal, but please give him this much: he left our aesthetic appreciation intact—albeit altered [594,2016]. The beauty we now admire is not in our anatomy but in the antiquity of its parts and in the epic stories they have to tell [2384]. Our parts have played many roles over the eons, like versatile actors in an itinerant repertory troupe [595,2384,2802]. In Darwin's own words,

> Thus throughout nature almost every part of each living being has probably served, in a slightly modified condition, for diverse purposes, and has acted in the living machinery of many ancient and distinct specific forms. [563] (p. 284)

The impetus for this project was seeing my father's delight whenever I shared my tales of life's curiosities, which I've collected ever since I fell in love with the idea of evolution in college. He was not a scientist, so I had to explain esoteric concepts in plain English—an approach that I tried to use here, albeit with limited success.

My mother, brother, and sister, all of whom are accomplished artists, provided encouragement during the writing process. I, in contrast, could never have attempted to draw the figures in this book without Adobe Illustrator as my crutch! George and Ann Asquith served as coach and cheerleader, respectively, and Sam Braudt shared books, blogs, and wisdom. Chris Curcio was my ardent advocate and supportive editor at Cambridge University Press.

Larry Blanton, Richard Campbell, John (Trey) Fondon, Joseph Frankel, and Jeff Thomas read the entire manuscript and improved it greatly. Drafts of chapters were critiqued by generous colleagues, including Robert Bradley, Sam Braudt, Jim Carr, Tom Cline, Jason Cooper, Barry Davis, Mark Hamrick, James Hutson, Thurston Lacalli, David Moury, Robert Paine, Julie Rosenheimer, Kent Rylander, and David Weisblat. For the record, Trey rejects my favoritism of *cis*-regulation, described in Chapter 4 (*cf.* [792,1598]); Tom disputes my portrayal of *dsx* as a switch in Chapter 5; and Sam disdains my reluctance to define cognition in Chapter 7 (*cf.* [1504]). Long ago, to celebrate my first articles, Barry (a college chum) gave me *The Encyclopædia of Ignorance: Everything You Ever Wanted to Know about the Unknown* (1977, Pocket Books, New York). That book's focus on what we *don't* know gave me the prism that I needed to see anatomy anew.

Citations of Aristotle's *Parts of Animals* are encoded in standard format as "{PoA: Book #: Part #: Page # of Bekker's 1831 edition: column # a or b: Line #}" sensu Barnes [137]. Darwin revered Aristotle. In an 1882 letter to William Ogle [565], who translated *Parts of Animals*, he wrote: "From quotations which I had seen, I had a high notion of Aristotle's merits, but I had not the most remote notion what a wonderful man he was. Linnaeus and Cuvier have been my two gods, though in very different ways, but they were mere schoolboys to old Aristotle." To many of us who admire Darwin, he eclipsed even Aristotle in his powers of observation, deduction, and exposition.

If Darwin were alive today, he would be thrilled to see how much his ideas have helped us to interpret the flood of findings from the goldmines of comparative genomics [358,604,1194,1628]. His noble legacy of innocent inquiry lives on today in countless laboratories around the world [1964]. Rank-and-file researchers who have drawn inspiration from the well of his insights owe it to him—especially during this jubilant commemorative year—to let at least a little of evolution's grandeur shine out from our ivory towers. Hence this book.

Lewis I. Held, Jr.
Lubbock, Texas
October 2008

CHAPTER 1

Background

If aliens had surveyed Earth 10 million years ago (MYA), they might have concluded that our planet not only lacked intelligent life but also had no obvious prospects for acquiring any. After all, how could they have known that East Africa would dry out, forcing some of its hairy tree dwellers to venture onto the plains, walk upright, utter sentences, make tools, and build skyscrapers?

What role did chance play in our origin?

It is sobering to think that our species might never have arisen without the timely uplift and desiccation of the Rift Valley [341], but that incident was only the latest in a long string of haphazard events that made humanity possible [515]. Indeed, the African continent itself only emerged by chance when Gondwana cracked randomly into half a dozen pieces ~140 MYA [1450,2662]. Our luckiest break was undoubtedly 65 MYA when a wayward meteor collided with Earth, killing off the ruling reptiles and making room for our mammalian ancestors to spread, colonize, multiply, and diversify.

If we ever do encounter aliens on our future voyages into outer space, they might be able to deduce our home planet just by studying our bodies and behavior. They could guess Earth's mass from the thickness of our bones, the depth of its atmosphere from the size of our diaphragm, the rate of its rotation from our sleep–wake cycle, and the spectrum of its sun by the optics of our eyes.

Humans are not alone in bearing the stamp of our planet in our anatomy. All Earthlings do [592]. However, mythology and science fiction are full of creatures that could not have evolved here because they would violate physical laws [81,2595]. Steven Spielberg's character "E.T." (the Extra-Terrestrial) is a case in point. His cantilevered head was simply too heavy to be supported by his thin neck—assuming that his spine contained the same sort of bony vertebrae as ours does [242].

Other species could have evolved but never did because of the vagaries of how history happened to unfold [1050]. For example, babies that look like mermaids are occasionally born to normal parents [1841,2665], as are one-eyed cyclopes [488], but neither of these "monsters" typically survives to adulthood due to side effects of their respective syndromes [1047,1524,1599]. However, if having a propulsive tail fin or a median eye had proven adaptive for hominins, then such traits might have been refined incrementally by natural selection [142], with their associated afflictions subsiding as felicitous mutations accrued over time [2589].

Physics thus restrains the conceivable to the possible [2137,2700], whereas history confines the possible to the actual (Fig. 1.1) [967,1266,1493,1792]. The beings that have populated Earth constitute a single, rather limited experiment in biological evolution, albeit a long one, spanning ~3.5 billion years. Only in the last 0.01% of that history did our species arise, spread, and come to rule the planet. If the movie of life were rerun, the likelihood of humans ever evolving again is nearly nil considering how many chance events chiseled our anatomy as our ancestors were nudged from niche to niche [341,729,970,2007,2400]. We are, in short, "a glorious accident" [978].

One historical constraint that profoundly affected the shape of our body was the molecular machinery that moves our muscles [830,2699]. A minimum of two opposing muscles is necessary to maneuver any given bone because muscles can only exert force by pulling [48,2213]. Muscles cannot push because myosin ratchets along actin fibers in only one direction [1563,2448]. If the gadgetry had instead relied on microtubule (*vs.* actin) fibers [328], then muscles could both pull and push because the motor proteins kinesin and dynein walk in opposite directions [2786].

Had we evolved push–pull muscles instead of pull-only ones, we would obviously look quite different [50,655]. Some of our organs have nevertheless managed to achieve some impressive feats of gymnastic flexibility despite the pull-only limitation [1490]. Our tongue, for example, extends or retracts without any bony support [2502] by alternately contracting transverse versus longitudinal muscles [50,2419,2806], our irises dilate or constrict our pupils by alternately contracting radial versus circular muscles [1478], and the penis of adult males rises hydraulically without any muscles at all by valving the flow of blood through spongy tissue [1362,2698].

How much did internal factors influence our evolution?

We have realized how natural selection drives anatomical evolution ever since Darwin put forward the argument in his magnum opus *On the Origin of Species*

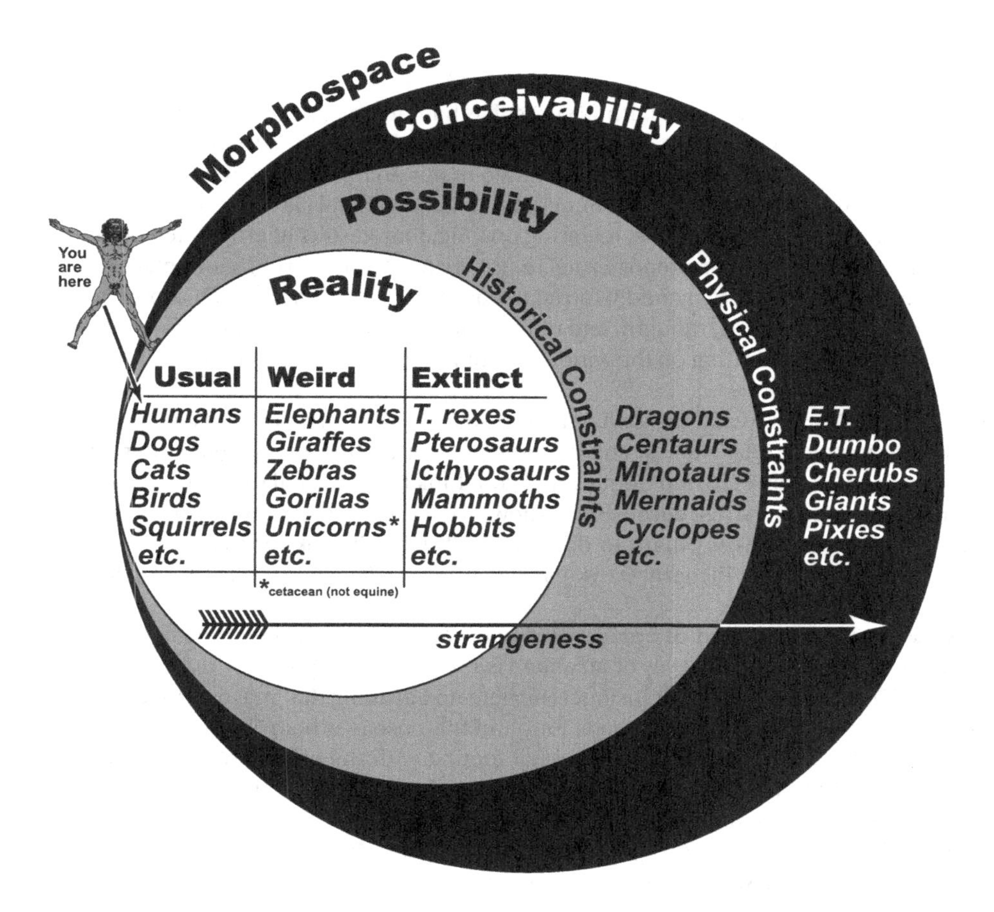

Figure 1.1. Real, possible, and conceivable subsets of vertebrate "Morphospace" [959,2600] in Venn diagram format. The arrow points from the familiar to the fantastic. City dwellers routinely see only a tiny part of the animal world: pets, birds, and the occasional squirrel. As children, we first met exotic animals (elephants, giraffes, etc.) at the zoo or circus and extinct dinosaurs at the museum. The thrill we felt at the novelty of those beasts has faded, but we can still get a similar frisson when we see science-fiction monsters in movie theaters. Some of those fabulous creatures could have evolved if Earth's history had unfolded differently [970], whereas others could not because they violate the laws of physics. For instance, centaurs could have evolved if the first fish to come on land had possessed three pairs of fins instead of two [823,959,2544], as some other groups of fish did at that time [2417]. Examples of conceivable, but impossible, animals include (1) Steven Spielberg's E.T. (the Extra-Terrestrial), whose neck was too thin to support his cantilevered head [2305], and (2) Walt Disney's Dumbo (the flying elephant), whose ears were too small (despite frantic flapping) to lift him into the air [49,1335]. The same is true for cherubs with their impotent wings. Mythical giants like Paul Bunyan could never stand because their proportionally scaled legs would not support their overly massive torso [955], nor could Disney's pixies like Tinkerbell exist because their brains would be too tiny to afford intellect. On the other hand, hobbits (~1 m tall à la Tolkien) not only *could* have evolved but *did* [92,2611], at least on one small island [407,2212]. Unicorns also evolved, albeit in aquatic form as narwhals [298,795], and, as noted by Aristotle {PoA:3:2:663a23} [137], the Indian rhino is technically a unicorn as well, given its median nasal horn [2085]. Heavier- or lighter-gravity planets may have fostered a rich assortment of alien faunas [556,1049], which we may someday encounter. Sadly, our Moon is lifeless, and although we like to think of it as colonizable, we are ill suited to walking there [1764,2685]. Indeed, the *Apollo* astronauts resorted to hopping and skipping to get around [50,1812].

REFLECTIONS ON FIGURE 1.1

Given how blasé we are today about the natural world, it is hard to imagine how amazed Europeans must have been when they first saw African wildlife in the 1700s [501] and dinosaur fossils in the 1800s [1472]. We take too much for granted in our anatomy (and behavior). One aim of this book is to rekindle our sense of wonder about life in general [594] and ourselves in particular. "The way we walk, for example, teetering on long, paired stilts of articulated bone, is unique among mammals, and as preposterous in its way as elephant trunks and platypus feet. We also communicate by tossing oddly intricate noises at one another, which somehow carry complex packages of feeling, thought, and information. We share and understand these sounds as if they were scents drifting on the wind, and our minds... sniff the fragrance of their meaning" [2737].

During the Middle Ages, narwhal tusks were marketed as "unicorn horns" and sold for 10 times their weight in gold [1876]! Now, of course, we know better: they're just teeth after all, so they sell nowadays for only $125/foot. But wait! If you think that narwhal tusks are any less mysterious, majestic, or magical just because we "know" what they really are, then you haven't thought about (1) what it takes for them to develop or (2) what it took for them to evolve (*cf.* Fig. 2.4). You should read Richard Dawkins's *Unweaving the Rainbow* [594]... or just read on.

This book was written as a kind of amusement park. Its thematic "pretend game" is to inspect each body part through the eyes of an alien visitor who asks, "Why is it *this* way and not *that*?" (*cf.* [2037]; his p. 523). Why, for instance, is there no Earthling that makes its skeleton out of metal, considering the ubiquity of metal ions and their use in other roles [2699]? In the face of such questions, the neophyte is on an equal footing with the expert. *No one* knows! Therein lies the *fun* of interrogating Nature. Asking the right questions is an art form unto itself, and some people are naturally gifted. Darwin was one of them.

For an even wilder ride through Fantasyland, see Dougal Dixon's *Man After Man: An Anthropology of the Future* (1990, St. Martin's Press, New York).

by Means of Natural Selection [832]. The essential idea is exquisitely simple [109,596,1336]. Individuals vary. When the environment changes, those best suited to the new conditions leave more offspring than those less suited, so that the population's average anatomy shifts in the next generation [664]. Over time, the population's gene pool may deviate enough to establish a new species [531]. Species thus manage to adapt to new ecological niches [2007,2460]. However, if the demands of the environment exceed the supply of useful variations, then a population or species can go extinct [330]. Surprisingly, extinction has been the fate of the vast majority ($\geq$97%) of all the species that have ever lived [1684,2461].

In short, the genome proposes, and nature disposes. Evolution is a groping, ratcheting, trial-and-error process fueled by hereditary variations [588,1234,1607]. The supply of those variations can thus limit its rate [118,375, 587,1065]. The possibility that variations might also constrain its direction was proposed by William Bateson (1861–1926) [1058], who actually coined the term "genetics" [1556]. His classic treatise on this subject was called *Materials for the Study of Variation, Treated with Especial Regard to the Discontinuity in the Origin of Species* (1894) [151]. As the title suggests, he concluded that traits can vary stepwise *before* nature winnows them, rather than smoothly in all directions [44,2262]. He stated his thesis clearly in his Introduction (boldface is his):

> If then all the individual ancestors of any given form were before us and were arranged in their order, we believe they would constitute a series. . . . In proportion as the transition from term to term is minimal and imperceptible we may speak of the series as being **continuous**, while in proportion as there appear in it lacunae, filled by no transitional form, we may describe it as **discontinuous**. . . . Variation has been supposed to be always continuous. . . . That this inference is a wrong one, the facts will show. [151] (p. 14*ff.*)

One drastic kind of discontinuity that Bateson analyzed was "homeosis" (another term he coined) [1535,1703]—the transformation of one body part into another [2689]. Most of the human homeoses he listed involved odd vertebrae—for example, a neck vertebra sprouting ribs as if it were a thoracic vertebra. Whether this phenomenon has any relevance to evolution is debatable, and it has been vigorously debated [661,957,1040,2475,2858]. The most extreme hypothesis—that homeosis is the main driving force behind macroevolution—was proposed by Richard Goldschmidt (1878–1958) [930,931]. He thought of homeotic mutants as "hopeful monsters" [662,2733] that could launch whole new species under the right circumstances [1163,2589]. Evolution may have followed this route in rare instances [1163,2179,2589], but its usual mode is decidedly more gradual [38,101,916,2790].

Bateson is famous not only for cataloguing homeoses but also for discovering an odd geometry of abnormally branched legs in animals as different as cockroaches and salamanders [819,1135]. Such legs, he found, always manifest new planes of mirror symmetry that obey what is now called "Bateson's Rule" [150,304]. Evidently, limb development is limited to a predictable subset of morphologies whenever disturbances occur. In a similar way, perhaps, homeoses tend to be confined to certain organs when they are induced by teratogens [1479,1958]. For example, fly embryos exposed to ether vapor show a nonheritable conversion of haltere to wing but few other abnormalities [1166].

Internal constraints apply not only to mutant individuals but to normal ones as well [943,960,1492]. One oft-cited example is the mollusk shell. The spectrum of naturally occurring shell shapes (from conch to nautilus to clam) is produced as a function of only three developmental parameters: (1) the rate of growth of the shell's mouth, (2) its rate of revolution around a vertical line, and (3) its rate of translation along that line [2132]. These internal variables define the axes of an imaginary cube in which most mollusk species can be plotted as single points (*x*, *y*, *z*) [465,2133]. Not all regions of this "Morphospace" are occupied [959,2600], mainly because of the contingent conditions (*i.e.*, *historical* constraints) that governed how the various lineages of mollusks happened to evolve [2325,2499,2729].

If shell growth, revolution, and translation are independent of one another in how they are controlled genetically, then mutations will tend to steer a species along a trajectory that is parallel to one axis of the Morphospace at a time [45,2498,2728]. The same applies to any species (mollusk or otherwise) with generative determinants that are likewise uncoupled. Under such conditions, which do indeed apply to many anatomical features [265,2109], certain paths through the pertinent Morphospace will tend to be well traveled, whereas others remain untrodden [793]. The evolutionary routes available to any given anatomy at any particular time will thus depend on its cellular program of development [44,1344,1687], which in turn will depend on the circuitry within its genome [1782,2816].

Here we have hit on the raison d'être for the hybrid field of "evo–devo" (evolutionary–developmental biology) [1143,2238]. Only by deciphering how genes control development can we hope to discern how evolution tinkers with anatomy [98,2110,2721].

Bateson was one of the first theorists to sense this connection. Another was his mentor Francis Galton (1822–1911), who is famous for advocating the use of fingerprints in forensics [2484] and *in*famous for advocating the use of eugenics in social engineering [704]. Galton devised a clever way to illustrate how development constrains evolution [854]. He imagined the organism as a polyhedron that rests on one facet at a time [959,960]. If pushed hard enough by natural selection, the polyhedron will topple onto one adjacent facet, and then another—thus following a trajectory governed by its geometry. The analogy stands in obvious contrast to the behavior of a billiard ball, which responds compliantly in both direction and degree to every nudge of natural selection [965,968]. Figure 1.2 sketches this metaphor and two others that have proven useful in this field. Galton actually fancied the polyhedron more for its tendency to stay put than for its ability to constrain evolution (boldface added) [916]:

> The mechanical conception would be that of a rough stone, having, in consequence of its roughness, a vast number of natural facets, on any one of which it might rest in "stable" equilibrium. That is to say, when pushed it would somewhat yield. . . . **On the pressure being withdrawn, it would fall back into its first position.** But, if by a powerful effort the stone is compelled to overpass the limits of the facet on which it has hitherto found rest, it will tumble over into a new

> position of stability, whence just the same proceedings must be gone through as before, before it can be dislodged and rolled another step onwards. [854] (p. 369)

Stability in the face of perturbing forces was also an issue for Conrad Waddington (1905–1975) [1054], but instead of facets on a solid, he imagined grooves in a landscape (Fig. 1.2b) [905,2283,2712]. The deeper the grooves, the greater the resilience of the pathways. If the embryo were a ball rolling down a groove, then mutations might alter its path by reshaping the landscape [99]. Waddington coined the term "canalization" to denote the stability of pathways in the face of mutations [2714,2813]. At the genetic level, it involves "buffering" [693,784,899,1999]. The fact that humans normally have five fingers is a manifestation of this phenomenon [844].

Waddington depicted his landscape (in later versions) as a corrugated canopy, much like a sagging circus tent. The shape of the canopy is set by a cobweb of guy wires beneath it (= gene circuitry). Those wires, in turn, are anchored by pegs (= genes) [2713]. Given this hierarchy, the effect of uprooting any single peg at the ground level (= null mutation) on the final anatomy (= fate of the rolling ball) is hard to predict because cross-links in the wire matrix (= *epi*genetic network) preclude a 1:1 relationship between specific pegs and parts of the landscape [1374,2195].

One upshot of this genetic cooperativity is that tweaks at the gene level can have complex, wide-ranging, and counterintuitive effects at the phenotypic level [1545,2379], and herein lies much of the fun in studying embryos. They often surprise you, and their mutant syndromes are typically as challenging as a Sherlock Holmes mystery or a Sunday *New York Times* crossword puzzle [414,1137].

Over the past century, the field of developmental genetics, which Waddington helped found [2711,2712], has deciphered thousands of etiologies and, in so doing, has uncovered a few "emergent properties" [576,2147] that characterize genomic control systems in general [190,331,571,1413,1741]. Those properties include robustness (resistance of pathways to perturbations) [133,846], pleiotropy (involvement of single genes in multiple circuits) [753,1175,1400,2720], nonlinearity (dosage-independent effects due to redundancy, etc.) [335,2773], thresholds (tipping points) [1536,1782,1798,1821], and feedback (dampers or auto-activators) [165,269,814,2018,2594]. However, the system property that has exerted the greatest constraint on *gross* anatomy has probably been allometry [479,882,1531,2421]—an inherent (canalized) divergence of growth rates in certain body parts relative to the body as a whole [323,871]. Wherever allometric

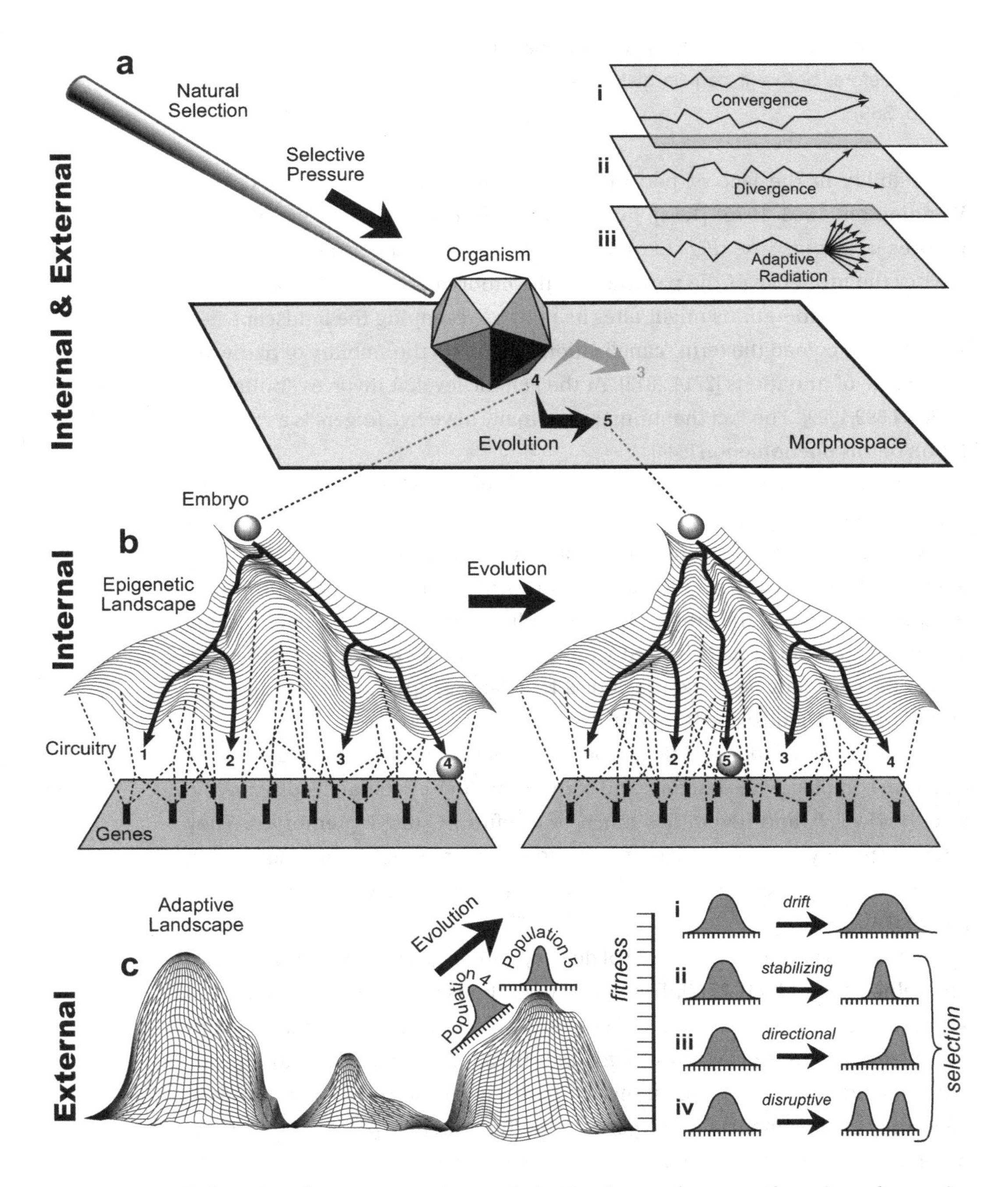

links exist, they cause a change in body *shape* whenever there is a change in body *size* [716,2305,2477]—for example, antlers getting disproportionately larger as deer grow bigger [952]. D'Arcy Thompson (1860–1948) was one of the first to show how startling such reshaping can be [953], and his grid transformations are iconic in the evo-devo literature [100,1702,2231]. Allometry is relevant to our own history because it may have boosted hominid brain size beyond the level

Figure 1.2. Metaphors for evo-devo that illustrate internal factors, external factors, or both. Modified from ref. [1135].

a. Galton's polyhedron analogy [960]. Francis Galton thought that species go from one stable anatomy to another when exposed to selective pressure [855]. He likened the stability to a polyhedron that stays poised on one facet unless pushed hard enough to topple it onto an adjacent facet. The organism would thus zigzag through its available "Morphospace" [2129]—the set of all possible shapes (*cf.* Fig. 1.1) [493,1702]. Insets (*upper right*) depict some trends in this context, where the *y* axis represents a one-dimensional Morphospace and the *x* axis is time. **a.i.** Convergence. Similar anatomies often evolve in separate lineages [1989,2821] because of physical constraints that allow few solutions for any given problem [1105,1176]. Similar *behavioral* traits can also evolve independently—for example, intelligence in humans and dolphins [1651]. **a.ii.** Divergence. Splitting of lineages is thought to occur mainly through geographic separation (*i.e.*, allopatric speciation), where different niches demand different adaptations [1154,2007,2580]. Initially, the splitting may only entail subspecific morphological specializations (*e.g.*, dog breeds), but the greater the differences become, the more likely they will lead to reproductive isolation (*i.e.*, separate species) [531]. **a.iii.** Adaptive radiation [870,2209]. Species can ramify rapidly into multiple offshoots [1523,2265] when they colonize ecosystems full of vacant niches [2108,2304,2446]. Mounting evidence suggests that the stress of habitat colonization itself may reveal hitherto-untapped potential for wild variability, on which natural selection can then act [1180,2525,2790].

b. Waddington's epigenetic landscape [2712,2713]. In his original metaphor, the ball was an embryonic cell that rolls to a differentiated state [905]. Here the analogy is broadened to the entire cohort of cells (*i.e.*, the embryo) and their collective destiny (*i.e.*, the adult). The canopy (~circus tent) is shaped by a network (circuit) of guy wires tied down by genes (pegs). Its grooves correspond to facets of the polyhedron in **a.** Initially, the species has anatomy #4 because this is the ball's path of least resistance (*i.e.*, lowest energy). If the environment changes, then any mutations (deviant tensions on the network) that happen to change anatomy to a more favorable state will be selected for [738]. As the landscape's shape changes (diagram to right of the thick arrow), the ball may fall into a prior groove (#1–3), or, as shown here, one that arises *de novo* (#5) [1064]. If the transition between grooves is sudden, then anatomy could change abruptly [151,930] (but see critiques [38,2858]).

c. Wright's adaptive landscape [869,2857]. In this imaginary example, height (scale at right) denotes fitness of genotypes (points in the mesh) [613]. If the environment changes (not shown), then contours would change accordingly. The three hills represent different ways to elevate fitness [593] (*cf.* [1304]). One interbreeding population is plotted as a bell curve, where the height of each point denotes the number of individuals having the genotype beneath it on the hill. The population is plotted before (Pop. 4) and after (Pop. 5) selection has exerted an effect, where 4 and 5 correspond to anatomies from the pathways shown in **b.** Panels at the right show how selection varies with topography [815]. **c.i.** On terrain that is relatively flat, there is negligible selection, so the span of genotypes can broaden via random mutation and recombination [118,738,1840]. In small populations, this "genetic drift" [1480,1606] can lead to fixation of neutral (or even deleterious) mutations [891,1224,2581,2632]. A case in point was the bottleneck ~5 MYA that led to fixation of our chromosome #2 from a harmless (but useless?) fusion of two ape chromosomes [1317]. **c.ii.** At an adaptive peak (hilltop) any deviations from the mean will be disfavored, thus narrowing ("stabilizing") the bell curve. **c.iii.** While a population is in the process of climbing a hill, individuals at the leading edge will leave more offspring, thus skewing the distribution ("directional" selection) [714]. Climbing can only occur by (1) mutational creation of new alleles or (2) recombination of old ones, both of which are random events. **c.iv.** If a population straddles a valley, then cohorts will be pulled apart ("disruptive" selection) [1702], and this divergence can lead to speciation [1407].

---

expected from adaptation alone—at least during the initial phases (*cf.* Ch. 7) [770,2181,2212,2309].

Historically, Waddington's metaphor of the "epigenetic landscape" was a derivative of Sewall Wright's notion of an "adaptive landscape" [2410], although

REFLECTIONS ON FIGURE 1.2

Such metaphors give us a feeling for how genes control anatomy. For more on how "Genospace" maps onto "Phenospace," see articles by Lieberman and Hall [1545] and Weiss [2778]. Ultimately, it was our ricocheting through "Ecospace" that caused us to land at the point we now occupy in Morphospace [482,493].

a. In Morphospace, a species would actually occupy an area, not a point, and the size of that area would reflect its span of variation [653]. Indeed, as mentioned in the text, a better metaphor for a species (≈ bell curves in **c**) might be an amoeba that moves over the terrain by extending and retracting pseudopodia. Species can retain their anatomy for eons [833,2750] (*i.e.*, stay put in Morphospace), although structural stasis need not imply genetic stasis [1124]. Darwin called such species "living fossils" and inferred that they must have "inhabited a confined area and... thus [were] exposed to less severe competition" [559] (p. 107). **a.i.** One clear example of convergence involves hooves. Any animal that runs *en pointe* (like a ballerina) risks injuring its toes, and the same protective devices evolved separately in even- and odd-toed mammals (*e.g.*, pigs *vs.* horses)—namely, fused toes and thicker toenails ("hooves") [2085]. Convergence also arises when lineages fill similar niches [482,2007,2692]. Thus, marsupials evolved species that look eerily like the placental wolf, cat, mole, squirrel, and anteater [631,2153,2446,2861]. **a.ii.** A classic example of divergence is Darwin's finches [1691,2770]. The molecular basis for finch beak divergence has finally been revealed by some recent evo-devo analyses [8]. **a.iii.** Explosive speciation can be sparked not only by access to uninhabited areas [811,2304] but also by the appearance of novel structures [237,1548,1890,2873]. For example, the debut of the neural crest created jaws that allowed a predatory (*vs.* filter-feeding) lifestyle [1624,1900] and hence spawned clades of carnivorous fish [472,796]. Radiation is aided by founder effects in small populations [1755] (*e.g.*, colonizing of the Galapagos by Darwin's finches [997,998]). Surprisingly, recent data refute the old cliché that mammals only radiated after dinosaurs disappeared [200,1597].

b. The uniformity of the depicted pegs may leave the impression that all genes are equal [2778]. Far from it! A tiny subset (the "toolkit") is most critical for building anatomy [392], and evolution results mainly from mutations therein [390].

the latter was devised for an entirely different purpose [1702,2035]. In Wright's original (1932) formulation [2857], unique combinations of alleles (= individuals) were represented by (x, y) coordinates in a plane, and their fitness for particular environments was denoted by altitude (z value) above the plane [613,1465]. To use a trite example, consider a population of giraffes with neck lengths that vary as a function of each animal's genetic constitution (= x–y plane) [53]. Those with longer necks will be more apt to survive (*i.e.*, have higher z values) when the only leaves that remain are high in the treetops (= natural selection). Such favored genotypes would thus constitute optima (= peaks) in the landscape.

In Figure 1.2c, a population is schematized as a bell-curve distribution that initially does not overlap any optimum in the landscape. If it resides on the side of an adaptive peak, then selective pressure will impel it to climb higher [815,1702]. For giraffes, this might mean evolving longer necks. If a population's breadth is relatively small, then it might ascend a lower peak from which it can't get down [1144,2054].

A more down-to-earth analogy might make Wright's world a bit more intuitive. If you imagine the population as an amoeba, then the cardinal rule is that it can only crawl uphill [593]. If the amoeba is small relative to the hills, then it must climb the hill where it currently resides (Fig. 1.2c.iii), regardless of whether other hills are higher. It will then be stuck there forever (Fig. 1.2c.ii) [2235] unless (1) the hill shrinks because of changes in the environment [493,1081,2692] or (2) the amoeba expands because of "genetic drift" (Fig. 1.2c.i) [1480,1606,2581,2632]. If either of these escape routes becomes available, then the amoeba may find itself straddling a valley [587] and hence be pulled in opposite directions toward adjacent hilltops (Fig. 1.2c.iv) [869]. Speciation is thought to happen under just such conditions [531,1407]. That is, populations can easily be ripped asunder (≈ the amoeba dividing) by the forces of disruptive selection [1702,2790] when those forces are exerted in conjunction with either geographic isolation or assortative mating [1955,2079].

Many suboptimal features of human anatomy are attributable to our ancestors having become trapped on low peaks in rugged landscapes [587,593, 2824]. Some of those flaws are examined in Chapter 6.

Are fruit flies really our kissing cousins?

The greatest surprise to come out of evo-devo research thus far is the universality of covert genetic circuitry, despite the overt diversity of external anatomy. For example, we look nothing like flies, but our genomes turn out to be wired in eerily similar ways [571]. (*N.B.*: For ease of comparison, human genes are henceforth written like fly genes—*e.g.*, *Sry*, not *SRY*.)

This "Circuitry Epiphany" has had a profound practical benefit. Most of what we have learned about how genes control development in the past 100 years has come from only a few nonhuman "model" species [195,1290], with the fruit fly *Drosophila melanogaster* chief among them [103,1137]. Because of our common circuitry, those insights turn out to pertain to us as well, so we can directly apply them to figuring out human development [1645] and to eventually curing or preventing birth defects [723,1024]. Such applications are being actively pursued, as are other spin-offs in the clinical realms of cancer, aging, and limb regeneration [195,1829,1906,2443].

Humans and fruit flies do share one aspect of external anatomy: we are both symmetrical. Nevertheless, the idea that we might have an anatomically complex ancestor in common seems preposterous because our nervous

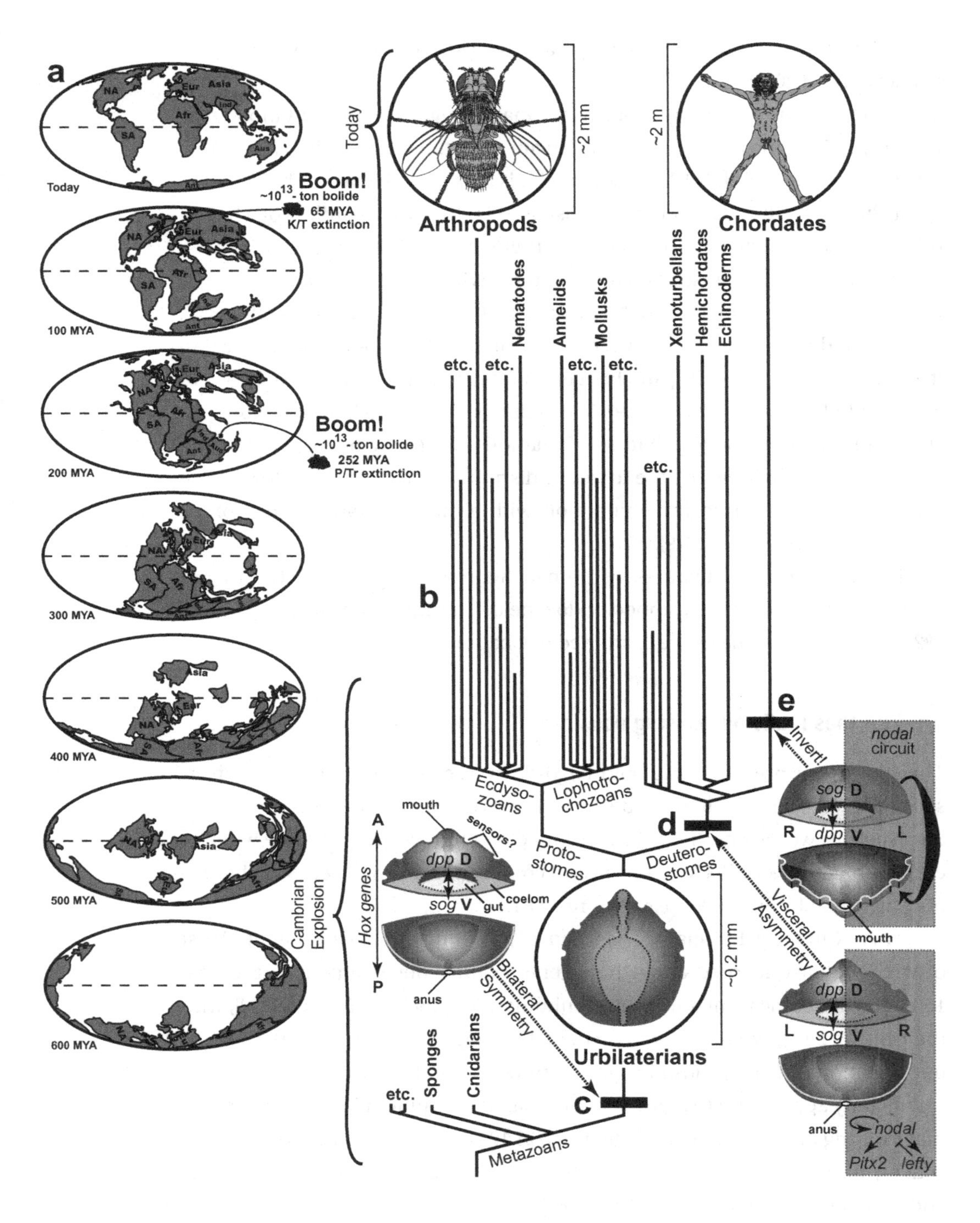

systems are on opposite sides of our bodies [977]: ours runs down our back, and theirs runs down their belly. Evo-devo studies have resolved this riddle with another shocking revelation. The back of a human and the belly of a fly are apparently—are you ready for this?—*homologous* [86,606,1457]! They use

Figure 1.3. Overview of bilaterian evolution. The intent of this diagram is to convey the vast amount of time that has been available for evolution to reshape the anatomies of metazoan phyla. In particular, the *Hox* genes that form the scaffolding for all bilaterian bodies (**c**) are >20 times older than the Himalayan Alps [604,2023]. Humans and flies shared a multicellular ancestor who lived more than half a billion years ago.

a. Timeline spanning one seventh of Earth's 4.55-billion-year history [620]. Modern continents or their pieces are traced back millions of years ago (MYA). The "Cambrian Explosion" [506] was an abrupt radiation of body plans ~543 MYA [1297,2023,2210,2663]. Back then, the continents looked nothing like they do today, and days were shorter (21 hours) because Earth spun faster (425 days per year) [2122].

b. Pedigree of metazoans, simplified from refs. [25,1048,1189]. Arthropods and chordates are singled out but are only 2 of ~37 crown phyla [2663]. (There are just 4 extant deuterostome phyla [252,2536].) Lines that end before today denote extinctions but are only meant to give a general impression. Black bars mark key events.

c. Urbilaterians are the ancestors of all phyla that have a medial plane of symmetry, a two-ended gut, and a coelom [195,886,2663]. Cross-section shows the A–P (anterior–posterior) axis [1189,1193] where *Hox* genes specify an "x" coordinate [1519,2316], and the D–V (dorsal–ventral) axis where the genes *decapentaplegic* (*dpp*; a.k.a. *BMP4*) and *short gastrulation* (*sog*; a.k.a. *chordin*) specify a "y" coordinate [39,1246,1779,2890]. Urbilaterians are depicted with the D–V polarity of protostomes for reasons explained in **e.** The tick-shaped animal (circled) belongs to the oldest known bilaterian species, *Vernanimalcula guizhouena* (580–600 MYA) [431], although it may not be our direct ancestor [2493,2675], which was probably more wormlike [119,1462]. Note its tiny size versus the fly and man. It has 3 pairs of concave pits (sense organs? [1190,1267,2301]), but no sign of segments [443,1226] or appendages [1761,2092,2096], which apparently evolved later in several descendant clades [195,604,1048]. The gut (*dotted outline*) is straight without even a hint of the left–right asymmetry seen in larger metazoans (*cf.* Fig. 2.1). The ventral mouth leads to a thin pharynx, a wide stomach, a narrow rectum, and a posterior anus.

d. All deuterostomes thus far examined share asymmetric expression of the gene *nodal* (see Ch. 2), which marks left versus right through the circuit shown (or parts thereof) [243,681], although the trigger upstream of *nodal* varies among classes of vertebrates [108,1528]. Arrows indicate activation; T-bars inhibition. Basal deuterostomes are drawn with this circuit on the right for reasons explained in **e.** How asymmetry arose in protostomes is unclear [528,1156,1216,2255,2429], nor is it known how often they recruited different signaling pathways for different asymmetric organs [13,1625,2430,2571].

e. An inversion of the bilaterian body must have occurred here because (1) chordates are upside-down versions of annelids and arthropods with regard to the *dpp-sog* axis [86,606,1457,1778], (2) the D–V polarity of hemichordates resembles arthropods rather than chordates [886,1583], and (3) the *nodal* circuit is on the right side in sea urchins but the left side in chordates [680,681,2367]. Our protochordate progenitor must have flipped over like a swimmer switching from breaststroke to backstroke at the end of the pool [629,2579]. What had been its belly became its back and what had been its left became its right [778,977]. The cross-section shows the *sog-dpp* axis in chordate orientation and the *nodal* circuit (*shaded rectangle*) on what has become our left side. Thus, the animals diagramed in **d** and **e** are facing nose-to-nose.

---

exactly the same genes to pattern their embryonic axes [195,778]. Comparative genetics has even allowed us to identify *which* of our ancestors flipped over and to pinpoint *when* the inversion happened (Fig. 1.3e) [629,2579]!

This "Inversion Epiphany" has allowed a more apt reclassification of animal life [1194,1289]. It has also spawned a lively guessing game among evo-devotees about what the last common ancestor of all bilaterally symmetrical phyla looked like [119,195,392,604,886,908,2663]—and especially whether it had segments [443,731,1226], limbs [1761,2092,2096], eyes [90], or separate genders

REFLECTIONS ON FIGURE 1.3

a. Global extinctions at the K/T (Cretaceous/Tertiary) [62,248,2322] and P/Tr (Permian/Triassic) boundaries [162,730] likely occurred as a consequence of bolide impacts (meteors or comets) [620], although other theories exist [1485]. Not shown here is the *really* bad day (~4500 MYA) when a Mars-sized object hit Earth so hard that we belched a molten bolus that became our moon [363,2619]. The moon's pockmarked face records the subsequent shower of smaller bolides that finally abated ~3800 MYA [1075]. Only then could life gain a foothold. For a pithy and poetic survey of evolution—both geological and biological—see Nigel Calder's *Timescale* [341]. Continents are redrawn from Scotese [2326] except that his artifactual gap along the Rocky Mountains is omitted [1834]. Dates, weights, and impact sites of bolides (asteroids, comets, or meteors) are from ref. [620].

c. *Hox* control of the anterior–posterior axis may actually predate cnidarians [646,764,2247].

d. Echinoderms somehow evolved radial, fivefold symmetry as adults [2415], but they retain bilateral symmetry as larvae [2536]. Strangest of all, they metamorphose by an asymmetric process as bizarre as having a wart on your left cheek that gradually expands to become an umbrella [681,1766].

e. One abiding mystery of the dorsal–ventral axis inversion is how the chordate mouth later migrated down the head to a ventral site [456,1463,1464]. If we can figure out this riddle, the answer might shed light on why our pituitary splits apart and rejoins [1348,2192] and why it evolved with our oral region in the first place (*cf.* Fig. 3.1) [356,1191,1637].

[737,2623]. Finally, it has sparked a worldwide hunt for the fossilized remains of that mythical beast [731,2293,2664]. Figure 1.3c depicts the likeliest candidate unearthed so far [431]. This figure also shows the vast depth of time spanned by this epic story. Earth's crust has split and shifted many times since our human–fly progenitor swam its seas or tunneled its sands—so much so that we would hardly recognize our own dear planet from the "Rubik's cube" configuration of its continents back then.

Yet another big surprise has been how few "words" cells actually need to communicate [138,392,604]. Animal anatomy is mainly patterned by protein signals that are transduced by only five signaling pathways: Hedgehog, Wnt, TGFβ (Transforming Growth Factor-β), RTK (Receptor Tyrosine Kinase), and Notch [604,883,907,2040]. For all but one of them, the signal is emitted by a group of cells and spreads throughout the rest of the developing organ, so that its intensity declines with distance [138]. (Notch is the exception: it is used for private conversations between adjacent cells [601].) Cells estimate their distance from the source by gauging the level of signal [2889] and then proceed to adopt identities that suit those locations [104,1429,1483]. Because these diffusible molecules ultimately create morphology, they are called "morphogens" [2542]. The theoretician who solved this problem abstractly long before it was confirmed biochemically was Lewis Wolpert (b. 1929) [2843]—a protégé of Waddington's and a mentor of many prominent researchers working in the field today.

This "Morphogen Epiphany" has allowed researchers to dissect development into a few orthogonal morphogen gradients per incipient organ [108,2380], and we now know the major signals that are used in many diverse systems. The knottier challenge has been to figure out *how* the transduction of those "(x, y)" signals evokes differentiation (nerve, muscle, bone, etc.) via downstream effects within the genome [2844].

The concepts introduced here are fleshed out in later chapters, and flies are used as a convenient counterpoint to contrast with humans. Chapter 2 shows how a diffusible signal is the root cause of our internal asymmetries, and Chapter 3 focuses on our sagittal plane of symmetry and its associated geometric riddles. Chapter 4 examines how morphogens create meristic patterns of repeated modules, such as fingers, teeth, and vertebrae. Chapter 5 delves into the genetic basis for sexual differences. Chapter 6 asks how our ancestors happened to get trapped on so many suboptimal peaks on their way to becoming *Homo sapiens*. Chapter 7 closes with the quirk that sets us apart from all other species on the planet: our sentient brain. The overall theme is that our anatomy can only be understood in terms of our ancestry. Our body is made of ancient parts with roles that have changed many times since they first evolved under dramatically different circumstances.

Throughout the book, a motley menagerie of other organisms (aside from humans and flies) are employed as a supporting cast of characters. The reason is simple. We cannot grasp how odd we are without noting the paths that our forebears could have taken but didn't [587,595,2824]. Besides, this multi-ring circus exudes a charm all its own, which is why many of us fell in love with biology when we were kids.

The diversity of anatomical gadgetry that has solved any given evolutionary problem is staggering [1890], but so is the paucity of parts from which the gadgetry arose [1265,1266]. That paucity is due to (1) the common ancestry we share with other phyla [382] and (2) the versatility of the parts themselves [309,558,1690]. In this sense, the transformations in any given lineage are like the tumbling mosaics in a turning kaleidoscope [2461]. Darwin sketched the broad outlines of this underlying unity. Now the details are becoming clear, thanks to breathtaking breakthroughs in evo-devo research [393].

Evo-devo gives us two lenses with which to examine anatomy. One is the ultimate lens of ancient history: how did what exists originate, and how has it changed over the intervening eons? The other is the proximate lens of antecedent architecture: how are structures built from the zygote and altered through intermediate embryonic stages? Combining these lenses has certainly provided a powerful compound perspective [1824], but what has really allowed us to crack so many age-old enigmas in the past few years has been our ability to ask evo-devo questions at the molecular–genetic level [392]. As new answers have been gushing from far-flung labs at an ever-faster pace, the sentiment in the research community has swelled from surprise to wonder to awe. How fitting that the conceptual seeds Darwin planted so long ago should be bearing such lovely fruit just now, on the occasion of his 200th birthday!

CHAPTER 2

Symmetry and Asymmetry

Humans look symmetric on the outside but have stark asymmetries on the inside (Fig. 2.1). Thus, we have a spleen on the left but not the right. Our left lung has two lobes, but our right lung has three. Our heart and stomach are shifted left of center, our liver is shifted right of center, and our intestines meander throughout our abdominal cavity with no regard for the midline at all.

Why are we symmetric outside but asymmetric inside?

The paradox of outer symmetry and inner asymmetry actually makes sense from an evolutionary perspective. Our outer shell of bones and muscles has always served a locomotory function [1871]. External symmetry allowed our aquatic ancestors to become streamlined for speed [1672,1871,2766], and it has been retained ever since because it is just as useful for walking as it was for swimming [517]. In contrast, the only locomotory restriction on our internal organs has been that their weight be distributed evenly relative to the midline.

If we trace the history of our anatomy back before the fish stage of evolution, we find that the inside of the body used to be as symmetric as the outside [512,1462]. Fossils of "Urbilaterians"—the earliest animals with bilateral symmetry [195,1393]—show a digestive tract that ran directly from the mouth to the anus, with no twists or turns along the way [431].

Why did the gut of vertebrates then become asymmetric? Part of the answer concerns size [242,955,1238]. The ocean is a fish-eat-fish world, and the bigger you are, the more likely you are to survive (all other factors being equal). However, a body with a straight gut cannot grow beyond a certain size because the supply of nutrients (proportional to gut *area*) cannot keep up with the increasing demand (proportional to body *volume*) [511,1049]. For Urbilaterians (~0.2 mm long) to reach our size (~2 m tall), their gut had to grow much longer than their body [512,2806]. Our intestines are ~10 times the length of our torso.

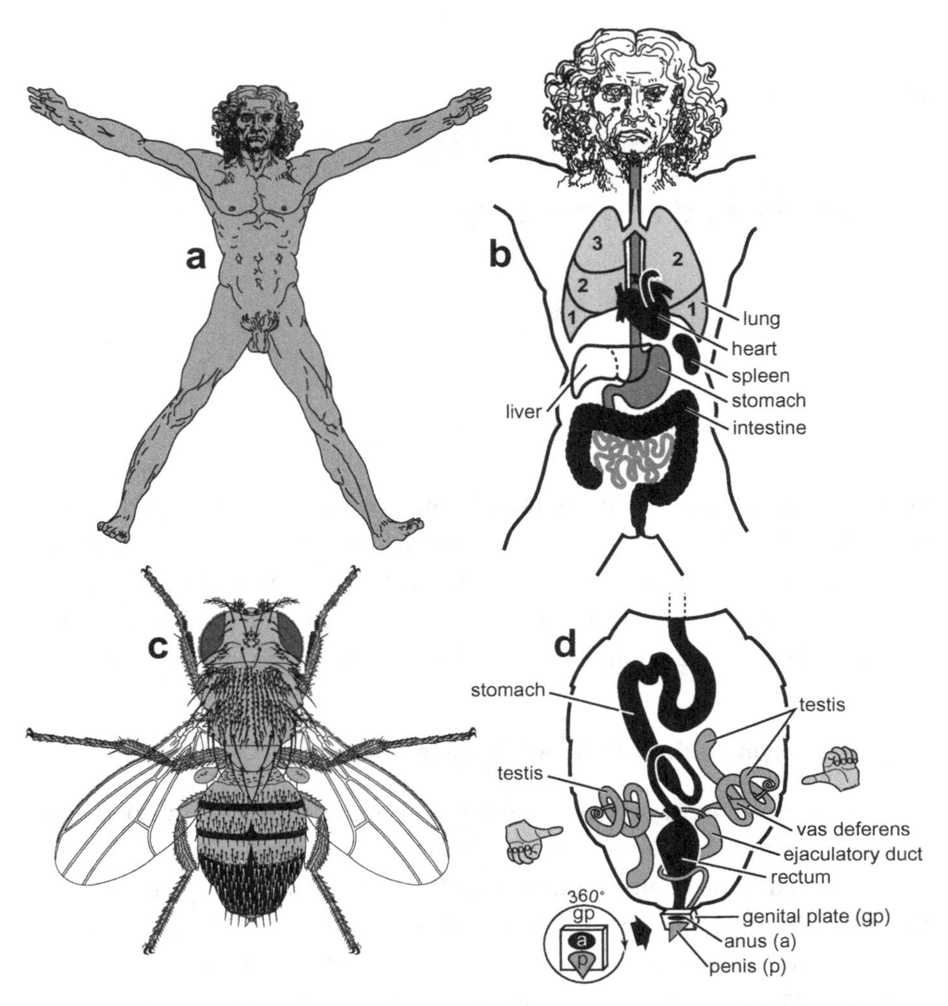

Figure 2.1. The paradox of external symmetry versus internal asymmetry, as exemplified by a human and a fly.

a. Vitruvian Man (by Leonardo da Vinci, *ca.* 1492), a member of the phylum Chordata.

b. Our right lung has three lobes but our left lung only two [256]. Our heart is shifted to left [308], our spleen is located on the left [322,2000], and our stomach bulges to the left, whereas our liver is shifted to the right [376,2311]. Our colon curls into a question mark, although its exact path can vary from person to person [187,1866,2828].

c. Vitruvian Fly (with apologies to Lenny). Male fruit fly (*Drosophila melanogaster*, ~2 mm long). Its phylum (Arthropoda) is only distantly related to ours (Chordata), but we share a common ancestor (*cf.* Fig. 1.3).

d. The fly's long stomach has constant loops (as does our gut) [528,1746]. What is really weird, though, are its testes, which form spirals. At first glance, the spirals look as if they might be symmetric, and in most fly species they are [2467], but in *D. melanogaster*, they both coil around their vas deferens like the fingers of a left hand curl around its thumb [1745,2467]. In other words, they are *not* mirror images! Rather, they're like a man with two left hands. No one knows how this asymmetry is generated or why it exists. Another quirk is a 360-degree rotation of the genitals during development, which twists the ejaculatory duct sinistrally around the rectum [919,1216]. (See text for discussion.) Males can still mate as long as their genitals are not >20 degrees off kilter [250]. Accessory organs (paragonia and sperm pump) have been omitted (*cf.* Fig. 5.2).

REFLECTIONS ON FIGURE 2.1

Mutations can completely reverse the asymmetry of the viscera to a create a "situs inversus" phenotype (not shown) in both humans [321,397,1258,1711,2441] and flies [1216,2429]. Ambiguous intermediate states are also commonly found [1430,1482]. Drawings of human viscera are adapted from refs. [353,1646] and those of fly viscera are simplified from refs. [1745,1746]. Although fly and human guts appear quite different anatomically, they turn out to use the same genetic circuitry and stem-cell gadgetry to replenish their epithelial cell cohorts [2041,2558]. This homology adds to the growing list of shared bilaterian devices uncovered by evo-devo research [884] (*cf.* "Study Aids" at Tom Brody's *Interactive Fly* Web site).

About 1 in 10 of us are left-handed [1715,1808,2022], a frequency that has not changed in at least 10,000 years [522,750]. Curiously, handedness varies independently of visceral asymmetry: only ~10% of people with situs inversus are left-handed [1713]. Hence, these traits must be under separate genetic control [511,512,1630]. Indeed, the circuitry for left–right asymmetry must have multiple branchpoints [208,441,732,1528].

b. Mouse lungs are even more asymmetric than ours: they have *four* lobes on the right and only *one* on the left [1181]. Other human asymmetries (not shown) include the sinistral coiling of our umbilical cord vessels and the dextral coiling of the ducts of our gall bladder and sweat glands [510,2024]. (Do Schwann cells wrap axons with the same chirality, regardless of whether they're on the left or right?) In some other mammals (*e.g.*, elk and lemurs) the colon forms a perfect spiral with 6 or more gyres [2806].

Our most dangerous asymmetry is inside the heart, where we make only one, *unilateral* pacemaker [1201]. We would be better off if we had a "backup" pacemaker on the other side that could take over in the event of a heart attack. The evolution of the vertebrate heart is an epic story that we are only now beginning to unravel morphologically [777,2062,2402] and genetically [438,481,582,1933,2280].

d. For clarity, the tip and coils of each testis are drawn on the same side of the abdomen (they really lie on opposite sides), the gyres of coils are spaced apart, and other organs (crop, Malpighian tubules, accessory glands, and ejaculatory bulb) are omitted. See Fig. 5.2 for a fuller rendering of the fly's reproductive system.

The discovery that myosin (*vs.* tubulin) is the key agent in fruit fly left–right asymmetry [1216,2429] offers hope that the upstream circuitry will yield its secrets [528]. However, our grasp of dipteran phylogeny is pathetically weak compared with what we know of chordates [1016], so it will be hard to decipher the order in which the genetic tinkering occurred.

To pack such a "fire hose" into the body cavity requires that it deviate from the midline. Historically, any mutations that let the gut become asymmetric (*e.g.*, those that created a coelom [462,1762]) would have proven profitable to their bearers because they would have permitted a virtually unlimited increase in size.

The deeper mystery is why our gut doesn't just coil haphazardly to fill the available space. Instead, it undergoes a ritualized sequence of twists and turns [2311] that normally culminates in a clockwise colon (ascending right, crossing, and descending left) [16,582]. Even if we could decipher the etiology of this intestinal origami, it would not explain (1) why our spleen and liver depart from the sagittal plane when our stomach gyrates about its axis [376,1866,2000,2311] or (2) why the stomach itself bulges leftward to form the fundus [685].

The epitome of asymmetric complexity, however, must be the heart, which arises through pretzel-like contortions of an initially symmetric tube [7,308]. The evolutionary issue here is whether these choreographed deviations from the midline are adaptive (physical constraints of organ packing?) [1976,1980,2678] or arbitrary (side effects of morphogenesis?) [1838]. The developmental issue is how they are programmed genetically and implemented mechanically.

If you think *our* asymmetries sound odd, consider the indelicate indignity of *Drosophila* males, who spin their genitals around their anus 360 degrees during metamorphosis [919,1617]. This pointless pirouette wraps their ejaculatory duct around their rectum internally (Fig. 2.1d) [1216] but has no net effect externally, where the angle of the penis really matters. Here at least, there may be a simple evolutionary explanation. In primitive dipterans (*e.g.*, mosquitoes), the penis only rotates 180 degrees, and it does so reversibly during sex [437]. If those early flies rotated their genitals in this way to suit a novel mating posture, but subsequent species (*e.g.*, *Drosophila*) "changed their mind" and reverted to the old posture for some reason (insects are kinky [1220]), then the latter flies could have restored the original penile angle in one of two ways. They could have (1) stifled the rotation at 0 degrees, which would have been the most efficient solution, or (2) extended the rotation an extra 180 degrees, which is what they actually did [1016,1871]. The lesson here is that in evolution, the ends justify the means—*any* means, no matter how silly they may seem in hindsight. This lesson can help us unravel illogical features of our anatomy as well (*cf.* Ch. 6).

Is visceral asymmetry due to molecular chirality?

To solve the mystery of why visceral asymmetry is orderly versus random, it is necessary to consider structures at a molecular level. Life is based on carbon atoms. When tetrahedral carbon binds four different partner atoms, they can be arranged in two alternative ways (Fig. 2.2) [183]. These "stereoisomers" cannot be superimposed any more than your left hand can be turned to look like a right one [2593]. Hence, organic molecules have an inherent handedness or "chirality" [495]. Amino acids, in particular, exist in either an "L" or "D" form [1995]—terms that denote whether their crystals rotate polarized light to the left or right (levo- *vs.* dextrorotatory) [83]. Living things universally use L amino acids in their proteins [1713,2848]. Because D amino acids should have worked equally well [827,1757], this story illustrates a type of contingency that Francis Crick called a "frozen accident" [535]. As a consequence of this arbitrary preference,

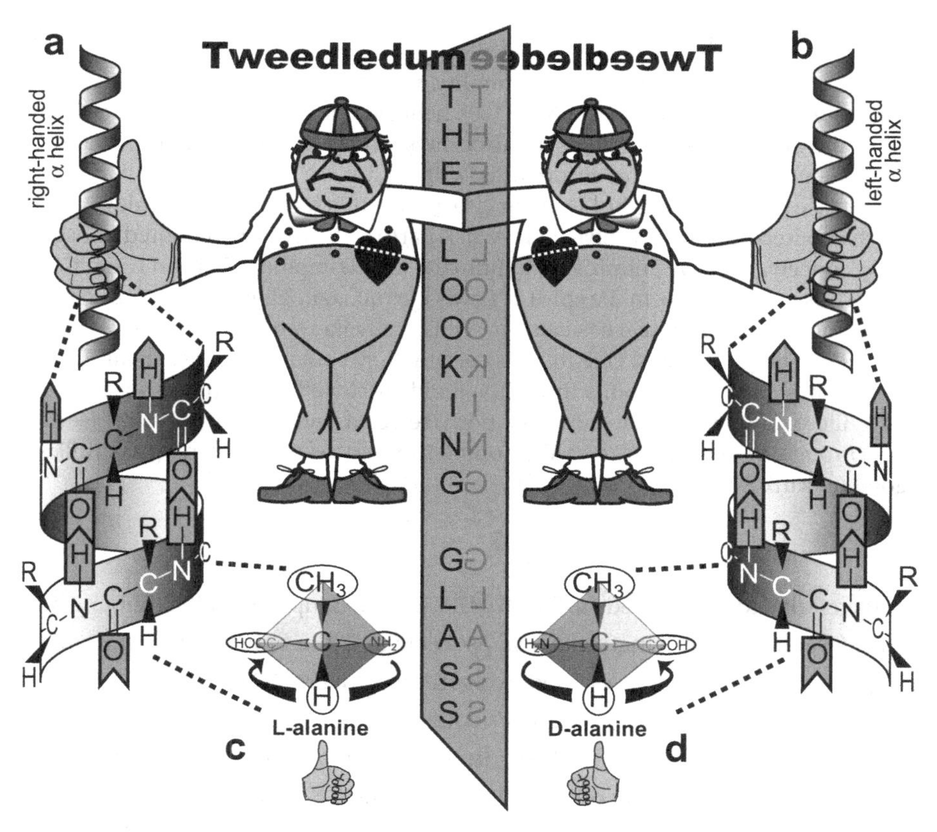

Figure 2.2. The geometric basis of molecular chirality. On the left of the mirror are molecules as they naturally occur in living things [495,1682]. On the right are their mirror images, which are rare in living things [827].

a, b. Proteins typically fold into right-handed α-helices (**a**), not left-handed ones (**b**) [852]. Handedness is conventionally defined by whichever thumb matches the ribbon direction traced by the curl of the fingers. Magnified section below shows the peptide backbone (N-C-C-N-C-C) and hydrogen bonds (between C=O and N–H groups) that bridge the gyres. Abbreviations: C = carbon; H = hydrogen; N = nitrogen; O = oxygen; R = reactive side group that varies among amino acids.

c, d. Mirror-image stereoisomers of the amino acid alanine (R = CH_3 group). Bonds are denoted by lines out of (*black triangles*) or into (*hollow triangles*) the page. The central carbon binds four constituents (vertices of tetrahedron) that can be arranged in either one of two ways [1682]. One isomer rotates polarized light to the left (L-alanine = levorotatory), the other to the right (D-alanine = dextrorotatory).

proteins, which are long chains of amino acids, have their own chirality: they tend to twist into right-handed helices [450,853].

Could the asymmetry of our viscera be due to the chirality of our proteins? It seems silly to think so given the huge difference in scale, but evidence has been accumulating in the last few years that this is indeed the case. The explanation rests on a marvel of molecular machinery called the cilium [568]. Cilia are hairlike structures that act like oars to propel protozoans through water [1937]. In multicellular organisms like ourselves, the cilia beat in a similar way to move

REFLECTIONS ON FIGURE 2.2

a, b. Other molecules that have a characteristic chirality, regardless of whether they are on the left or the right side of the body include DNA [1130], helicoid chitin fibers in the cuticle of insects [1870,1871], and helicoid collagen fibers in the corneal stroma of vertebrates [2629], and the same is true at a higher level for sister chromatids [254].

c, d. Living things use L-amino acids but D-sugars [495]. How this bias arose at the dawn of life has been much debated but never resolved [114,483]. If Earth's history were rebooted, those early events could conceivably have culminated in humans with D-amino acids and L-sugars instead [865,1681]. Thus, it is possible, in the spirit of Lewis Carroll, to imagine a counterpart of Alice from that mirror-image universe. Let's call her "Ecila." If Ecila were to encounter Alice, would they look and act alike? Perhaps. One thing is certain: they could not share a meal. The reason is clear. Enzymes catalyze reactions through a hand-in-glove fit with their substrates [287,1431,1731], so Alice's food would not suit Ecila, nor vice versa. Indeed, Carroll addressed this problem in *Through the Looking-Glass*: Alice opined to her cat that looking-glass milk might not be good to drink [380].

fluids relative to the ciliated surface [209,580]. For example, the cilia that line our lungs sweep the overlying mucus so as to cleanse the epithelium continually [529]. These same cilia are disabled in cigarette smokers [2686], and the resulting stagnation of mucus causes "smoker's cough" [1213].

Most cilia beat back and forth in a single plane, but one special type spins like a propeller in a cone-shaped path (Fig. 2.3) [1253,1768]. Mammalian embryos have a lawn of such cilia whirring away at our midline [1164,2376]. The "Gulf Stream" that they create wafts certain signaling molecules from right to left [1846,2548]. The rising intensity of those signals on the left side eventually turns ON a master gene called *nodal* [772,2352], which forces tissues on our left side to develop differently from the default condition (*nodal* = OFF) on our right side [1067,2136]. *Nodal* exerts its control through subordinate genes collectively called the "*nodal* circuit" [1620]. The terminal effector of that circuit is the transcription factor Pitx2, which directly turns downstream genes ON or OFF during the development of the heart [848] and other organs [554,1254,1559,2377] to produce the overt asymmetries alluded to earlier.

In summary, the causal chain is as follows: (1) the chirality of our amino acids determines (2) the handedness of our proteins, which dictates (3) the clockwise spin of our midline cilia, which forces (4) the leftward movement of the overlying fluid, which activates (5) the transcription of *nodal* on the left side of our body, which leads to (6) the delegation of Pitx2, which finally controls (7) the asymmetric development of our viscera. This basic mechanism must be ancient because lawns of rotating cilia are also seen in birds, amphibians, and fish. However, it turns out that the chain of command differs among classes

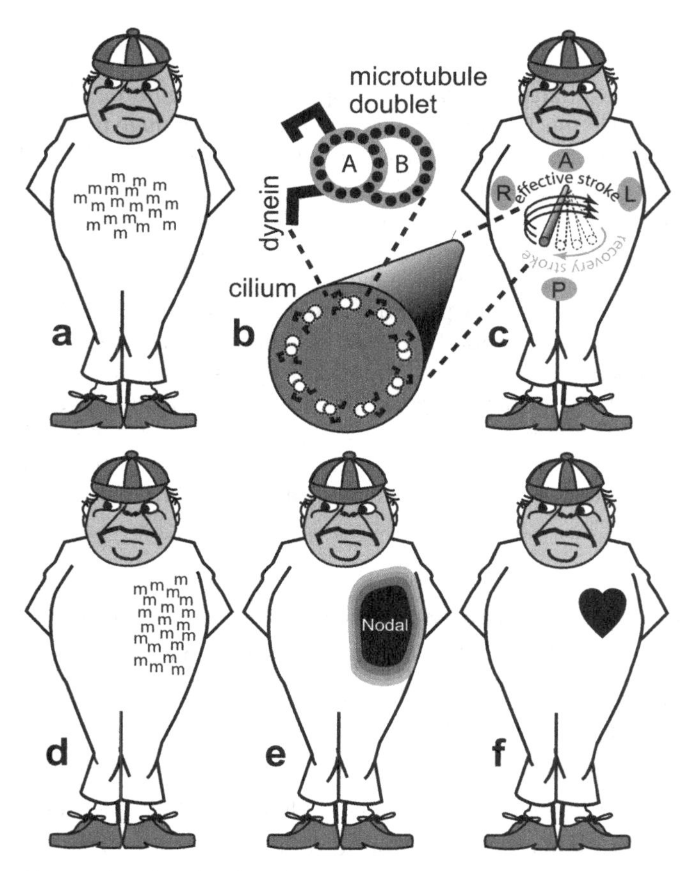

Figure 2.3. How molecular chirality dictates visceral asymmetry in mammals [1924]. Tweedledum represents a human embryo at the gastrula stage.

a. Signaling molecules termed "morphogens" (m) are secreted near the midline in an initially even distribution, left versus right. (Candidates for the actual signal, which is not yet known [1898,2569], include Sonic hedgehog [1966] and retinoic acid [2754].) In this same area is a lawn of ~250 hairlike cilia [1164,2376], although only one representative cilium is depicted henceforth.

b. In cross-section, each "9 + 0" cilium displays the 9 microtubule doublets (AB) typical of standard "9 + 2" cilia but lacks the central pair of singlet microtubules. Small black circles are protofilaments of the protein tubulin. (B tubules can have 10 or 11 protofilaments [1837,1892,2524].) The A tubule bears two dynein motor proteins that walk along the B tubule neighbor of the adjacent doublet [2140]. Slippage of adjacent doublets moves the cilium [1875].

c. Each 9 + 0 cilium (enlarged) spins like a propeller (*vs.* beating like a metronome) [1253], and other 9 + 2 and 9 + 4 cilia here do also [756]. A, P, L, and R denote anterior, posterior, left and right. Because each cilium is tilted ~40 degrees from the perpendicular [1924], the fluid fails to move (because of surface adhesion) as the cilium swings close to the surface (recovery stoke), so it only flows (right to left) during the effective stroke [394]. The leftward flow is thus caused by (1) the clockwise spin of each cilium [773,2529], which is due to (2) the handedness of the tubulin and dynein proteins, which ultimately comes from (3) the chirality of their L-amino acids (*cf.* Fig. 2.2).

d. Gradually, "m" molecules intensify on the left [1898,2569].

e. Eventually, they activate the gene *nodal* [2257,2691], whose protein product (Nodal) then suffuses the area.

f. The heart and other organs develop asymmetrically [308,1201] because of Nodal's influence on their cellular processes [2119].

REFLECTIONS ON FIGURE 2.3

b. Single-celled protists called "ciliates" use cilia for propulsion and put more of these "oars" into the water than a Roman trireme. For them, ciliary chirality poses enough puzzles of polarity, asymmetry, and geometry to fill a book [851,1861], and Joseph Frankel has indeed written extensively about their looking-glass world [804–806].

f. Other classes of vertebrates use different tricks to create a *nodal* bias [1528]. Recent evidence indicates that protostomes may set up left–right asymmetry using *nodal* as well [1156]. The "Nodal Cilia" Model illustrated here is named for Hensen's "node" (*cf.* Fig. 3.2a) [166,569]. The fluid flow actually occurs in the posterior notochordal region [1164,1924], which is on the dorsal side in a human gastrula (not the ventral side depicted here for convenience). A variant of this model invokes mechanical (*vs.* chemical) signals to activate leftness genes [1705,2548,2885]. Evidence that seems inconsistent with both versions includes (1) the failure of *inv* mutant mice, which show 100% situs inversus [2882], to exhibit reversed flow [511,1923] and (2) a prior asymmetric bias in the expression of *Lefty1* (a component of the *nodal* circuit) in the mouse blastocyst [2557]. Other models are still being debated [2101].

because evolution has added symmetry-breaking gadgets to the *nodal* circuit in series or in parallel since the classes diverged from one another [1959,1976]. The developmental sequence of the contraptions in each class is now being ascertained [1528], after which it should be possible to deduce the evolutionary order in which they arose [1574].

Although the *nodal* circuit governs the left side of all the members of our phylum (the chordates) [427], it is found on the right side of our echinoderm relatives [680,681]. The astounding explanation, based on comparative genetics, is that chordates are anatomically upside-down relative to all other bilaterian phyla [1583]. That is, when our chordate ancestors diverged from their deuterostome relatives, they inverted their body (see Fig. 1.3e), like a swimmer flipping from breaststroke to backstroke upon reaching the end of the pool [778]. What used to be our left became our right and vice versa. In this sense, our entire phylum is quirky.

No mutations have ever been found that undo this inversion, but mutations do exist that reverse our viscera. About one in every 10,000 humans exhibits "situs inversus" [1499,2617] in which the heart shifts right instead of left, the colon arches counterclockwise, and all other asymmetries (spleen, liver, etc.) are reversed as well [1941]. Twenty-five percent of these "mirror-image" people suffer from Kartagener Syndrome [110]. This disease is also known as Primary Ciliary Dyskinesia because its symptoms arise from a primary defect in ciliary motion. Affected individuals also manifest bronchial congestion ($\approx$ smoker's cough) due to malfunctioning of the cilia in the lung. This peculiar syndrome confirms the key role of our cilia in establishing the asymmetry of our viscera during normal development.

Why do fingerprints violate mirror symmetry?

Despite the general impression that our body surface is symmetric, there are exceptions. One is fingerprints. If you compare the ridges on corresponding fingers of your left and right hands, you'll notice that their patterns are not the same. In theory, these discrepancies could be due to *nodal*, but in fact our fingerprints are largely beyond the control of our genome [754,916]. Proof of this assertion comes from the Dionne quintuplets, each of whom had unique fingerprints despite being genetically identical [1614]. The same is true for identical twins [1269,1551].

The ridges on our fingertips help us grip objects [307,453]. They are a legacy from our primate forebears [543], whose ability to grasp branches securely was literally a matter of life or death, given their risky acrobatics so high above the jungle floor [222,2077]. The added friction afforded by the ridges is further enhanced by the presence of sweat glands and the absence of oil glands [1848]. The feature of the ridges that is most critical is their spacing—not their contours, which merely must avoid aligning along potential slippage axes (parallel to the fingers). Hence, ridge patterns have been able to vary relatively freely "under the radar" of natural selection [916].

The most common types of swirl patterns are arches, loops, and whorls [543]. What creates them? The process may be the same as that which forms asymmetric whorls of hair on the back of the head [35,1338,1371,1394]. In both cases, the streams appear to be an emergent property of polarized skin cells [1291,2518,2747] that bind one another [703,1009]. Just as a herd of cattle is easily spooked by the least disturbance, the contours of these streams are susceptible to random perturbations that arise during development [1784]. Hence, the outcome on any given fingertip is unpredictable, and the final design is a "frozen accident" of development [2747]. In this sense, each of us bears a miniature painting of our prenatal history on each of our finger pads.

Why do blood vessels violate mirror symmetry?

Blood vessels are yet another trait that has evolved "under the radar" of natural selection [907]. (For an overview of stochastic indeterminacy, see [2781].) The smaller trees of arteries and veins vary greatly from person to person [187,884,1866,2828], and even the larger ones may occasionally meander, as Darwin noted in his *Descent of Man*:

> The chief arteries so frequently run in abnormal courses that it has been found useful for surgical purposes to calculate from 12,000 corpses how often each course prevails. [561] (Vol. 1, p. 108)

Notice, for example, that the veins on the undersides of your wrists branch differently, left versus right [1241]. Similar idiosyncrasies in the vessels of our eyes make it possible to use retinal scans for identification screening [1648,2061]. Despite these obvious vagaries at the small- and mid-scale levels, the gross layout of everyone's circulatory system is relatively constant [50,2032].

During embryonic development, each incipient network of vessels matches its blood supply to the needs of its target tissues by sprouting more than enough branches and then pruning the excess [2187,2227] in response to local cues such as oxygen tension [2202,2226,2645]. Ultimately, what matters is the volume of blood flow through the vascular tree, not the exact shape of the tree itself [2651].

Indeed, tubular branching processes in general rely more on ad-libbed local signals [1145,2744] than on hard-wired genetic blueprints [20,575,924,1734]. Other instances of the genome's laissez-faire attitude toward the routing of our plumbing include bronchiole branching in our lungs [1733] and ductal branching in our mammary glands [1511,1863,2199], salivary glands [2260], kidneys [577,2341,2755], prostate [669], and lymphatic system [1866,2828].

Pruning is also used as a fine-tuning trick throughout our nervous system [463,637,2829]. Successive stages of sprouting and culling dictate how batteries of neurons achieve parity relative not only to one another but also to their muscle targets [1343,1940]. Surprisingly, blood vessels and nerves use many of the same chemical cues to navigate throughout the body [709,1310,2772]. Hence, their paths actually coincide to a startling extent [149,2697,2896], despite the inherent vagaries of local conditions. Because nerves evolved before vessels [1830], this congruence must be due to a "Johnny-come-lately" vascular co-option of previously established neural guidance cues [377]. ("Co-option" entails a new function for an old feature [2631].)

Despite the fact that sprouting is messy and pruning is wasteful, this overall strategy is remarkably reliable [2519]. It is also ironically Darwinian because it uses a "survival of the fittest" filter to sift an initially random population [884]. The automatic ability of vessels and nerves to tailor their supply to the demand of their targets has meant that natural selection could alter virtually any body part (*e.g.*, lengthen our legs) without having to "wait" for other mutations to adjust the neural or vascular input every time it did so [885]. Exploratory processes like these have therefore made anatomy much more "evolvable" than anyone had ever guessed [759,1392,2716,2790]. They have also made our anatomy more variable from person to person (*e.g.*, muscle insertions) than most medical textbooks would ever lead one to presume [187,1866,2828].

One vascular asymmetry that all of us manifest is potentially lethal. Our abdominal aorta, which transports oxygenated blood to our legs, runs alongside the inferior vena cava, which drains deoxygenated blood from them [1866].

Each of these trunk lines branches to form a "Y" shape as it enters the legs, but the Ys are not superimposed congruently, one atop the other (*cf.* Fig. 6.1). Rather, they are offset, like two lovers lying side-by-side on their backs, one of whose legs (the right common iliac artery) is draped over a leg of the other (the left common iliac vein). As a result of this peculiar geometry, the left common iliac vein is more prone to compression and occlusion than its counterpart on the right, especially during the late stages of pregnancy when a fetus puts pressure on the juncture, squeezing it between the baby and the spine [196]. This asymmetry explains the otherwise enigmatic fact that pelvic vein clots occur *eight times more often* on the left than on the right [320,1683]!

How have animals evolved external asymmetries?

Although our chief asymmetries are hidden inside, this is not true of all other vertebrates. Many breeds of dogs and cats have variegated coats that are highly asymmetric (*e.g.*, dalmatians and tabbies) [2307], and the same is true for a few breeds of horses and cows (*e.g.*, pintos and Guernseys) [177]. Some of their patterns seem as random as cloud formations, and the chemistry that creates them may actually be governed by the same mathematical rules [125,942,1724,1836]. The prevalence of dappling among mammals suggests that humans could have also evolved leopard- or giraffe-like coloration [1177], and piebald people have indeed been documented throughout history [748,1502,1524]. In fact, *all* women would be as blotchy as calico cats (due to X inactivation) were it not for the fortuitous fact that there are no pigment genes on our X chromosome [1738] (*cf.* [1618]).

The random asymmetry of fur or skin color, however, is a far cry from the *directional* asymmetry of the viscera discussed earlier [992,1521,1787], in which the spleen is *consistently* on the left and the liver is *consistently* on the right [243,1975,1978]. Directional asymmetries do exist in the external anatomy of certain species, although they are rare [1974,1976,1977]. Three of the oddest cases are depicted in Figure 2.4 (narwhal tusks, owl ears, and flatfish eyes), along with what humans would look like if we were ever to evolve similar left–right disparities. We know virtually nothing about the genetic basis of external asymmetries [1980] (except for handedness in snail shells [97,586]), but it may be instructive to think about the issue in terms of what we do know about visceral asymmetry.

To recap the highlights of the earlier argument, we concluded that the left side of our body starts out the same as the right side. At an early stage, a lawn of ciliary propellers at our midline wafts a stream of chemicals leftward, and their rising concentration activates *nodal,* which then steers the fate of the left

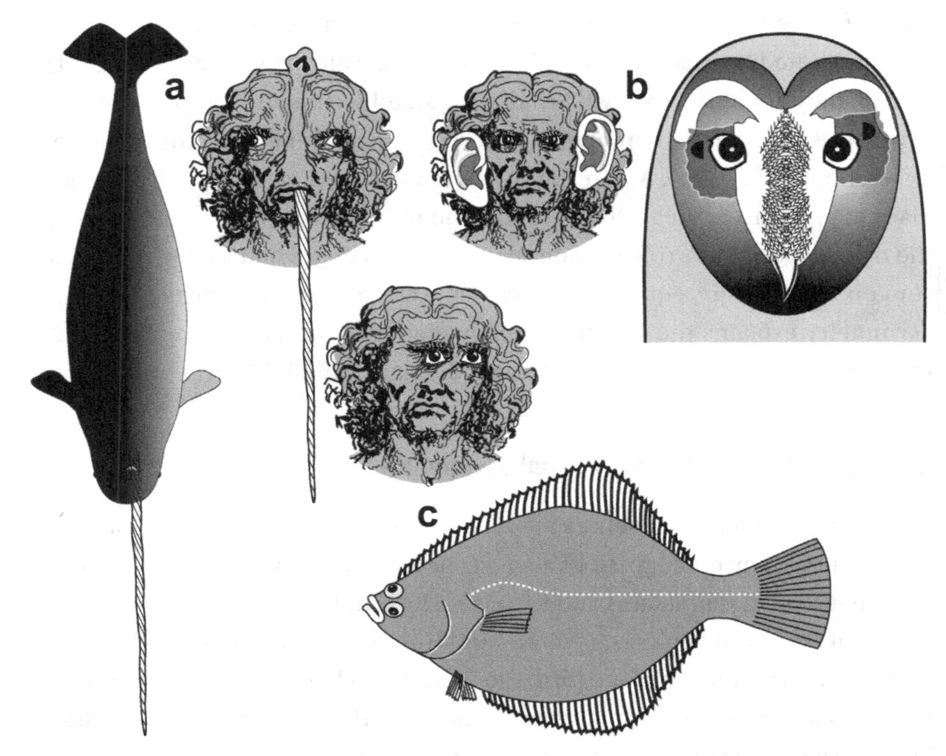

Figure 2.4. External asymmetries of three vertebrate species and what we would look like if we evolved like them.

a. The narwhal (a whale), *Monodon monoceros* [1876]. Male narwhals have a tusk up to 2.6 m long (possibly used for jousting [2399]) that is a modified upper left incisor. It is cone-shaped with helical grooves that always twist sinistrally [1388]. Rarely, the right incisor grows out to form a second tusk (not shown), and in every such case its grooves are also sinistral [2146]—a situation as strange as a man with two left hands (*cf.* Fig. 2.1d legend). Another asymmetry in narwhals (both sexes) is a blowhole left of center [1718,1987], which evolved from fused nostrils with differing roles: respiration (left) versus vocalization (right) [532,1155].

b. Face of the barn owl, *Tyto alba*, with most feathers removed to reveal preaural skin flaps (gray areas flanking the eyes) that cover the ear openings (D-shaped holes; adapted from ref. [1408]). The ears' height disparity allows localization of prey in a vertical plane [1893,2703]. The system is fine-tuned visually during adolescence [1409], and it is aided by the barn owl's having the longest cochlea of any bird [1421]. (Insects use other tricks for directional hearing [2194].) *N.B.:* Ear asymmetry occurs in a third of all owl species. In the great gray owl, the asymmetry is opposite (*i.e.*, right ear higher), and the shift is seen not just in the skin but also in the skull [1612].

c. Starry flounder, *Platichthys stellatus*, redrawn from ref. [2057]. Juvenile flatfish look normal, but then one eye migrates across the midline so that both eyes look up when the adult lies flat [2250]. The trajectory (left *vs.* right) varies among species and is independent of visceral asymmetry [1111]. Goldschmidt cited flatfish to support his "hopeful monster" idea, because it was admittedly hard to imagine any utility for a partial migration [930]. Nevertheless, fossils of just such missing links (early flatfish with partly shifted eyes) have now been found [822,1276]. Having both eyes on one side should make the fish dizzy when swimming [2315], but evolution also rewired the vestibular system to compensate [988].

REFLECTIONS ON FIGURE 2.4

All of the depicted traits are "directionally asymmetric" insofar as they appear consistently on one side only. In contrast, traits that are "randomly asymmetric" (a.k.a. "antisymmetric") appear as often on the left as on the right [243,929,1871,1976,1980] (*cf.* "fluctuating asymmetry" [992,1521,1787]). Another bizarre dental bias (aside from the narwhal) exists in a snake species that eats dextrally coiled snails: it has ~25 teeth on the right side of each jaw but only ~17 teeth on the left [2143]!

An oddity as baffling as any of those shown here (OK, excluding the Picasso fish) occurs in the long-legged fly, *Erebomyia exalloptera* [2242]. In males of this species, the right wing is always concave at its tip, whereas the left is always convex—the most extreme wing asymmetry of any flying animal, although the males can still fly! Why such a lopsided trait when symmetry is typically seen as ideal [996]? Consider that flies in general are notorious troubadours (second only to birds), who use their wings as musical instruments [2438]. *E. exalloptera* are no exception: the males serenade females by fanning their wings, so their odd scalloping might possibly be for retuning their love songs [898]. Other fly species have negligible wing asymmetries [1401], and repeated attempts to select for directional asymmetry in various traits have failed in *Drosophila* [706,1688,1978].

One bias that has been figured out occurs in manatees, which lack hindlimbs but retain pelvic remnants that are larger on the left side [2344]. They lost their hindlegs because of reduced Pitx1 (which sets hindlimb identity [2193]; *cf.* Fig. 4.7), but they still make Pitx2 on their left side (as part of the *nodal* circuit; see text). Thus, Pitx2 is evidently compensating for the loss of its Pitx1 paralog [2344] as in sticklebacks [174,175]. Darwin cited manatees in *Origin* (p. 454) because they have silly vestiges of fingernails on their flippers!

Other notable asymmetries, some of which were pondered by Darwin [1979], include unilateral lungs in snakes [16], unilateral ovaries in the platypus [966] and in birds [1028, 1254,2053,2082] (except kiwis for some unknown reason [969]), lopsided syrinxes (vocal cords) in birds [399,2532], twisted beaks in crossbills [180,1438], corkscrew mega-penises in waterfowl [273], skewed nasal passages in whales [188,1335,2085], helical partitions ("spiral valves") in shark intestines [16,2417], analogous partitions in the rabbit appendix [2806], asymmetric claws in crustaceans [985,2004,2871], twisted abdomens in hermit crabs [281,1107,1264,1712] (a trait utterly undone by king crabs [544]), and gnarled genitals in male water bugs [1236], not to mention another whole world of spiral and helical traits in plants [674,1112,1301,2592] (*cf.* reviews [510,1871,1974,2024]). Perhaps the strangest asymmetry ever described is a flatworm larva whose eyes rely on different types of photoreceptors—rhabdomeric type for the right eye versus a combination of rhabdomeric and ciliary types for the left one (*cf.* Ch. 6) [695].

Aristotle was especially intrigued with claw asymmetries {PoA:4:8:684a26*ff.*} [1941,1975], which are common in fiddler crabs [2871,2902], hermit crabs [1107,1264,1712], and shrimps [1725,2004]. In lobsters, as he noted, either the left or the right claw can dominate. Which claw becomes the larger crusher (*vs.* smaller cutter) is fixed before adulthood based on usage [987,1564]: both claws become cutters if they are underused, but an unknown neural circuit prevents them both from becoming crushers [985,986]. The crusher–cutter asymmetry in crustaceans is eerily similar to our molar–incisor dichotomy, and the convergence gets even spookier in the molarlike "teeth" that line their crusher. Both gadgets (claws and jaws) are lovely illustrations of physical constraints because they evolved as simple levers and are subject to the same formulae as scissors and pliers.

The only asymmetry that has been investigated genetically to any useful extent is helical coiling in snail shells [97,1156,2297]. Dextral shells are the norm and sinistral ones are rare in most gastropod species [976,984], although a few whole species are sinistral [97,2196,2297].

REFLECTIONS ON FIGURE 2.4 (*continued*)

Why dextrality prevails is unknown [538,850,1305,2684]. Mating between mirror-image snails can occur [102,1210], and chirality turns out to be due to a single gene that acts before fertilization [586,1031,2196,2520,2658]! How the maternal gene product (a cytoplasmic RNA or protein) biases the spiral cleavage of the zygote [813,2363,2741] is unclear [1085,1156].

Our annoying inability to use both hands with equal grace can be blamed on our hemispheric lateralization [79,369,872]. Brain lateralities of some kind [79,261,574,872,920] must also dictate left–right behavioral preferences in fish [603,1203,1554,1714], frogs [207,1630,1845], snakes [2142], birds [659,933,1229,2666], rats [1474], humans [112,519,520,750,2879], pre-human hominins [856,914], and nonhuman primates [361,1068,1117,1517,1803]. Indeed, lateralized brain function has even been found in honeybees [1526]! It remains a mystery [369,790, 1945,2001].

Charles Dodgson (a.k.a. Lewis Carroll)'s fascination with asymmetry, which is so apparent in his book *Through the Looking-Glass*, may have been sparked by his own asymmetric peculiarities: one shoulder was higher, his eyes were at different levels, and his smile was crooked [380].

side differently through Pitx2, culminating in the asymmetries of our viscera. We also noted that the use of *nodal* to distinguish the two sides of the body (left–right, or L/R) from one another predates the divergence of chordates (our phylum) and echinoderms (sea urchins, etc.) ~500 MYA.

The great antiquity of the *nodal* L/R switch means that it has been around a lot longer than some of the organs that it now controls. Our lungs are a good example, because they evolved in our fish ancestors ~400 MYA (*cf.* Ch. 6) [1498]. How did the number of lobes in our left lung come to be ruled by *nodal* and, more immediately, by its subordinate Pitx2?

Our lungs start as a balloonlike outgrowth of the throat that splits into two bronchial buds [1540]. The left bud forms two branches, the right bud forms three, and each branch then goes on to form a separate lobe [256]. Interestingly, the growth of the left bud lags behind the right one [1430,1998], so one plausible scenario is that *nodal* gained dominion over the *growth* of lung tissue, rather than the number of lobes per se. By slowing down the left bud, it could have reduced the number of its branches, assuming that they arise at regular time intervals (or size thresholds) during a finite period [1181,1733,2748]. At some point in the ancestry of our species, therefore, a mutation may have caused *nodal* to inhibit a growth gene in the lung (*cf.* Fig. 4.2), and that gene henceforth limited the rate of bronchial bud growth on the left. If having only two lobes on the left instead of three proved helpful (by making more room for our heart? [1181]), then individuals carrying this mutation would have survived better and produced more offspring. Eventually, the mutation would have spread through our whole species.

Using similar logic, it is easy to see how differential growth rates along the left versus right side of the gut could have caused the twists and turns that occur at certain points in gut development [1204,1826]. The initial handedness of that folding has recently been traced to *nodal*-dependent differences in cell shapes within the dorsal mesentery [582]. Indeed, the use of *nodal* to fold the gut neatly as it lengthened evolutionarily was probably a safer alternative than letting it meander randomly, because haphazard looping can tie the gut in knots, leading to lethal sigmoid volvulus [459,1801]. Here, finally, we may have a partial answer to the riddle raised at the outset of this chapter—namely, why our viscera look so orderly despite being so asymmetric.

What this mental exercise has shown is that *any* gene could have come under the influence of *nodal*, including, perhaps, one that affects *external* anatomy. Having one leg shorter than the other would have obviously been maladaptive, as would most lopsided traits we might think of, not to mention that finding a mate might be hard because symmetry is a key aspect of beauty [996,1558,2165,2601]. The conclusion, in short, is that humans *could* have evolved external asymmetries (and might still do so someday), but whatever change arose would have to enhance—not impede—our chance of survival. The details of how genes such as *nodal* enslave anatomy-affecting genes during evolution are explored later (*cf.* Ch. 4).

Finally, it is worth mentioning a model organism in which the etiology of external asymmetry should be approachable experimentally. The protochordate amphioxus is asymmetric as a larva but symmetric as an adult—the converse of the flatfish life-history story [2250] (Fig. 2.4). As a tadpole, its mouth is on the left, and it has a single row of gill slits on the right [1990]. During metamorphosis, these asymmetries disappear [632]. What makes this transformation even more intriguing is its relevance to one of the greatest mysteries in evo-devo—namely, how the body flipped upside-down near the base of the chordate clade (*cf.* Fig. 1.3) [1463]. It is conceivable that the migration of the mouth during amphioxus metamorphosis recapitulates the inversion itself [1464]. The recent sequencing of the amphioxus genome [874,2100] gives us a new playground where we can explore our ancient past. The fun is just beginning!

CHAPTER 3

Mysteries of the Midline

If you move one arm, the other does not automatically follow suit. In contrast, our eyes are yoked together whenever we shift our gaze [767,1562]. Eye coordination makes sense because seeing objects from two points of view lets us compute distances based on parallax (Fig. 3.1) [1215,2065,2783]. Depth perception was critical for our primate ancestors, who cavorted from branch to branch high in the jungle canopy [1354,2515]. The fact that our eyes blink in sync, however, makes no sense because it makes us blind, albeit for only an instant [2824].

Our lenses refract photons from each point in visual space onto the "nasal" half of one retina (near the nose) and the "temporal" half of the other (near the temple). For us to compare the images from our left and right eyes, nasal axons cross the midline and join temporal axons from the other eye before being bundled and sent to higher centers for stereoscopic analysis [59,503]. It is astounding that we do not perceive even the slightest hint of a seam where our visual field is constantly being split right down the middle [647].

Why do optic nerves cross the midline?

Retinal axons cross at a juncture called the "chiasm," which is named for the X shape of the Greek letter *chi* [728,2061]. Rare mutations have been found in humans that obliterate crossing completely [1150,2266], and, as expected, the affected individuals have negligible depth perception [289].

A more common type of miswiring is seen in albinos [1027]. Their depth perception is impaired [1507] because of *excessive* (>50%) crossing of axons at the chiasm [1282,2306]. Albino humans can compensate for this error (at least partially) by crossing their eyes [67], and albino mammals in general are notoriously cross-eyed [1026,2465]. The mutations that cause albinism entail defects in an enzyme called "tyrosinase" that is critical for pigment synthesis [1911,2614].

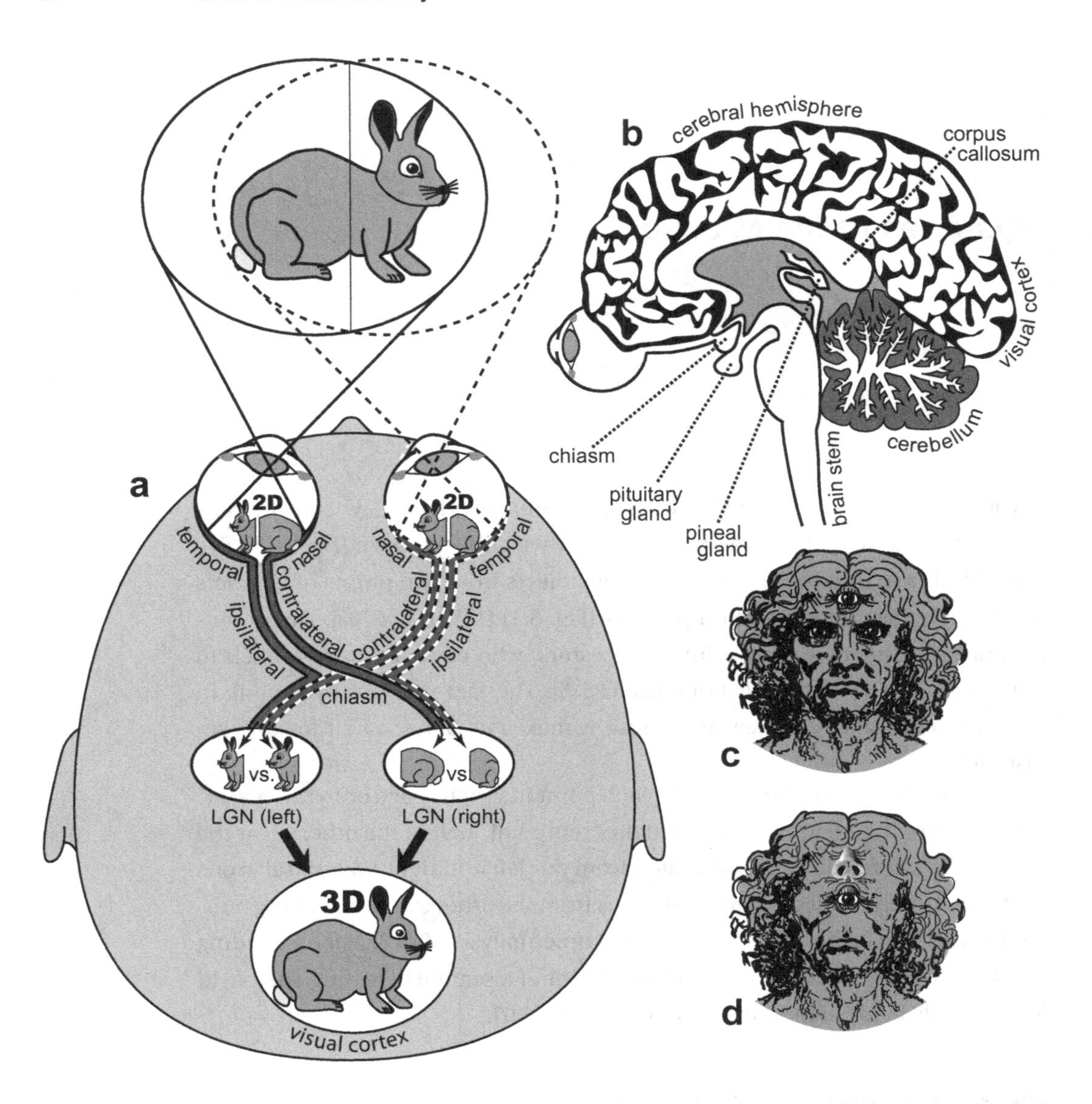

Why a primary loss of pigment should cause a secondary misrouting of optic axons is unclear [1284,1387,2464], and the etiology of this peculiar pleiotropy is currently under investigation [1022,1495,1833].

Given that overlapping visual fields like ours involve 50% crossing, we might naïvely expect walleyed animals like fish to have 0% crossing, because their eyes have no field overlap whatsoever. On the contrary, fish—and, indeed, most vertebrates aside from primates—exhibit 100% crossing [1283]! Why should the ancestral condition have been a crossed ("contralateral") projection [999] when a same-side ("ipsilateral") path would have been patently shorter and simpler [950]? The answer appears to be that the visual system was merely obeying a rule that already governed the entire vertebrate central nervous

Figure 3.1. How our eyes perceive depth, and how we might look if we reverted to a 3-eyed (reptile) or 1-eyed (protochordate) stage of evolution.

a. Our eyes point frontally, and their fields overlap extensively. When we look at something like a rabbit, each eye sees it from a distinct angle. With the rabbit nearby and facing right, our left eye sees more of its rump, while our right eye sees more of its face (although the disparities are grossly exaggerated). Our lenses focus the inverted images onto our retinas, and each image is then bisected. "Nasal" axons (those near the nose) cross over (at the chiasm) to the lateral geniculate nucleus (LGN) on the contralateral (opposite) side, whereas "temporal" axons (those near the temple) go to the LGN on the ipsilateral (same) side. The half-images are stitched back together in our visual cortex [691], and depths are computed from differences between left and right images [2651,2783]. The farther away the rabbit, the more we rely on parallax disparities of the background (*vs.* those of the rabbit itself) [1037]. It is a testament to the virtuosity of our visual system that we are oblivious to our visual field being cut in half [6,647].

b. Sagittal section showing the visual pathway and pineal gland relative to other brain structures.

c. What we might look like if our pineal gland regained its ability to form a third eye, assuming it could somehow grow through our cortex and skull (*cf.* [694]). Our fish ancestors may have actually had *two* such eyes at the midline [496]. The extra eye should function adequately because surgically implanted third eyes can weave their nerves into existing pathways [504,1895]. How a median eye might steer its axons at the chiasm, however, is unclear [752].

d. Holoprosencephaly (cyclops) syndrome in its most extreme manifestation. The nose lies above the eye, perhaps because the neural crest cells that normally migrate down between the eyes to make the nose can't do so because they're blocked by the median eye [2573,2811]. However, the culprit could instead be (1) absence of instructive signals from the underlying (abnormal) brain [712] or (2) blockage of other (permissive?) tissue shifts (T. C. Lacalli, personal communication).

---

system [2278]. As for why that crisscross rule arose in the first place, some possible rationales are offered in the next section.

The first hints of ipsilateral routing evolved in primitive mammals [1147] in environments where some degree of binocular vision proved advantageous [710,2135,2223]. This neuronal rewiring could have been implemented by a simple change in the cell-surface receptors that help to steer growing axons [1473]. In our own primate lineage, the extent of the ipsilateral projection increased as our eyes converged to an even more frontal gaze [1147,1476,2135]. However, these two shifts—in eye angle and axon routing—were probably ratcheted by *separate* stepwise mutations [1207] because eye angle does not dictate axon pathfinding in development, nor does the latter cause the former [2025].

Why do pre-motor nerves cross the midline?

Contralateral wiring turns out to be an integral feature of vertebrate nervous systems in general [2708]. As shown most starkly by human stroke victims [790], the left side of our brain controls the right side of our body and vice versa [120]. This diagonal control is mediated by a crossing of corticospinal nerves in the brain stem [1674,2832]. Why do we possess what in essence is a chiasm

REFLECTIONS ON FIGURE 3.1

a. Neural yoking devices force our eyes to (1) move together [767,1562], (2) blink together [2824], and even (3) constrict their pupils together [1588,2671,2810], although only the first reflex has any obvious utility—*viz.* the aiming of both foveae on the same objects for maximal acuity [2287] and depth perception [1147,2026,2452].

The cleavage plane that splits our visual field is shifted in albinos because of a defective tyrosinase enzyme that misroutes axons at the chiasm [1027,1282]. To compensate, albinos cross their eyes [67]. A similar syndrome affects Siamese cats [1323,2306] due to a milder mutation in the same gene [1912]. Interestingly, their black nose, ears, paws, and tail are also due to the disabled tyrosinase, which only makes pigment at the colder tips [2447]. Amazingly, a "Siamese cat" mutation has even been found in humans [2614].

Our eyes arise on the sides of our head like a fish and gradually move frontally [1998], an obvious validation of the largely discredited dictum that "ontogeny recapitulates phylogeny" [956]. At 6 weeks postconception, they point 160 degrees relative to one another (no binocular overlap). By 7 weeks, the interocular angle is ~120 degrees, and by 10 weeks, it is down to ~70 degrees, nearly reaching the adult angle of ~60 degrees [1998].

In contrast to vertebrates, arthropods only process their visual inputs ipsilaterally, a strategy more sparing of axon length. Nevertheless, they have their own quirky (unilateral) chiasms [312,2681]: the visual image gets inverted twice on its way back into the brain (180 degrees from lamina to medulla and another 180 degrees from medulla to lobula), thus restoring its original orientation [1946,1948]—a seemingly pointless exercise [1723].

b. Expression of *Pax6* in the mouse pineal gland is further evidence for its homology with the lateral eyes [2030,2740]. The pituitary is another midline gland with at least as many plot twists in its history as the pineal [599], if not more of them [420,1350]. During human development, its front and rear halves fuse from separate parts that travel weird paths [1348,2192], and these parts are traceable to separate precursors in our protochordate [356,1191,1637,2361] and bilaterian [2588] ancestors.

c. In hindsight, it might have been better if we had evolved the proverbial extra "eye(s) in the back of our head" [975]. Regardless of whether the pineal rudiment becomes an eye or a gland, it arises like an optic vesicle (*cf.* Fig. 6.2) insofar as it starts as a diverticulum from the diencephalon [376,539,1801,2278]. Because the lenses of the lateral eyes come from invaginations of head ectoderm [1084], one might expect the lenses of pineal eyes to arise likewise, but they actually develop within the vesicle wall itself (in reptiles) [694]. This distinction may not mean much, however, because lateral eyes in various vertebrates can regenerate a lens (from the iris! [1020,1125,1126]) through an *intra*planar route [1273,2350,2639].

d. The proboscis is actually more tubular than depicted here [485,486], presumably because its shape depends on signals from the underlying forebrain [700,712,1217,1642], which itself is deformed in the syndrome [2811]. Cyclopic babies typically do not survive beyond a week after birth [140,1047]. Cyclopia is also seen in the fused heads of conjoined twins (*cf.* Fig. 3.2g).

writ large? It serves no obvious function—either anatomically or physiologically [120]—but it may have proven handy long ago in our pre-fish ancestors.

One idea is that crisscrossing arose as part of a "coiling" reflex [1340]. Aversive stimuli detected on one side of the body—say, a predator's shadow—would have instantly caused the muscles on the opposite side to contract [120], thereby turning the body away from potential harm (*cf.* Fig. 6.2) [368]. Another theory holds that diagonal wiring evolved in conjunction with undulatory locomotion to coordinate sinusoidal swimming [2278]. We don't know which idea, if either, is correct.

Axonal crossing at the midline is regulated by a "traffic signal" that must date back ~600 MYA to the Urbilaterians [893,1571,2106] because it is shared by animals as diverse as humans [1288], nematodes [828], insects [2919], and flatworms [405]. All of these groups use the gatekeeper molecule "Slit" and its axonal receptor "Robo" to (1) steer axons toward the midline, (2) allow them to cross once, but (3) prevent any backtracking by henceforth repelling them [192,658]. In mice, where the details of the mechanism have now been worked out, the switch from attraction to repulsion involves differential splicing of Robo isoforms before (Robo3.1) and after (Robo3.2) crossing [212,433]. In humans, defects in the Slit-Robo gadgetry may be responsible for dyslexia [1080], a reading disorder that affects ~1 out of every 10 people [2842].

Why do symmetric organs fuse at the midline?

A number of our body parts arise separately during development as bilateral halves that then fuse together along the midline. Among them are our heart [2453], palate [1017], sternum [1560], uterus [91], urethra [2332], and much of our face [485]. From an engineering standpoint, it would have been much safer to sculpt such shapes *in situ* because assembling them piecemeal is risky. To wit, the parts must be in the right place at the right time in the right state of readiness for fusion to proceed successfully. Failures present themselves clinically as cardia bifida [34], cleft palate [2371], cleft sternum [3], duplex uterus [91], hypospadias [459,1780], and a variety of facial dysplasias [485,1319,2656], including cleft nose [91,2479].

Bilateral fusions are also critical for the assembly of our gross anatomy. They close the neural tube along our dorsal midline (*cf.* Fig. 6.2) [514] and the abdominal wall along our ventral midline [278]. If mistakes occur during these large-scale mergers, then the clinical consequences are profound—namely, anencephaly [1433], spina bifida [685,1965], and umbilical herniations [91,1644].

The explanation for at least some of these fusions appears to be that they form *tubes*, and tubes tend to be built by the invagination (*e.g.*, rolling) of cell sheets [1182,2107], rather than by the (safer?) cavitation of solid cylinders [1447,1586]. Wherever such an invagination occurs, fusion of the two epithelia must follow in order to (1) separate the tube from its generative layer and (2) seal the zone of contact as seamlessly as possible.

Here we encounter an "Origami Epiphany." We tend to think of ourselves as sculpted statues, but we are really more like rubber sheets that have been folded to make a hollow shell plus an internal labyrinth of zigzag pipes [576]. Traces of this 2D-to-3D strategy are evident not just in our tubes but also in some structures that form spheres—for example, our inner ear [2256] and our lens [548]. Even our brain, which looks like a solid cauliflower, is actually a crinkled balloon that has risen like dough to resemble a rumpled baguette [2113].

This reliance on folding instead of sculpting turns out to be widespread across metazoans [576] and may have arisen in Urbilaterians [341,2651]. Conceivably, we still use this antiquated approach because it is so ingrained in our genome that it has stymied any mutations that could have launched us in a different direction. In Waddington's lingo, folding and fusion would be "canalized" traits [2262]. The canalization cannot be *total*, however, because it was overcome in the case of gastrulation [427,605]. To wit, gut morphogenesis was drastically reconfigured in reptiles (ingression) relative to their amphibian forebears (invagination) [87,2408], and we have retained the reptilian mode ever since.

From the standpoint of development, the fusion strategy raises deep mechanistic issues [576]. Topologically, the process is as eerie as clapping your hands together, only to discover that they will not come apart because your palms have vanished, leaving only a seamless sheath of skin around both hands. How two epithelia become one is best understood for the palate [1855], where both cell migration and cell death are known to mediate fusion in response to the growth factor TGFβ3 [29]. We are still left to wonder how the "Merge with me?" overtures and the "OK, let's do it!" responses evolved in the various organs where fusions occur.

A separate circuit appears to govern fusions where epithelia join side by side in wound healing (*cf.* eyelid fusion [2893]), rather than apex to apex as in the cases noted earlier [1272,1668]. One of the latest surprises to emerge from evo-devo research is that wounds are repaired by the very same genetic gadgetry in mice as in flies [1089,1817,2508,2606].

Where did the neural crest come from?

Closure of the neural tube differs from fusions elsewhere in the body insofar as it unleashes a swarm of cells that disperse throughout the embryo [1099,1405]. These "neural crest" cells form nerves [1219] plus a hodgepodge of non-neural tissues [671,1486], including jaw cartilage [1293], teeth [1002], and pigment cells [1235,2868]. Indeed, the neural crest plays so many key roles that it is deemed to be a fourth germ layer [1055]. Moreover, it has been credited with jumpstarting vertebrate evolution ~500 MYA [286,634,1900] and with empowering our subsequent adaptive radiation [884].

Embryologists are so accustomed to the crest's wild antics that we lecture on this topic as if it were no big deal. On the contrary, any architect presented with such a building plan would laugh you out of his or her office. How did such a wacky innovation arise? Antecedents of the crest have been found in pre-vertebrate chordates [1286] but not in other phyla that are our nearest relatives (other deuterostomes) [2500]. In one of those pre-vertebrates—the urochordate *E. turbinata*—certain cells scatter from the neural tube to become pigment cells [1287], and in another—the cephalochordate amphioxus—sheets of cells migrate from the edges of the neural plate toward the midline and form epidermis [122]. Thus, the "migrant worker" strategy predates the origin of the vertebrates, but the itinerants were not nearly as versatile as their vertebrate counterparts.

The pithier question, therefore, is: how did these "proto-crest" cells acquire so many more duties in vertebrates [1900]? One idea is that the neural border in basal chordates was primed—or "preadapted" [442]—for further roles [1191,2302] because amphioxus expresses many crest-specific genes there already [951]. Border cells could have been released from the epithelium through subsequent mutations that allowed delamination, migration, and differentiation [1356,1952,2365]. They may have gained access to diverse histological pathways (*e.g.*, cartilage formation) by co-opting existing genetic circuits (*cf.* Fig. 4.2) [1735,2282].

The ability of the liberated gypsy cells to reshape anatomy (*e.g.*, teeth [1414,2427]) stemmed from a rainbow palette of new cell types [2687] that began to emerge as these nomads started to gossip with their newfound neighbors wherever they settled [671,1393,1501]. Ultimately, the migrants' responses (differentiated states) to the provincial dialects (paracrine signals) must have been fine-tuned by a pervasive rewiring of the genome [671,2366,2710].

To trace the genealogy of our crest, we need to learn a lot more about the urochordates [1060,1516] because they turn out to be closer to us phyletically than the cephalochordates [627,874,1192]. The quest for the crest's ancestry remains a lively subfield of evo-devo research [141].

How did our third eye become our pineal gland?

Deep inside our brain, at the midline, is a gland that used to function as a third eye in our vertebrate ancestors (Fig. 3.1b) [59,694,1114]. A *third* eye? Yes!

Our pineal gland controls our sleep–wake cycle by secreting the hormone melatonin [1380,1925], but it evolved from an eye that can still be found in certain fish [2454], amphibians [665], and reptiles [2618]. In those species, the median eye typically resides in a hole in the skull and uses the same kinds of photoreceptors as the lateral eyes [2521], although it is smaller and immovable. In lizards it serves as a solar compass [810].

The evolutionary transformation from an eye to a gland can be detected anatomically as far back as turtles and snakes [2278]. As the pineal organ submerged over the eons, its window in the roof of the skull closed, and it lost its ability to sense light–dark cycles directly. Instead, it came to depend on indirect neural inputs from the lateral eyes [1029]. Meanwhile, its cells transformed histologically from a photoreceptor cytoarchitecture to an endocrine configuration [711]. At the same time, their secretory output switched from a neurotransmitter to a hormone [1396,2115].

Intriguingly, the hormone melatonin is made from the neurotransmitter serotonin [203], and the two enzymes that are needed for the conversion had actually evolved much earlier in the cephalochordates [1396]. Hence, the third eye was already preadapted physiologically to becoming a gland. We do not yet know (1) how the release of melatonin came to depend on day–night rhythms [1426,1727] nor (2) how the hormone eventually entrained an apt set of target tissues to respond accordingly [2704].

Amazingly, the immature pineal gland of a frog can be spurred to grow as big as an eye—and to resemble one anatomically—by merely (1) transplanting it from a younger to an older tadpole [1275] or (2) removing the lateral eyes [1274]. Evidently, the pineal Sleeping Beauty is only a princely kiss away from awakening to assume its former role. If a series of mutations were to someday turn our pineal back into an eye, then we would have an extra eye buried in the center of our head. For that eye to resume its rightful place on our crown, however, it would have to somehow penetrate the overlying mass of cerebral cortex, poke its way through our skull to reach the surface, and then part the skin to peek out (Fig. 3.1c). Don't bet on it!

Are human cyclopes evolutionary throwbacks?

Nevertheless, people are sometimes born with an eye in the middle of their face (Fig. 3.1d). This median eye does not develop from the pineal gland but rather comes from a fusion of the two normal eyes. They are brought together whenever the intervening tissue fails to form. Midline tissue is extensively missing in "holoprosencephaly" (HPE) syndrome [488]. This syndrome manifests cyclopia as its most extreme symptom, but it can be so mild as to only cause a median incisor in the upper jaw [857,2811]. It is named for a midline fusion of the prosencephalon (cerebral hemispheres) [1389] and is seen in as many as one out of every ~250 fetuses from induced abortions [625,2372]. Cyclopic babies usually die within a week after birth because of other midline problems with the hypothalamus, pituitary, or brain stem [140,1047].

The earliest known vertebrates had a head with paired eyes [1454], but the "eyes" of their protochordate forebears were situated along the midline [1460]. Urochordates have a single eyespot in the larva [2188,2368], whereas cephalochordates have four kinds of light detectors [1461,2800]. At best, such lensless organs perceive only shadows but cannot form images. How one or more of these devices became the vertebrate eye is an unsolved riddle [1463].

The midline location of these primitive light-sensing organs suggests an affiliation with our pineal rather than with our lateral eyes, but the latter link is equally plausible because vertebrate retinas actually begin development as a single, median eye field [720]. This field later splits in half when it receives signals from the underlying cell layer [2215]. Artificially removing that layer (by surgery) or genetically blocking its signal (by HPE mutations) causes cyclopia [215].

If the lateral eyes of vertebrates actually did evolve from a median eyespot by splitting and subsequent separation [1463], then the cyclopic eye of mutant humans might indeed be an "atavism" (an evolutionary throwback) [2811,2831]. Further speculation along these lines is moot, however, because (1) cyclopia only restores one aspect of the putative ancestral condition [487], and (2) the gulf in complexity between our eyes and those ancient sensors makes any congruence hard to assess [1079,2500].

Why do conjoined twins tend to be symmetric?

One out of every ~100,000 births results in conjoined twins [2440]. Most pairs are remarkably symmetrical. They look as if someone got stuck while trying to enter a mirror, but, to continue the analogy, there is a perplexing spectrum in (1) the *angle* at which they entered the mirror and (2) the *extent* to which they traversed it before getting stuck (Fig. 3.2). A similar spectrum of symmetric twinning occurs in other vertebrates as well [1345].

Conjoined twins are monozygotic, so they must start life as a single embryo that splits in two [2440]. It would thus seem reasonable a priori to suppose that the anomaly is due to incomplete splitting [1345,1616]. However, that idea fails to account for (1) why heads can meet at skewed angles, (2) why "Janus" faces can form perpendicular to the body axis, or (3) why certain contact sites predominate (Fig. 3.2) [2435]. These trends are more easily explained by secondary fusions *after* initial splitting and total separation [231].

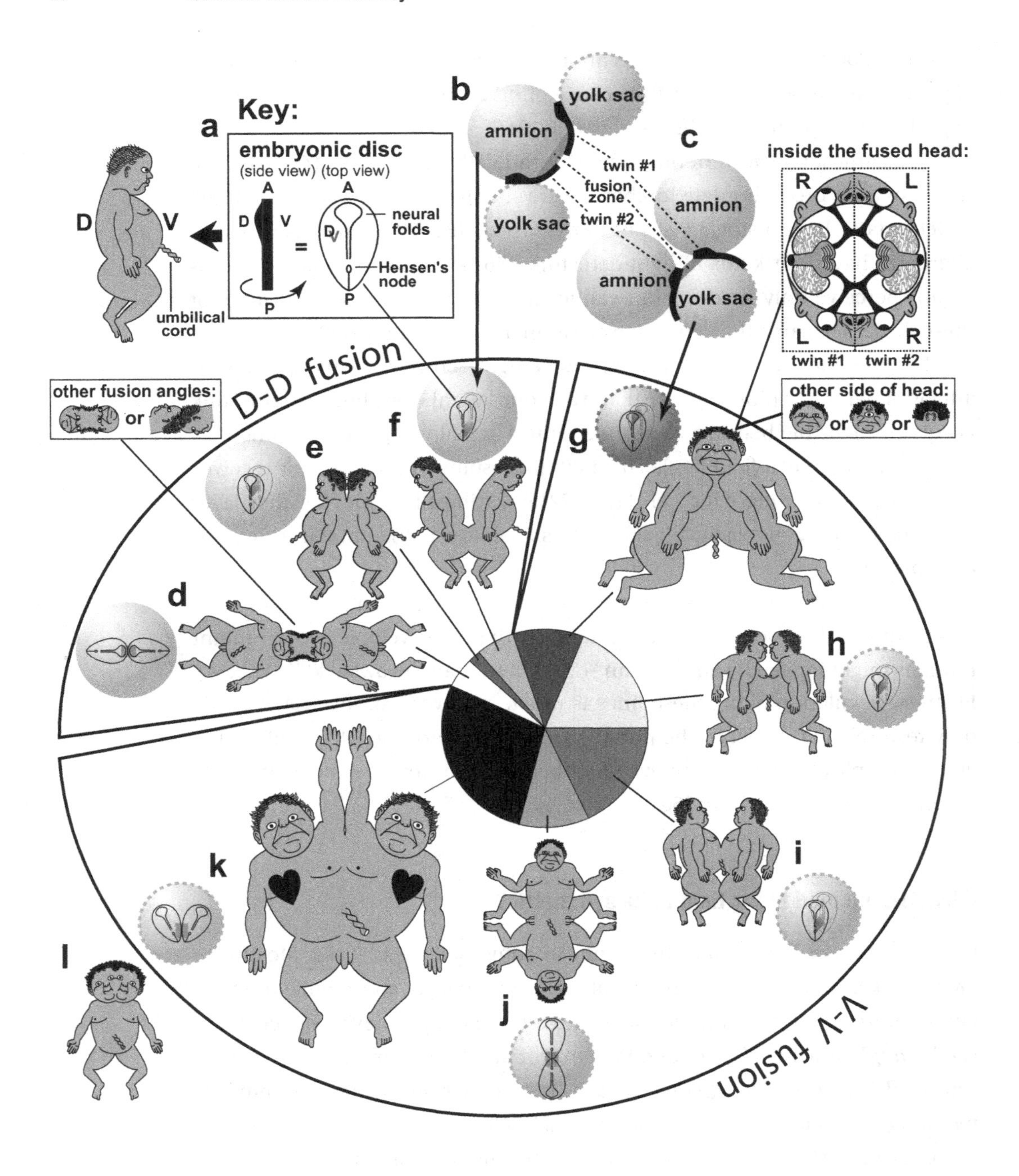

If conjoined twins arise when identical twin embryos bump into one another, then why can't collisions happen at *any* angle and produce fusion geometries that are markedly *asymmetric*? Collisions may indeed be random at first [1036], but any resulting disparities are evidently eliminated by death of the mismatched tissues or by the stunting of their growth [1616,1771].

Figure 3.2. Extra planes of symmetry in conjoined twins [2435].

a. Axes of a normal embryo (A, anterior; P, posterior; D, dorsal; V, ventral). The human "embryonic disc" stage (~19 days after fertilization [1801]) is shaped like a guitar pick, as shown at the right, where the disc is seen D-side up (V-side dimmed below). The neural folds flanking the anterior neuropore (≈ keyhole) will later meet at the midline to seal the roof of our brain [91]. Hensen's node is where superficial cells went inside (to form the notochord) during the previous gastrulation stage. In the sketch at the left, the disc is drawn in side view as a bar that is swollen at the head end to denote the elevated neural folds there [1801,2311]. (The baby alongside is provided to clarify the D-V polarity.) This "swollen-bar" icon is used in **b** and **c** to represent the tail-to-tail (**b**) or head-to-head (**c**) orientation of adjacent twin embryos.

b, c. Inferred geometries of fusing discs [2433]. Each disc inflates two balloons made of spare embryonic tissues: an amnion dorsally and a yolk sac ventrally [91]. Depending on how the original zygote splits, the twins can either fuse dorsally ("D-D," **d–f**) across a shared amnion (**b**) or ventrally ("V-V," **g–k**) across a shared yolk sac (**c**) [1339], although in the latter case, the amnions must also fuse at a later time [1345]. Fusions are depicted as gray bands between the two embryos (black bars). There are eight categories of conjoined geometries, which occur at different frequencies (pie chart below; parasitic twins excluded) [2435].

d–f. Twins with two umbilical cords (13% total; yolk sacs omitted). **d.** Head fusions due to neural fold contact. Heads are often twisted at odd angles (inset), an asymmetry that rules out fission alone as the basis for joining. **e.** Back fusions due to medial contact. **f.** Sacral fusions due to posterior contact.

g–l. Twins with one umbilical cord (87% total; amnions omitted) [278]. **g.** Frontal fusions due to contact of oral membranes [1152,1616]. The head has two faces that point sideways [1151,2436]. Structures often vanish at the fusion plane (small inset) [1616], resulting in cyclopia or fused ears [1345,2411]. Each face comes from the left half of one twin and the right half of the other, and the optic nerves detour to form inter-twin chiasms (large inset) [2688]. Darwin was fascinated with skewed fusions (small inset): "In one instance of two heads united almost face to face, but a little obliquely, four ears were developed, and on one side a perfect face, which was manifestly formed by the fusion of two half-faces" [560] (Vol. 2, p. 333). **h.** Thoracic fusions, in which the heart can have as many as seven chambers [911]. **i.** Abdominal fusions [1035], the most famous case of which were the original "Siamese" twins Chang and Eng Bunker (1811–1874) [678]. **j.** Posterior fusions, in which genitalia are diverted laterally like the half-faces in **g**. **k.** Side-to-side fusions with loss of tissue at the fusion plane, resulting in one pair of legs. The right twin often has situs inversus (heart on wrong side, etc.) due to trans-twin leakage of signals that stifle the *nodal* gene (*cf.* Fig. 2.3) [1529,2884]. Walking is difficult because of separate brain control [678,1751]. This Y-shaped anatomy (2 heads, 1 rear) can be artificially induced at high frequency (80%) in frogs by centrifuging fertilized eggs [214]. **l.** An anomaly not included in the pie chart is a single body with two heads (same etiology as **k**?) [348,911]. The heads can be fused, as sketched here for Lali Singh, a girl born in India in March 2008 (*cf.* Wikipedia: Diprosopus), or they can be separate, as for Syafirtri, a girl born in Indonesia in August 2006 (not shown; *cf.* Wikipedia: Polycephaly). Lali, who died two months after birth, was able to feed from either of her two mouths, and when she blinked, all four eyes blinked together.

The two-headed animals that grab the media spotlight from time to time presumably develop in this way (*cf.* [348,911,1095]).

A comparable phenomenon has been studied in *Drosophila*, where randomly induced cell death causes the fly's two forelegs to fuse. The resulting "mermaid" leg is missing varying amounts of tissue at the midline, but it is invariably mirror-symmetric [2069]. Interestingly, this tendency for global (organ) symmetry to emerge from local (cellular) disparities may also explain Bateson's Rule [1135], which governs the symmetry planes of branched (≈ conjoined)

REFLECTIONS ON FIGURE 3.2

Given that conjoined twinning has so little genetic basis [1991], any of us could have inadvertently sculpted a doppelgänger from part of our own body. Indeed, all vertebrate embryos are as malleable as putty at early stages [91,570]. This reality actually intruded into my brother's family. His son Zachary had an identical twin who dueled with him *in utero* in a "Twin-Reversed Arterial Perfusion" Syndrome [2335]. When twins share a single placenta, one typically perfuses the other with venous blood [1073], dooming it to death at birth. Zack survived; his twin died. Indeed, half of such contests kill *both* twins [1631], so Zack was lucky to have survived at all.

We are used to thinking of identical twins as being *genetically* identical, but this is certainly not true for females. Female mammals inactivate one X chromosome per cell during embryogenesis and hence are patchworks of paternal-X and maternal-X territories [1738]. Because inactivation occurs randomly (after the twinning stage), twin girls are unlikely to have same pattern of patches. Other epigenetic changes occur stochastically in both males and females after conception. Those changes include histone acetylation, DNA methylation, and DNA copy number variation—all of which must be taken into account in clinical twin studies [297,800].

As for how the splitting that causes twinning occurs in the first place, an early embryo might be bisected by a zone of cell death. An apt analogy may be the manner in which extra legs arise when parasites tunnel through the leg buds in frogs, creating fragments, each of which reorganizes "regulatively" [570] to form a whole leg [2339,2340]. Nevertheless, the splitting and rejoining of embryo fragments can also produce twins of unequal size [238,2434]—the hideous "parasitic twin" phenomenon [678,1524].

appendages [304]. The nature of the algorithm that cells use to iron out those initial disparities remains to be elucidated.

Some categories of conjoined twins can survive to adulthood despite pervasive changes in their internal anatomy [677]. Clearly, human anatomy is more malleable than most medical textbooks would lead us to think [2790]. Nevertheless, two-headed people would be at a severe selective disadvantage in any imaginable environment [2824]. No species has ever resorted to conjoined twinning as an evolutionary strategy [2814], although a few species do employ twinning—sans conjoining—as discussed in the final section.

What causes identical twinning?

Once in every ~250 pregnancies a human embryo splits completely (without rejoining) to form identical twins [1345]. Why can embryos survive bisection when babies cannot? (Recall King Solomon's resolution of the maternity dispute?) And why does each piece go on to form a *whole* human instead of just a *half*?

The answer to both questions is that vertebrate embryos are "regulative" in early stages [1799]. This term signifies that cells know where they are along the body axes, and that they can revise these "area codes" whenever the axes are altered [376]. Dividing such an embryo is like cutting a magnet in half to yield two smaller, but fully functional, magnets [480] (*cf.* the surreal polarities of acoel worms [2395]). Young embryos manage to survive the trauma because the fates of their cells are so robustly flexible. Moreover, if some cells die, it doesn't matter. Others can replace them.

The plasticity of mammal embryos is so great that three mouse blastulae can be scrambled together to make a single mosaic (hexa-parental) mouse [1655]. Moreover, even embryos of two different species can be intermixed. The most extreme case was reported in 1984 [755]: researchers combined the cells of a goat embryo with those of a sheep embryo to make—are you ready for this?—a crazy-quilt goat-sheep [755], or, if you will, a "geep."

Such mix-and-match animals are as fabulous as any of the composite chimeras ever imagined by the Greeks or Romans in their wildest dreams, such as centaurs, minotaurs, and satyrs [351]. Indeed, if Shakespeare were alive today, he might write a happier ending for *Romeo and Juliet* in which stem cells are salvaged from the dying lovers [760], scrambled into a mosaic, implanted in a surrogate mother, carried to term, and named "Romiet" (or "Juleo"), so that the star-crossed pair could be reunited (*literally*). Then again, he might not.

Given the numbers of chorions (C) and amnions (A) in monozygotic human twins, 31% of them (2C2A) must split before ~day 4, 65% (1C2A) between ~days 4 and 7, and 3% (1C1A) after ~day 7 [230]. At the ~5-day peak of this distribution, the embryo itself has only about 20 cells [2859], so the deaths of just a few cells in the middle of the array could theoretically bisect the cluster. If each half restores its axes (like a half-magnet replacing a missing pole), then it could go on to make a whole baby.

A superficially similar transect scenario has been seen in a seemingly unrelated context. When limb buds of tadpoles are infected by trematodes, the parasites carve the leg buds into islands, and each fragment then goes on to make an *entire* leg, so that the resulting frogs end up with as many as 12 hindlegs [220,1306,2339]! Extra legs can also be induced in healthy frogs by implanting inert beads as partitions [2340].

What could cause a subset of cells to die in a human embryo at 5 days after fertilization? One possibility is X-chromosome inactivation [1738,1791]. Females inactivate one of their two Xs at around this time in development [2522]. Half the cells turn OFF the paternal X, and the other half turn OFF the maternal X. If the maternal X, say, were to carry a recessive cell-lethal mutation (*cf.* [917]), then half the cells would die, and it is a matter of chance how they'd be arranged [2863]. If, furthermore, the paternal-X-OFF (dying) cells happened to line up on a tic-tac-toe diagonal, then they would bisect the cluster. This hypothesis predicts that there should be more female than male identical twins. Are there? No! The monozygotic sex ratio is about 1:1 [1616]. Nevertheless, there is a 3:1 ratio of female:male *conjoined* twins, so this scenario might pertain to that situation [1616].

The most extreme known case of monozygotic multiple offspring occurred in Canada in 1934 when Mrs. Elzire Dionne gave birth to identical quintuplet girls [1614]. The original embryo in this case must have undergone several rounds of splitting. Conceivably, it kept disintegrating like a crumbly cookie because of some deficiency in intercellular adhesion.

Is twinning ever adaptive?

There is little evidence that monozygotic twinning is hereditary in humans [1991], so it would seem safe to assume that splitting is beyond the reach of natural selection [1593]. However, that assumption is disproved by the armadillo genus *Dasypus* [1345]. Believe it or not, armadillo litters typically consist of identical twins in *D. kappleri*, identical quadruplets in *D. novemcinctus*, and identical octuplets in *D. hybridus* [839]! These powers of two imply that the splitting mechanism here—whatever it is—must be more regular than the one that led to the Dionne quints. How does the partitioning happen? We don't yet know.

Why did embryo splitting become the norm in the *Dasypus* genus and nowhere else in our entire phylum [2510]? The adaptive explanation that springs to mind is that the littermates must be helping one another survive and therefore increasing the chance of passing along their own genes. But, no! The pups treat kin and nonkin alike [1575].

The actual reason, it turns out, is a geometric peculiarity of reproductive anatomy. The armadillo uterus, it so happens, cannot hold more than one egg at a time. If the one-egg constraint evolved in the ancestral stock but ecological conditions later changed so that more pups per litter conferred a survival advantage (*cf.* [1120]), then embryo splitting might have been the only option available to increase litter size [839].

The human uterus has no such quirk, so we have never found ourselves in such a bind. Nonetheless, a similar harnessing of splitting is theoretically possible [417], and it is amusing to think how different our society would be if everyone came in 2, 4, or 8 clonal copies.

Stranger still is the notion of a species in which every individual comes in ~2000 clonal copies, but such is the case for the wasp *Copidosoma floridanum*. When these wasps lay an egg inside a host caterpillar, the embryo splits repeatedly to form ~2000 cell clusters that make identical adults [2912]. Because the totals involved are not obligate powers of two, the partitioning is probably random (*i.e.*, the "crumbly cookie" scenario).

A similar strategy evolved independently in other insect groups, each of which is an endoparasite in a much larger host [2510]. (Human embryos resemble endoparasites insofar as we sap nutrients from our mother [1008].) Evidently, the transient availability of nutrients rewards mutations that maximize the number of offspring during that brief window of time.

Generally speaking, asexual reproduction (*e.g.*, the embryo "budding" just described) is an easy way to exploit ephemeral, unpredictable resources. This strategy is epitomized by aphid parthenogenesis [282,1685,2823]. If evo-devotees were to adopt this sort of strategy in their own lives, then they would feverishly try to clone themselves every time they were awarded a big research grant!

CHAPTER 4

Merism and Modularity

The first part of your body you saw as a baby was a hand as it flitted in front of your face. You may not have noticed it at the time, but those five digits are quite similar to one another. Later, you discovered that five comparable things protrude from each foot. Our 20 fingers and toes are all variations on the theme of a digit module. Indeed, modularity itself is a theme throughout our body [277,2300,2303]. Other examples include our spine, which has 33 vertebrae (12 with ribs), and our mouth, which sprouts 32 (adult) teeth. In each case, the units are arranged serially and spaced regularly. In short, they are meristic [151].

Spatial periodicity is a ubiquitous feature of living things [383], and it has been studied extensively in embryology [2777]. Periodic patterns can be made in various ways, depending on the species and the body part [1102,1135,1724].

What makes the thumb our only opposable digit?

Our fingers are virtually indistinguishable when they are first detectable as protrusions from the edge of our palm [452,1638]. Later, differences are imposed on them [1872,1950,2605] by two "morphogens" (molecules that inform cells about their positions [2542]): "Sonic hedgehog" (Shh) [1985] and "Bone Morphogenetic Protein" (BMP: types 2, 4, and 7) [1889]. Shh is produced near the pinkie and diffuses as far as the forefinger [1244,2252], so that its concentration tapers to form a gradient (Fig. 4.1a) [148,1483]. One effect of this gradient is to reliably make five digits [844,1557].

The cells between the digits assess their level of Shh [555,2621,2910]—apparently by integrating their exposure over time [1704,2193,2292]—and secrete a proportional dose of BMP [2535]. Each digit then turns ON a subset of *Hox* genes [555] in accord with the local concentration of BMP [104] (or a related ligand [131,2534]). The *Hox* genes *d12* and *d13* are turned ON in all digits except

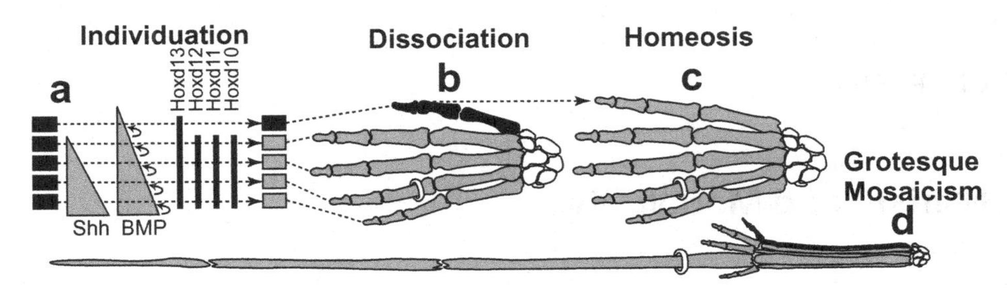

Figure 4.1. Development (a, b), transformation (c), and fanciful evolution (d) of human fingers.

a. Digits are first detectable as similar condensations of chondrogenic cells (rectangles) [1796] that then become different by "individuation" [833,1638]. This process begins when a signal—Shh (Sonic hedgehog)—diffuses from the pinkie region to form a concentration gradient (triangle) [2193,2252]. Interdigital tissue apparently measures how long it is exposed to Shh [645,1094] and responds by emitting a proportional dose of bone morphogenetic protein (BMP) [1045]. BMP may then activate *Hox* genes *d10–d12* in adjacent digits at a higher threshold than *d13*. Vertical bars indicate the spatial extent of gene expression along the anterior–posterior axis (not amounts!). Data are from mice [1404].

b. Left hand of a normal adult. The thumb's unique identity allows it to develop differently ("dissociate") from the other digits. Amazingly, the thumb's code (*d13* ON; *d12* OFF) dates back ~400 million years to our fish ancestors [581], who had no bona fide hands or feet [1302]. How the code works is unclear (see text) [379,841,2825,2904], but we do know that it must act indirectly via genes that affect anatomy directly (*cf.* Fig. 4.2). Embryologically, the thumb is the last digit to form in all land vertebrates except salamanders, where, strangely, it arises first [825]. Our thumb is longer than that of any other primate relative to the index finger [1676], a quirk that enables us to handle tools more adeptly [1848].

c. Left hand of a 35-year-old woman seen at Women's Clinic University in Graz, Austria, in 1957 [1133], which resembles a bear paw [16]. Her thumb was transformed into a forefinger [2676]—a phenomenon termed "homeosis" (see text)—and the same was true for her right hand, although it also displayed ankylosis of the first metacarpal and trapezium. *N.B.:* Despite superficial similarities, this case differs significantly from the typical presentation of "triphalangeal thumb" [1033,2582] insofar as (1) all of the muscles and tendons specific to the thumb were missing and (2) opposability was patently lacking [1133].

d. Human left hand redrawn to look like the hand of the pterosaur *Anhanguera piscator* (which actually lacks our pinkie) [426]. Note the long ring finger, which served as a strut for the pterosaur's wing membrane [52]. The ability of one member of a meristic series to undergo such drastic divergence ("grotesque mosaicism") should only be possible if it has its own genetic identity (*cf.* **a**) [2461].

the thumb [941,2676]. The thumb does not activate *d12*, presumably because its BMP level fails to attain the needed threshold [104].

Our thumb's uniqueness is thus attributable to a distinctive "*Hox* code" [1557]—namely, "*d13* ON; *d12* OFF." Consistent with this idea, mouse thumbs can be converted to forefingers by forcing them to express *d12* [1406]. Amazingly, a woman with this same thumbless, 5-finger trait in both hands was seen at a birth clinic in Austria in 1957 (Fig. 4.1c) [1133]. She did not consider her anomaly a disability. On the contrary, she said that it helped her play piano more easily! Her newborn baby looked exactly the same, so the trait is probably genetic. The general term for a transformation of one body part into another is "homeosis" [151]. Homeosis is often caused by mutations in *Hox* genes.

REFLECTIONS ON FIGURE 4.1

a. Models of digit patterning are compared in ref. [2547], and models of digit identity are critiqued in refs. [1797,2534]. One unsettled issue concerns thumb identity. The depicted model assumes that thumbness is specified digitally (no pun intended) by a combinatorial code of ON/OFF states ("*d13* ON; *d10–d12* OFF" [1406,1557]), but an analog mechanism is also possible [641,1794] (*i.e.*, thumbness could be dictated by a lower anterior dose of total Hox protein). Thumb identity may be the universal default state [439,2216] because removal of Shh converts all digits to thumbs [1557]. However, removing Gli3, an Shh transducer that binds Hox proteins [432], evokes odd shapes and more digits [2742]. Another debate concerns the role of *BMP* paralogs. Lowering the dose of several *BMP* genes fails to show the threshold effects predicted by the depicted model [131], but the basic idea is still viable because various other Noggin-sensitive (*BMP*, etc.) genes could be playing a morphogen role [2534]. Finally, there is the nagging question how Shh is transported and interpreted, a broader issue where progress is being made [644,645].

c. The woman's baby (born later at the same clinic) had the same thumbless 5-finger trait on both hands as she did. Homeosis of their thumbs might be due to broader *Hoxd12* expression [1406]. (*Hoxd13* mutations in humans can cause syndactyly [1831].) Strangely, however, the feet of mother and baby exhibited preaxial polydactyly (partly duplicated big toes) instead of homeosis. A possibly related syndrome of preaxial polydactyly coupled with triphalangeal thumbs has recently been traced to a defective enhancer of *Shh* [1033,1525].

d. The white-striped possum also elongated the fourth finger, albeit to a lesser extent [2669] (*cf.* the fourth toe of the eastern barred bandicoot [758]). The aye-aye converted its *middle* finger into a long, thin needle for retrieving grubs [52,758], and the slow loris shrank its *index* finger to a mere stub [1849,1881], while reverting that finger's nail (atavistically) into a claw for grooming [1260,1371].

Pterosaurs have another remarkable quirk in their hand. In their wrist is an elongated carpal ("pteroid") bone (not shown) that forms a fingerlike strut for the leading edge of the wing [2819,2820]. Another example of such a "wristdigit" is the panda's "thumb" [578,718] that was so deftly popularized by Stephen Jay Gould [958,961]. Quirks, of course, are in the eye of the beholder. Syndactyly, for example, is considered a dysplasia in humans, but it is the norm for hindfeet in some possum species [758].

A brief primer here might prove helpful (*cf.* [1673]). Hox is short for "homeobox," a conserved stretch of about 180 base pairs within homeotic genes [875]. The *Hox* acronym per se is reserved for members of the *Hox* complex [334,863], an ancient cluster of homeobox genes that patterns the anterior–posterior axis of bilaterians (*cf.* Fig. 1.3c) [2010]. The homeobox encodes a ~60 amino-acid, DNA-binding section (homeo*domain*) of a Hox protein [877,879]. Gene names are italicized, whereas protein names are not [1137]. Thus, for example, the *Hoxd13* gene makes the Hoxd13 protein whenever it is ON. "ON" means that *Hoxd13* gets transcribed into RNA, which, in turn, is translated into Hoxd13 protein. RNA polymerase cannot transcribe a gene adequately unless "transcription factors" bind DNA nearby at sites called "enhancers" [1996,2836].

Hox proteins constitute one class of transcription factor [866], so Hoxd13 can turn ON "downstream" genes by binding enhancer sites (through its homeodomain) adjacent to those genes. Different Hox proteins (*e.g.*, d13 *vs.* d12 [379]) recognize different enhancer targets [22,186,1903].

Genetic circuits consist of many such intergenic interactions. One example was discussed in Chapter 2: the *nodal* circuit uses Pitx2 (a homeodomain transcription factor [1559]) to control downstream genes that then create asymmetries in our heart [2119], lungs [1552], intestines [223,355], and other organs [511,835,1585,2246]. However, the circuitry is considerably more complex than any précis could possibly convey [1164,2136,2376]. For example, (1) different isoforms of Pitx2 play different roles [732,2324], (2) *nodal* bypasses Pitx2 entirely (using Foxh1 instead; *cf.* [2883]) to position the stomach and spleen [2000,2378], and (3) there are additional branches in the control hierarchy that are not yet understood [208,441,1129]. Most regulatory networks are at least as complicated as the *nodal* circuit, and some are much crazier [571]. Luckily for our discussion, the hand's *Hox* circuit is comparatively quite simple.

To see how a *Hox* code could steer the thumb to a unique morphology—specifically two instead of three phalanges—consider an imaginary gene "*Ph3*" that is needed for a finger to form a third phalanx [1730,2534]. If *Ph3* requires that Hoxd12 bind nearby for it to be turned ON, then it must be OFF in the thumb. Similarly, other genes must exist that elicit the muscles for opposability [475]. They would be activated only when Hoxd13 is present without Hoxd12. (The logic is made even simpler by the law of "posterior prevalence" [2878], in which one *Hox* gene overrules all others wherever coexpression occurs [604,2825].)

Our thumb differs qualitatively from the other four digits in its anatomy and mobility [9,1848], so it makes sense that it would have its own code. In contrast, the four fingers only differ quantitatively from one another [51], so their relative lengths could easily be tweaked by growth factors alone (*e.g.*, Shh [1953] and BMP [2049,2217]) [1071].

Among monkeys and apes, there is a wide spectrum of hand shapes [2076], but there is one consistent trend: when the lengths of the four fingers change, they tend to do so coordinately, as if they were a coherent module apart from the thumb [2161]. The interdependence of those fingers (and their independence from the thumb) is understandable in terms of the *Hox* code. That code also explains how evolution could truncate the thumb in a few primates without altering any other digits [2317,2552] (*cf.* thumb loss in cloven-hoofed artiodactyls [2085]). The only mystery, given the dexterity of the patient mentioned earlier, is why a thumbless, 5-finger hand never evolved in any simian.

What about other vertebrates with unique digits aside from the thumb? The most striking example is the huge ring finger of the pterodactyl [2661], which was so long and strong that it served as a strut for the entire wing ("ptero" = wing; "dactyl" = finger; Fig. 4.1d) [1894]. For it to grow ~10 times longer than the other fingers (beyond the range of any conceivable modulation in graded growth factors), it must have had a separate identity.

Could we evolve a giant digit like pterosaurs?

Consider what it would take for humans to evolve a finger as long as the pterodactyl's without sacrificing the uniqueness of our opposable thumb. We would need to somehow augment our *d12/d13 Hox* code [641,1582]. The code could theoretically be expanded by tapping *Hoxd10* or *Hoxd11*, which are also transcribed in mammalian digits, but both are expressed congruently with *Hoxd12* [2574], so they could only distinguish another finger if they shifted their domains [1404,2722].

Figure 4.2 uses a fishing analogy to show how this process of "individuation" [2178,2717] could happen. Suppose that a gene—call it "*Fng4*"—is expressed only in our ring finger (*cf.* [1611,2726]) and that its protein product, Fng4, is a transcription factor, which binds the DNA sequence "GACT." For *Fng4* to enlarge the ring finger, it must gain control over growth [2072,2835]. Mutations are randomly changing nucleotides all the time to create new motifs [1814,2638], so GACT might easily arise near a growth-stimulating gene [1591,2836]. At that point, Fng4 could (1) bind there [572], (2) prod transcription through *cis*-activation [288,2853], and (3) cause the ring finger to grow excessively [668]. In fact, most binding sites exceed four nucleotides [22,866], and the larger the site that is needed, the longer a species has to wait (on average) for its anatomy to change [118,391].

The broader lesson here is that body parts must acquire distinct identities genetically before they can diverge developmentally ("dissociate" [1857,2109]) to any great extent [1537,2111]. Another example was discussed in Chapter 2, where *Pitx2* played the role of *Fng4*, and the left side of the body played the role of the ring finger. Our left lung was presumably able to evolve fewer lobes than our right because (1) it expresses *Pitx2*, whereas our right lung does not, and (2) at some point a link was forged between *Pitx2* and a growth-controlling gene in the lung. That link must have been inhibitory, rather than stimulatory.

Notice also the fickleness of these rewiring events [939,970,2762]. The *nodal* circuit happened to capture lung, heart, and spleen genes in our ancestors, but it evidently harnessed tooth genes in narwhals, ear genes in owls, and eye genes in flounders (*cf.* Fig. 2.4). The *Hox* code happened to give primates an opposable digit that proved useful for climbing trees [1790,1813,2076], but

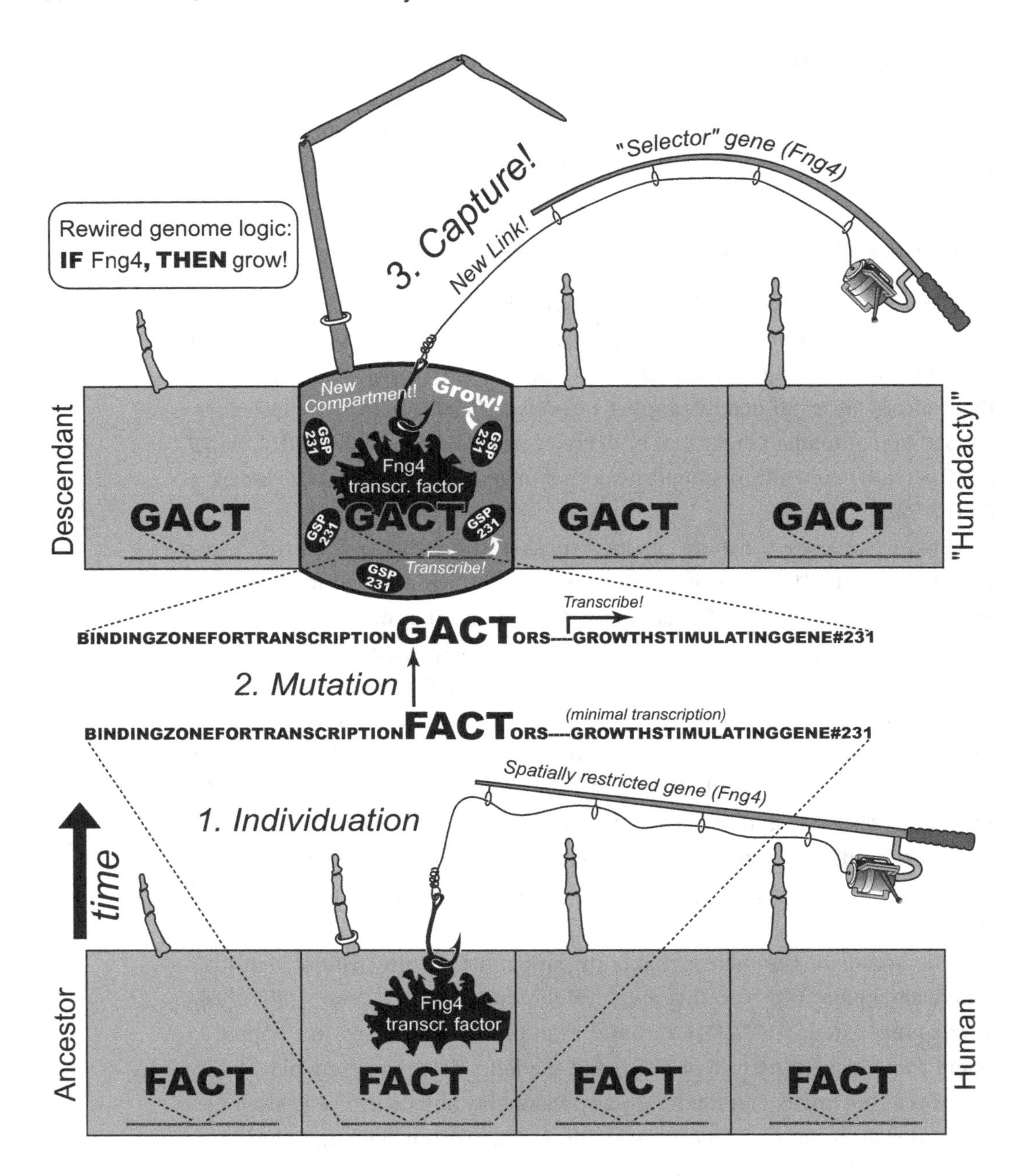

there is nothing sacred about a single thumb or its site of origin [844]: koalas have two [619,758], chameleons have three [2214], and pandas have a pseudo-thumb that evolved from a bone in their wrist [961]. Humans, it turns out, are no more quirky than other species. All species stumble through Morphospace via random mutations (*cf.* Fig. 1.2a). Every now and then, the resulting reconfigurations prove useful and may even look elegant in their final form, but the rewiring process itself is clumsy and blind [1605,2424].

Figure 4.2. A mental exercise in how a human hand could evolve to resemble a pterodactyl wing. One major mode of anatomical evolution—possibly the main one [241,401,2204,2478,2838]—involves changes in how genes respond to regional transcription factors [390,2093,2853]. This illustration is based on that mechanism. Each square designates a finger primordium ("phalanx-forming region") [2534], and the bulging square in the upper row denotes a finger growing excessively. Within each square is the DNA sequence (written here in English) of an imaginary gene, *#231*, that can provoke growth whenever it is transcribed. Normally, however, it is OFF.

Step 1. For one finger to grow 10 times longer than the others, it must have a separate identity ("individuation") [833,1582]. In this fanciful scenario, that identity arises when the hypothetical gene *Fng4* (fishing pole) has its expression spatially restricted to the ring finger (*cf.* [1611,2726]), perhaps in response to graded Shh or BMP (*cf.* Fig. 4.1a). *Fng4* encodes the transcription factor Fng4 (bait on the hook). Despite having a unique "area code" at this stage, the finger would look exactly the same because Fng4's shape doesn't fit any sequence of letters adjacent to gene *#231.*

Step 2. Over time, sporadic mutations alter the nucleotide sequences in the regulatory zones where transcription factors such as Fng4 can bind [2501,2855]. Those zones are typically a short distance (dashes) from the coding region [1530]. Here, only one letter changes (from F to G).

Step 3. If such a change happens to create a binding site for Fng4 (GACT) near a gene that spurs growth like gene *#231* (*cf.* [668]), then *Fng4* would "capture" it (*e.g.*, see [2347]) and become a "selector gene" [1862,2488,2762]. To the extent that Fng4 abets transcription (bent arrow), gene *#231* will be "ON," and its product, Growth Stimulating Protein 231 (GSP231), will cause the finger to enlarge. (Cases of "macrodactyly" have indeed been described in humans [2582].) Henceforth, *Fng4* can recruit more genes elsewhere (not shown), whereupon its links will fan out like a spiderweb into the genome [676,1933,2719], and the ring finger will continue to evolve as an autonomous "compartment" [884,2303]. Such compartments tend to acquire new features in the same way that Christmas trees "collect" ornaments [2726].

Once a spatially restricted gene recruits an organ-shaping gene, the former is termed a "selector" [859,1862], although it need not select an identity for *every* cell in its dominion [37,398]. As the selector gene commandeers more and more anatomy-affecting pathways [897,1636], the body part that it governs will evolve independently as a separate "compartment" within the genome [861,1393]. Compartments usually coincide with conventional anatomical subdivisions, but occasionally they do not [1137].

This process of "mosaic evolution" [2461] has been studied most intensely in insects [78]. The best understood case concerns how the *Hox* gene *Ubx* converted an ancestral hindwing into a tiny gyroscope in flies [2474,2763]. It did so by insinuating itself into pathways that control (1) wing size [536,537,1785], (2) vein pattern [1153], (3) bristle initiation [1973], and (4) sensilla formation [2203]. *Ubx* exerts its influence at various echelons within these different chains of command. It gives orders to a colonel in one, a lieutenant in another, and a sergeant in a third. The moral of this story is that we shouldn't think of "master genes" such as *nodal* and *Ubx* as serenely aloof executives [875]; they're actually neurotically intrusive micromanagers [37,1137].

Data from organisms aside from flies [527,571,2299,2764] confirm that anatomical divergence is typically a tediously gradual process [387]. Ironically,

REFLECTIONS ON FIGURE 4.2

This diagram is a gross oversimplification insofar as (1) proteins can bind DNA not only via a jigsaw-puzzle fit but also by electrical attraction [866], (2) genes can affect other genes at many levels besides transcription [60,61,1178], (3) gene inputs can be negative rather than positive [2620,2817], (4) links can be combinatorial (many to 1) rather than direct (1 to 1) [2414,2637], and (5) fingers are shown as existing before *Fng4* expression became restricted, but it could be the other way around [2245,2723]. Indeed, regarding this last point, the tetrapod pattern of *Hoxd12/d13* expression is evident in shark fins [816], suggesting that it predated the dawn of digits by ~100 million years (*cf.* [30]).

Step 1 might be rate-limiting, although enhancer-trapping studies show that it can occur simply [401,1639]. Step 2 could also be slow [118,391] because actual binding sites are typically ~5 to 10 letters long [22,866,2642], but mutations need not affect just one letter at a time [2814,2851]. Cut-and-paste shuffling of "word" and "sentence" fragments [329,2130] through transposition, gene conversion, and unequal crossing over [676] refutes the old "monkeys-on-typewriters" canard that evolution can't create complexity [588,960,1167]. After such rewiring occurs in an individual, the novelty can linger as a polymorphism and, if sufficiently adaptive, spread and eventually saturate the population [185,2851,2855].

Case studies that illuminate various aspects of the depicted scenario include (1) the accrual of binding sites throughout the genome after a selector gene arises (mammalian limb [2701]); (2) the evolutionary flux of transcription factor binding sites (yeast [246], *Drosophila* [1814,2632], and in general [1605,2130,2501]); (3) the tolerance of genomic networks for tinkering with their links (bacteria [181,1249]); and (4) the instrumental involvement of rewiring in evolution (reviews [1255,1605]). The most relevant case may be the amazing tale of how the swordtail fish got its sword, a question Darwin himself contemplated in the second half of *The Descent of Man, and Selection in Relation to Sex* [708]. Recent revelations from evo-devo research have essentially solved the mystery of how one element of a meristic series can be singled out for extraordinary growth [390,2263].

If, at some future date, a mutation (not shown) were to disable *Fng4*, then it would, in this analogy, break the pole and release all the "fish" that had been hooked over the eons [2815]. In that case, the ring finger would homeotically revert to its mundane morphology. Reversions to prior evolutionary states are termed "atavisms" [1056,2067,2482]. Atavisms can involve reappearance of lost structures in rare individuals or in whole species [151,1053,2178,2482]. Atavisms in humans putatively include apelike fur [153,765,862], doglike nipples along the milk lines [408], and fishlike "gill slits" on the neck [666]. Instances in nonhuman species include bird teeth [1100], snake fingers [268], horse toes [963], frog tails [1077], whale hindlegs [1053], and fly hindwings [818,860]. Traits that have reestablished themselves in whole species include lizard toes [1415], larval stages in salamanders [444], shell-coiling in gastropods [492], wings in stick insects [2497,2799], spots on fly wings [2094,2852], and sex combs on fly legs [1425]. Taxon-level reversals could theoretically start with "hopeful monster" mutants [661,662,957,2179,2733] or arise through a gradual awakening of dormant pathways [1040,2475]. However, dormant circuits are difficult to resuscitate after more than ~10 million years [310,311,1825,2584,2585] because of deterioration from disuse [1659,2922].

however, complexity that took eons to elaborate can be undone in the blink of an eye by a homeotic mutation that disables the selector gene [2407,2816]. Such regressive "atavisms" [2482] have deluded some researchers into thinking that evolution progresses in equally dramatic saltations [38,916]. It usually doesn't [2589].

Where did our tail go?

Humans differ from most other primates in that we lack a tail. Actually, each of us had a tail when we were embryos (Fig. 4.3a) [719], but the vertebrae that it contained stopped growing and fused to form our coccyx and sacrum before we were born [376]. (Rarely, babies are born with an external tail [91,561].) Darwin realized that this stunted vestige is a tangible legacy from our simian ancestors:

> The human embryo likewise resembles in various points of structure certain low forms when adult. For instance, . . . the os coccyx projects like a true tail "extending considerably beyond the rudimentary legs." [561] (Vol. 1, p. 16)

The ability of a subset of vertebrae to develop differently (*e.g.*, become vestigial in this case) depends on their having separate identities [2784], just as our thumb has a unique *Hox* code [797]. In fact, vertebrae had their own intricate *Hox* code long before our amphibian ancestors ever evolved digits [1059,1230]. Amazingly, the *Hox* code for the hand was co-opted *from* the *Hox* code for the body by a *de novo* limb-specific control site [640,816,940,1794]!

Mammals have four clusters of *Hox* genes (*a*, *b*, *c*, and *d*) [1011], which evolved through two successive duplications of an Urbilaterian *Hox* complex [64,1519,2906]. Within each cluster, the genes are numbered sequentially [682]. Thus, *Hoxd13* is the last (13th) gene in the *d* cluster. One unsolved mystery of the *Hox* clusters is their colinearity [347,416]: the zones of *Hox* expression along our spine (Fig. 4.3a) correspond to the order of the genes within each cluster—as if there were a little man (dubbed the "homeobox homunculus" [1135]) reclining along each of the respective complexes [318,642].

How do different *Hox* genes get turned ON at successive levels along our spine [791]? One clue is the ability of exogenous retinoic acid (RA) to shift vertebral identities when embryos are exposed to it [1367]. Depending on the time of treatment, identities can shift uniformly either anteriorly or posteriorly [1349]. Chordate embryos make their own RA [499,2225], and its usage along the anterior–posterior axis dates back at least to when chordates diverged from other bilaterians [59,357,829,1656] and possibly even farther [639,2401]. Thus, an RA gradient appears to govern the spine in the same way that an Shh gradient governs the hand [690].

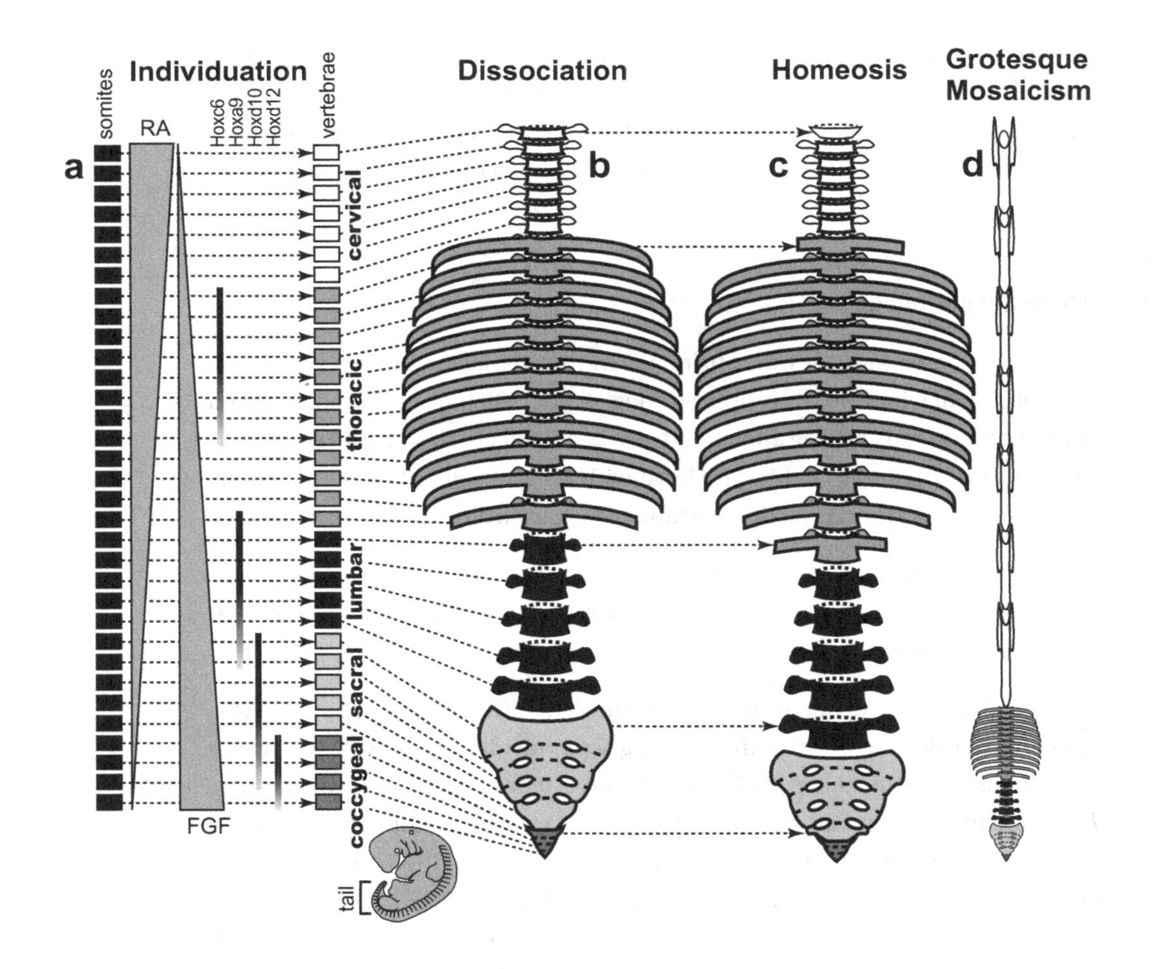

Different *Hox* genes have RA response elements that may turn ON at different levels of RA [1627]. Only four thresholds are needed to delimit the five types of vertebrae (cervical, thoracic, lumbar, sacral, and coccygeal; Fig. 4.3b) [104,376]. On the basis of where RA is synthesized, degraded, and transduced *in vivo*, the idea of a single gradient seems too simplistic [2149]. The preponderance of evidence implies *two* opposed gradients [1878]: a head-to-tail gradient of RA [1621,1805] and a tail-to-head gradient of FGF (fibroblast growth factor) [684,1570]. Together, they apparently assign segmental identities [622,925] (via *cdx* genes? [422,1212,2746]) in sync with an oscillator that works like the escapement device of a grandfather clock [1012,1915,1960,2652]. Parts of that device may date back to Urbilateria [557].

Codes for the five vertebral types have been deciphered in mice by deleting *Hox* genes or by turning them ON where they are not normally expressed and then studying the resulting homeoses [371,1707,2785]. Similar homeoses

Figure 4.3. Development (a, b), transformation (c), and fanciful evolution (d) of human vertebrae.

a. Vertebrae (and ribs [1218,1928]) develop from blocks of tissue called "somites" (rectangles). Somites and vertebrae are slightly out of phase [82,1810] (not shown; *cf.* [23] for an even stranger shift in frogs). Occipital somites are omitted because they form part of the skull [1211,1307,1592]. Somites arise in a wave from head to tail and initially look alike [123]. They acquire separate identities ("individuation") through a *Hox* code [521,791,2785] that uses a subset of our 39 *Hox* genes [1011]. Gradients (triangles) of retinoic acid (RA) [1621] and fibroblast growth factor (FGF) [684] specify where *Hox* genes get turned ON [1483], and these ON/OFF states endure [1379]. Graded vertical bars denote expression zones for genes whose anterior limits coincide with type transitions [318,376,1230], although expression in *pre*somitic mesoderm is more critical [371]. Limb buds (not shown) may also be positioned by RA/FGF because tadpole tails grow legs (and pelvic girdles) when exposed to RA [2234]. The ground state for vertebrae is thought to be "thoracic" (*i.e.*, rib bearing) [2785]. Below is a 31-day-old human embryo [2311]. Note the somites (along the spine) and the tail (which stops growing later to form the coccyx and sacrum) [1076]. Rarely, humans are born with short, external tails [91,561,2178,2802], but these cases may not be true atavisms [2683]. The specific vertebrae that comprise the sacrum vary among individuals [2806].

b. Axial skeleton of a normal adult (ventral perspective; *i.e.*, ribs curving toward viewer). Different types of vertebrae are indicated by different shadings.

c. Axial skeleton (ventral perspective) from a man's body donated *ca.* 1882 to Amsterdam's Vrolik Museum [1938]. The most anterior vertebra of every region appears to be shifted in identity by one unit. (The depicted shift of first cervical to an occipital identity (*i.e.*, base of the skull) is conjectural because his skull was not retained.) Similar shifts are seen in *Gbx2*, *Gdf11*, or *Cdx* mutant mice [371,834] or in embryos treated with RA [499,1368]. ("Frame shifting" of identity relative to merism also occurs in digits [2718].) Asymmetries in rib formation (here first thoracic and first lumbar) are surprisingly common in human axial homeoses [845] and may be due to insufficient RA [276]. More drastic homeoses (*e.g.*, ribs eliminated [371] or ribs everywhere [2785]) occur when *Hox* genes are broadly misexpressed or collectively disabled [1707].

d. Axial skeleton of a human redrawn to resemble that of the pterosaur *Quetzalcoatlus northropi* [426]. Note the incredibly elongated cervical vertebrae (dwarfing a giraffe's neck [2425]). The opposite trend (cervical compression) is seen in whales [188]. We do not know why the number of cervical vertebrae is so constrained in some groups of animals (*e.g.*, pterosaurs [426] and mammals [1850]) but not in others (*e.g.*, sauropods [746] and birds [1873]) [840]. Revamped vertebrae are also seen in turtles, whose ribs fuse with their shell [1842,2179], and in extinct gliding lizards, whose ribs formed wings that did not evolve from arms [2373,2752]. The pterosaur's long neck is counterbalanced by long legs (not shown).

---

have been seen in rare humans (Fig. 4.3c). The different identities of the vertebral types have enabled them to evolve independently as separate compartments (Fig. 4.3d) [306,766]. In each instance, *Hox* selector combinations appear to have harnessed downstream genes that affect growth (*cf.* Fig. 4.2). Examples include (1) our own ape ancestors, who probably lost their tail when *Hox* genes that govern caudal vertebrae accidentally captured genes that curtail growth [702,2668]; (2) frogs, who likewise lost their tail, perhaps by a similar mechanism [1077,1078]; (3) giraffes, who acquired longer necks by specifically lengthening their cervical vertebrae [2425,2824], although they were surpassed in this regard by the pterosaurs [426]; and (4) snakes, who converted their cervical vertebrae into thoracic (rib-bearing) vertebrae [490] and elongated their bodies by making more of the same. The trick that enabled snakes to make an unlimited number of thoracic vertebrae has recently been revealed: they sped up the aforementioned oscillator that slices their mesoderm into somites (*i.e.*, vertebral precursors) [935,2706].

Why do our teeth have different shapes?

Humans have about as many types of teeth (four) as vertebrae (five): incisors, canines, premolars, and molars (Fig. 4.4) [1300]. The individuation process is similar, too [1696]. Opposing gradients (of BMP4 and FGF8) establish zones of expression for homeobox genes (albeit outside the *Hox* complexes per se [2346]) [1770], and the resulting code dictates different identities [476,2021]. Incisors arise in response to high levels of BMP4 (via *Msx1* and *Msx2*), whereas molars are elicited by high levels of FGF8 (via *Barx1* and *Dlx2*) [2643].

Consistent with this logic, incisors can be transformed into molars by blocking BMP4, which results in activation of *Barx1* [2644]. These data come from mice, but humans develop similarly [1553]. Indeed, a girl with molars replacing incisors was seen at a dental clinic in Minnesota *ca.* 1990 (Fig. 4.4c). Codes for our premolars and canines, which mice lack, remain to be determined [1695,2488].

As with vertebrae, the separate selector genes have allowed evolution to modify the classes of teeth independently of one another [1590,2486]. Examples include conversion of premolars to molars in horses, transformation of molars to premolars in bats, and loss of premolars and canines entirely in rodents [1397,2020,2780].

How did the different classes of mammalian teeth originate? Reptiles already had all the signals needed for tooth diversification in the right areas of their jaw (BMP in front, FGF in rear, etc.) [476], but their teeth were uniformly peg-shaped [2027]. In the lineage that led to mammals, mutations must have somehow enabled tooth buds to (1) "hear" those preexisting signals [2488,2780] and (2) respond to them by turning ON cusp-patterning genes [340]. After this co-option by teeth of jaw signals [1695], further shape changes were probably relatively easy [333,1298].

Why are some features easier to evolve than others?

The ease or difficulty of specific anatomical changes has long been a mystery in biology [1143,2716]. Darwin was astounded at the plumage varieties attainable by artificial selection using the ordinary rock pigeon [559,560,974,1141]. Dog breeds likewise dazzle us by their wide array of shapes and sizes [560,792,1951]. Evidently, feather patterns and body shapes are easy to alter in these respective

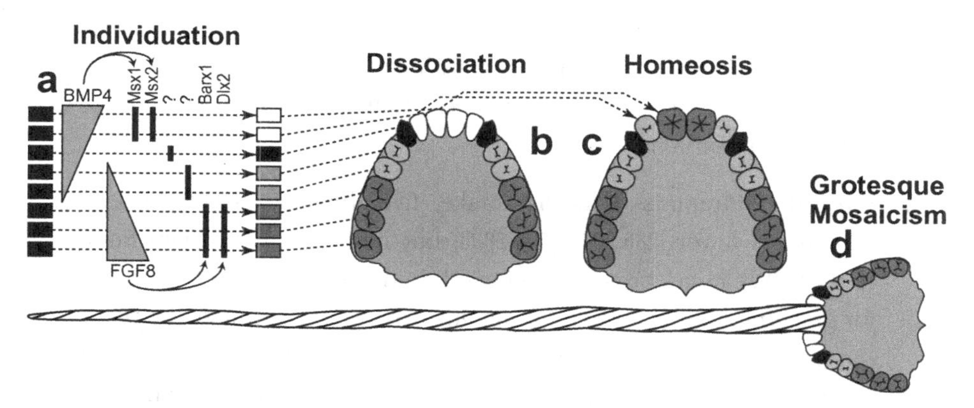

Figure 4.4. Development (a, b), transformation (c), and fanciful evolution (d) of human teeth.
a. Our teeth come from tooth buds (rectangles) [1917,2643]. Identities are dictated by antagonistic gradients of morphogens (BMP4, FGF8) [1561,1770] that turn ON genes for homeobox transcription factors (Msx1 and 2, Barx1, Dlx2, etc.) at different thresholds [2643]. Vertical bars denote expression zones. We do not yet know the codes for premolars and incisors [1346,1695]. Tooth positions may be set by similar circuitry [1867,2059]. The core of each tooth comes from the neural crest [412,1743], which imposes its final identity on the overlying, enamel-making epithelium [476,567] through a reciprocal dialogue [413]. Our reptile ancestors had only peglike teeth [1944], so the human default dentition may likewise be a mouthful of conically shaped teeth resembling our canines.
b. Upper dentition of a normal adult [2027,2374]. (As children, we have no premolars, and our juvenile molars are replaced inexplicably by adult *pre*molars [1866].) All of our teeth flank the midline, and the same is true for other mammals [151], except for one odd bat species (on an island off the African coast) that has three incisors in its lower jaw, the middle one straddling the midline [1321]. Median incisors are rare in other species [151], and they are diagnostic of a mild form of human holoprosencephaly syndrome (*cf.* Ch. 3) [857].
c. Upper dentition of a 13-year-old girl examined at the University of Minnesota *ca.* 1990 [1332]. Her central incisors looked like molars (with extra cusps), and her lateral incisors like premolars. Her bottom teeth (not shown) were relatively normal (as was her face), but one premolar was missing. She had no family history of dental defects, so the genetic basis is unclear, although she did show congenital hearing loss.
d. Upper teeth of a human redrawn to simulate that of the narwhal *Monodon monoceros*, in which the inner left incisor elongates enormously (in males) to form a tusk [2146]. Overgrowth might occur through targeted overproduction of Shh [566]. Adult narwhals actually lack all teeth except the tusk.

species, and body size knew no bounds in either sauropods [2272] or whales [188].

At the other end of the spectrum of "evolvability" [589,2422] are traits such as reptile teeth, which remained conical (like those of crocodiles) for more than 100 million years [1357], despite the fact that dental heterogeneity (molars, incisors, etc.) would have offered obvious advantages for chewing different types of food [1587]. Did the right mutations simply never occur [1309,1484], or was it just too hard to concoct the right binding site combinations [1687,2325]? No one knows.

We simply don't yet understand enough about the rules governing genetic circuitry to explain such limitations [1330,2264] nor to fathom how they were overcome in certain lineages [1689,2095,2262]. All we can say with confidence is that some (*intra*module) traits are strongly linked within the genome, whereas other (*inter*module) traits are weakly linked [1393,1400]. Other similar puzzles were touched on earlier:

1. Why do tetrapods typically make five fingers [148,844,1848,2362]—sometimes fewer [268,2345,2504,2805] but never more? Darwin famously mused: "What can be more curious than that the hand of a man, formed for grasping, that of a mole for digging, the leg of the horse, the paddle of the porpoise, and the wing of the bat, should all be constructed on the same pattern, and should include the same bones, in the same relative positions?" [559] (p. 434).
2. Why do mammals typically make seven cervical vertebrae despite a wide range of lifestyles [840,2824], whereas reptiles and birds managed to escape any such restriction [843,1850,2131]? Darwin also wondered about "the same number of vertebrae forming the neck of the giraffe and of the elephant" [559] (p. 479). He guessed that constancy could evolve if "natural selection... should have seized on a certain number of the primordially similar elements, many times repeated, and have adapted them to the most diverse purposes" (*ibid.*, pp. 437*ff*). Those "diverse purposes" are certainly manifest in the innervation map of the human neck, forelimb, and thorax [1674].

Many other cases of stasis are just as intriguing [959,2824,2861], especially with regard to why certain crannies in "niche space" have stubbornly resisted invasion for eons immemorial [227,482]. Ecologists have toyed with these irritating riddles using game theory [2692], but the ultimate answers are hiding in the genome [793,1303].

One key insight gleaned from the foregoing cases of fingers, vertebrae, and teeth is that merism itself enhances evolvability because it contains inherent redundancy [843,2776]. That is, the availability of multiple elements allows a subset to assume new roles without jeopardizing the original role for which the series evolved [2775,2824]. Another insight, mentioned earlier vis-à-vis selector genes, is that it is easier to lose a complex trait (via single regulatory mutations) than to gain it in the first place (via multiple rewiring events) [2093].

How do upper and lower teeth achieve quasi-symmetry?

Our upper and lower jaws are similar enough to be considered members of a meristic series, but one is upside-down relative to the other. How did this inversion arise? William Bateson (1861–1926), who wrote the definitive work on merism (*Materials for the Study of Variation*; 1894), wondered whether the jaw might have co-opted the body's midline gadgetry to create its mirror symmetry:

> In view of the fact that the teeth in the upper and lower jaws may vary simultaneously and similarly, just as the two halves of the body may do, it seems likely that the division of the tissues to form the mouth-slit must be a process in this respect comparable with a cleavage along the future middle line of the body. [151] (p. 197)

Molecular–genetic data refute this notion [2172], as does our current view of how the oral region evolved in vertebrates [1936,2171]. Rather, the oral plane of symmetry appears to have been superimposed on a prior architecture [2364], much like a new plane of symmetry was shoehorned into arthropod legs to create the mirror-image (upper–lower) surfaces of insect wings [1448].

If the latter case is any guide, then the new symmetry might depend on a diffusible signal [1137] from the boundary between the two halves [2780] that dictates inverted polarities [2898] (*cf.* also the boundary between the mes- and met-encephalon [2488]). Indeed, FGF8 appears to serve this very role [761]. It diffuses from the maxillary-mandibular boundary (Fig. 4.5) [477,1561], but whether it tells each half of the jaw which way to face is not known.

Perpendicular to the FGF gradients are BMP4 gradients [2172,2364] that appear to enforce a "*Dlx* code" for upper versus lower jaw [636,1292]: the homeobox genes *Dlx5* and *Dlx6* are ON in the lower jaw but OFF in the upper jaw, presumably because the upper jaw's BMP gradient never attains the threshold needed to turn them ON. If both genes are inactivated, then the lower jaw develops like an upper jaw that is upside-down [189,633]. The inversion confirms the notion of a "zone of polarizing activity" (ZPA [2193]) at the boundary [635].

Given their unique identities, the upper and lower jaws have been able to acquire various differences in their teeth [2486]. In humans, the disparities include (1) a slight shift ("occlusional offset" [1573,2154]) that allows a tight fit between our upper and lower teeth [50,2768], (2) bigger upper incisors, and (3) extra cusps on our lower molars [1004]. More extreme upper–lower inequalities are seen in a few mammal jaws (*e.g.*, elephants) [2178] and many bird beaks [1864], including those of the Galapagos finches [8,2770].

A revealing study in this regard was done on dog muzzles by Charles Stockard, who found overbites and underbites in various hybrids [2490]. These mismatches probably stem from disparate downstream genes (linked to the *Dlx* code by artificial selection [792]) for jaw shape in different breeds [2406,2556]. Indeed, hybridization wreaks havoc on coadapted gene complexes in general [11], which may be why discrete species (and their quirky isolating mechanisms) evolved on earth in the first place—that is, to protect the functional integrity of

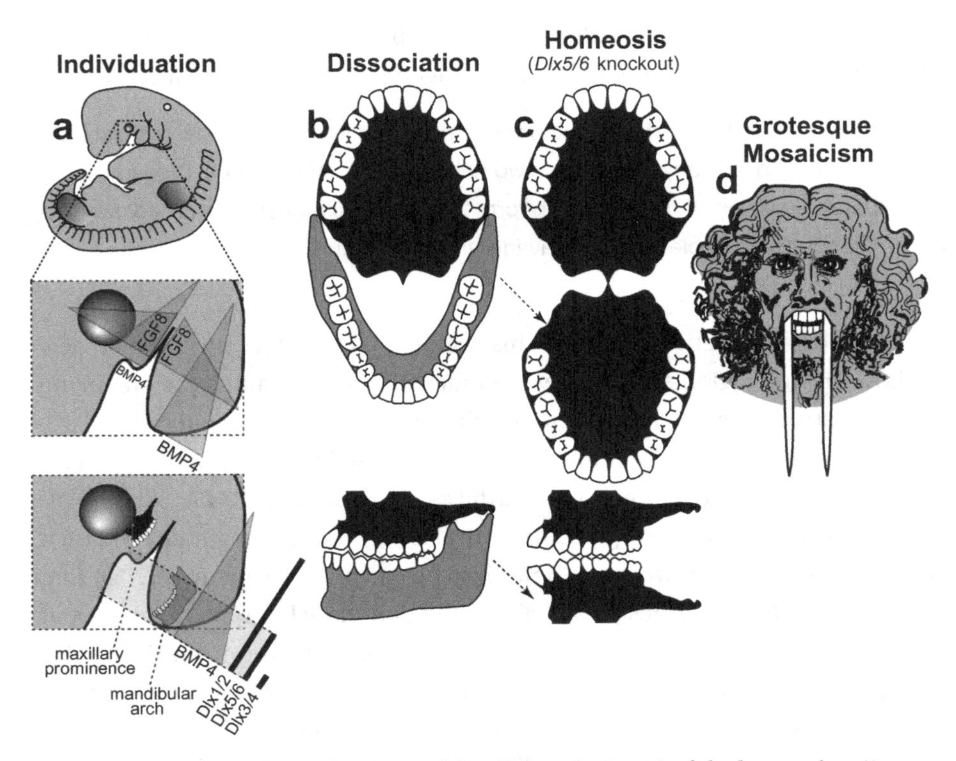

Figure 4.5. Development (a, b), transformation (c), and fanciful evolution (d) of the human jaw (upper *vs.* lower).

a. Above is a 31-day-old human embryo [2311]. Below are enlarged cartoons of the oral region. The large circle denotes the incipient eye. Our upper jaw comes from the "maxillary prominence" [409,1509], whereas our lower jaw comes from the "mandibular arch" [1142]. The oral area is criss-crossed by gradients of the same morphogens that cause teeth individuation (*cf.* Fig. 4.4). FGF8 spreads outward from the gap between the maxillary and mandibular areas [477,1123], whereas BMP4 diffuses inward from the tips of the two bumps [2172,2364]. Together they coax *Dlx* genes into a nested pattern, although *Dlx* enhancers studied thus far respond counterintuitively (positively to FGF8 and negatively to BMP4) [1992]. The nested domains (slanted bars) act as a "*Dlx* code" that distinguishes upper (*Dlx1/2* ON; *Dlx5/6* OFF) versus lower (*Dlx1/2* ON; *Dlx5/6* ON) jaw [636], where the shorthand "*Dlx1/2*" means "*Dlx1* and *Dlx2*." The maxillary is defined by *Dlx5/6* OFF, whereas the mandibular is defined by *Dlx5/6* ON [1292]. This *Dlx* code may have existed for eons in our fish forebears [2487] before fortuitous mutations caused it to capture anatomy-affecting genes (*cf.* Fig. 4.2), which led to visible dissociation of upper versus lower dentitions (**b**) [299].

b. Dentition of an adult human (upper jaw above, lower below). Note the differences (upper *vs.* lower) in sizes of incisors and the cusps of molars [1004]. Note also the occlusion of alternating cusps from the two surfaces (side view below) [1573,2154], which enables us to chew more effectively [50,2768]. Such a precise fit may only have been possible after our mammal ancestors stopped replacing teeth continuously (like reptiles) and started making two sets: deciduous (baby) teeth and permanent (adult) teeth [59].

c. Transformation of lower into upper jaw due to inactivation of *Dlx5/6* in mice [189,633], rendered here in human form. The same phenotype is obtained by blocking Endothelin-1 signaling [1961,2239], which acts upstream of *Dlx5/6* [2171]. (Disabling of Activin signaling also targets the lower jaw but does not transform it [1695].)

d. Teeth of a human redrawn to mimic the saber-toothed cat *Smilodon lethalis* [458,1098], where the upper jaw has huge canines, but the lower jaw does not.

REFLECTIONS ON FIGURE 4.5

a. Man's likeness to a fish at this stage may have been why Darwin decided to use such a drawing as his first figure in *Descent of Man* because no other image so clearly conveys his book's theme as stated at its end: "Man still bears in his bodily frame the indelible stamp of his lowly origin." Gill slits persist atavistically as cervical fistulae in rare humans [91,666,2178], although such "humafish" would presumably still need scuba gear to respire effectively underwater.

b. How mammals make a certain number of teeth [11] and how each tooth makes a definite number of cusps [1299] are active areas of investigation in evo-devo [1346,1770].

d. Other examples of upper–lower dental divergence include the tusks of elephants, walruses, boars, and narwhals (*cf.* Fig. 2.4) [52,2178] and the fangs of snakes [2897]. The evo-devo story of how the snake got its fangs has recently been deciphered [2705]. One knotty question that evo-devo is starting to address is why certain teeth grow continuously (*e.g.*, rodent incisors) but most do not [1769,2590].

finely tuned genomes [531]. Underbites are also seen in some dolphin species [1669]. A few genes are known to affect upper versus lower teeth differently within the same shape class [4,2486].

Despite the salience of these upper–lower discrepancies, they are the exception, not the rule in terms of evolutionary trends [635]. The two dentitions have remained similar in most mammal lineages even as dental formulae changed [2027]. Is this coevolution of upper and lower teeth due to some sort of yoke that makes it easier to alter both jaws at once [1400]? Or does natural selection enforce similarity by weeding out any large deviations with no help at all from any covariation constraints? Unfortunately, we don't yet know how the "jaw module" is wired in the mammalian genome [635], nor what genes therein have been captured by the *Dlx5/6* selectors in different lineages over time [1936,2780].

Why do our ears start out on our neck?

One month after fertilization, we look strikingly like a fish (Fig. 4.6a) [719]. Our eyes are on the sides of our head, we have what look like fins (our limb buds), and our neck has grooves that, in a fish, would become gill slits [1498]. In humans, these grooves never deepen enough to perforate, and all of them, except the first one, disappear. It forms the tube that funnels sound to our eardrum. In rare cases, remnants of the other grooves persist into adulthood as cervical or pharyngeal fistulae [685], and such people may drool through these residual holes in their necks [91,182].

The grooves alternate with bumps called "branchial arches" (BAs) [2295,2806]. In fish, the BAs form gills [989,2296], but in humans, they make a potpourri of other structures, including our outer ear (BA1 and 2), which is visible

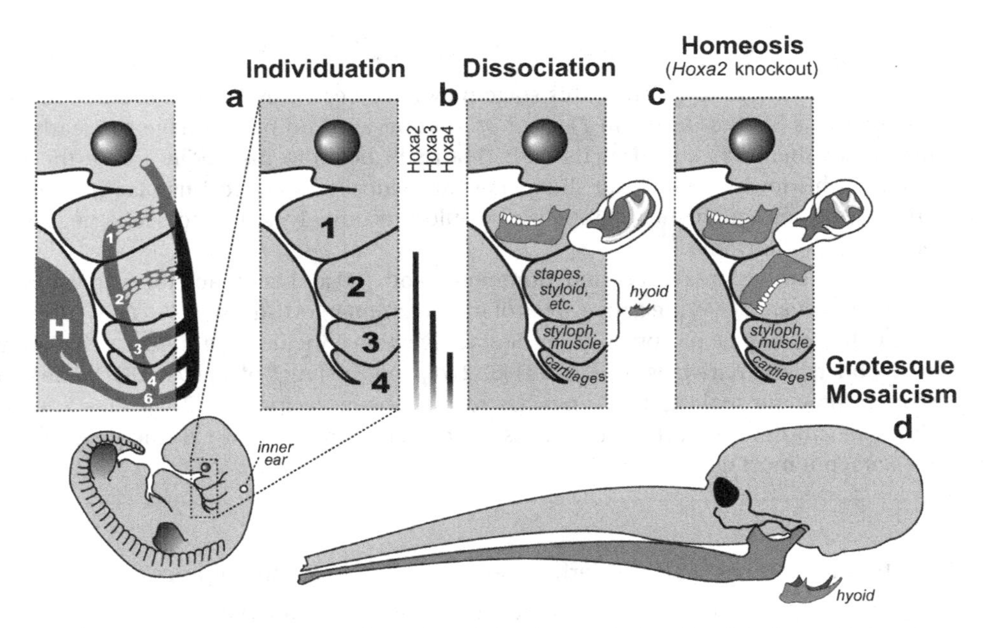

Figure 4.6. Development (a, b), transformation (c), and fanciful evolution (d) of the branchial arches.

a. Human embryo at 31 days postfertilization with neck region magnified to show branchial arches (BA1-4), branchial clefts (grooves between the arches), and underlying aortic arches (cartoon at left) that convey blood (*arrow*) from the heart (H) [2311]. Numbering obeys phyletic convention. BA5 never forms in humans [143], nor do the clefts perforate as in fish, where they would form gill slits. Aortic arches 1 and 2 disintegrate (schematized as Swiss-cheese holes) shortly after they arise and individuate (*cf.* [43] for variations). This strategy of making structures and then destroying them is idiotic from an engineering standpoint but makes sense evolutionarily, given that plumbing, in general, is easier to revamp than to reinvent [1838]. Darkly shaded protrusions are limb buds. The large circle is the eye primordium.

b. Some BA derivatives [2295,2311]. BA1 makes our lower jaw, BA2 our styloid process, BA3 our stylopharyngeal muscle, and BA4 our laryngeal cartilages. Our outer ear arises from BA1 and BA2, as do our middle ear bones (stapes from BA2; others from BA1) [91,757]. Our inner ear starts dorsally (**a**) as an otic pit that sinks to form a separately developing vesicle [229] in accordance with its own (ancient) ancestry [2513], although, interestingly, it also requires *Dlx5/6* [2201]. Our hyoid bone comes from BA2&3.

c. Transformation of BA2 to BA1 due to inactivation of *Hoxa2* in mice [880,2180], rendered here in human form. A second lower jaw forms in mirror image to the normal one [2420]. Mirror duplication of the front part of the outer ear is inferred, given that approximately the front one third comes from BA1 and the hind two thirds from BA2 [719]. Absence of hyoid (part of which comes from BA2) has been documented in frogs [130].

d. Head of a human redrawn to approximate the oversized snout of the giant anteater *Myrmecophaga tridactyla* [1669]—a feature also found in cetaceans (*cf.* [908]; their Fig. 23.18). The opposite disjunction (hyoid *vs.* jaw size) occurs in male howler monkeys [1161,1162], monitor lizards [2527], and predatory frog tadpoles [615].

on the side of our neck by 2 months [376,719]. How strange to think that our ears start out below our jaw and have to migrate up the sides of our face to reach their final position!

Here we have a classic case of an evolutionary change in function (breathing via water, then hearing via air) of a structure (first branchial groove) that led to its relocation [2802,2844] (*cf.* mammalian middle ear bones [1670,2168]).

REFLECTIONS ON FIGURE 4.6

a. Development of the aorta (not shown) involves unilateral destruction of an aortic arch [2876]. Interestingly, we make an aorta from our left fourth arch and erase our right one [2311], whereas birds do the opposite [1838]. Because mammals and birds both evolved from reptiles, in which arches stay nearly symmetrical, we are left to ponder why the left was commandeered in one clade and the right in another [1488,1838].

It is not known whether the nested domains of *Hox* gene expression (vertical bars [1936]) are delimited by a morphogen gradient(s) [990]. The domains may provide a code like that for vertebrae (*cf.* Fig. 4.3) [995].

Internally, the evidence of our aquatic history is even more compelling. Human embryos go to the trouble of making five pairs of aortic arches (which once sent blood to five pairs of gills) but then destroy two of them completely [2311,2806]. This Sisyphean stupidity only makes sense as a historical constraint [959]: it must have been genetically easier to reconfigure the existing plumbing than to scrap it altogether and start afresh [2876].

The uniform gill precursors of our fish ancestors were only able to diverge developmentally after they acquired different identities genetically [1626,1694]. Those identities were furnished by a version of the same *Hox* code that specifies vertebrae (*cf.* Fig. 4.3) [1936,2561]. Proof of this circuitry was adduced in 1993 when two teams of researchers knocked out the *Hoxa2* gene in mice [880,2180]. The resulting mutants had an extra lower jaw (albeit partial) below the normal one, and it was upside-down! Thus, *Hoxa2* must not only encode BA2, but there must also be a ZPA at the BA1–BA2 boundary, just as there is between BA1 (the mandibular arch) and the maxillary prominence [283]. The generality of these results was confirmed when *Hoxa2* was disabled in frogs, producing the same effect [130], and when it was pervasively misexpressed in chicks, where it yielded a reciprocal transformation (BA1 to BA2) [995].

The *Hoxa2* loss-of-function syndrome entails disappearance of the hyoid cartilage, which normally arises partly from (the now-transformed) BA2. Evidently, this code enabled the hyoid to grow to huge size (*cf.* Fig. 4.2) in male howler monkeys, where it amplifies their haunting screams [1161,1162]. An opposite trend in hyoid versus jaw size is seen in cetaceans and anteaters, where the jaw enlarged tremendously (Fig. 4.6d) [1669].

Why is our arm built like our leg?

As discussed earlier, pterosaurs had a huge fourth finger, but their fourth toe was not mentioned because it was not unusual [426]. The lack of a sympathetic change in the hindlimb is surprising, given the situation in the panda. In that case, the wrist bone that became the pseudo-thumb (radial sesamoid) enlarged along with a similar change (albeit less striking) in its ankle counterpart (tibial sesamoid) [578,1904], although the latter is not used at all for stripping bamboo [579,718] (*cf.* moles [1399]).

Stephen Jay Gould (1941–2002) [384], who popularized the panda's thumb as the quintessential quirk, also proposed (along with Richard Lewontin) the architectural term "spandrel" (a nook between two arches [1654]) for traits such as the panda's pseudo-toe that confer no advantage but evolve nonetheless because of a genetic link with traits that do [982,2105]. Gould imagined that *all* meristic modules tend to evolve together and hence evoke spandrels easily (boldface added) [958]:

> Repeated parts are coordinated in development; **selection for a change in one element causes a corresponding modification in others**. It may be genetically more complex to enlarge a thumb and *not* to modify a big toe, than to increase both together. (In the first instance, a general coordination must be broken, the thumb favored separately and correlated increase of related structures suppressed. In the second, a single gene may increase the rate of growth in a field regulating the development of digits.)

It is worth noting that Darwin himself expressed the same basic insight in *Origin of Species* (boldface added).

> We should remember that . . . **correlation of growth** will have had a most important influence in modifying various structures. . . . Moreover when a modification of structure has primarily arisen from the above or other unknown causes, **it may at first have been of no advantage** to the species, but may subsequently have been taken advantage of by the descendants of the species under new conditions of life and with newly acquired habits. [559] (p. 196)

Fore- and hindlimbs develop as serial homologs [2245] and, as such, are thought to use a shared "limb module" that dates back to their origin as paired fins [1034,1059]. Darwin was aware of arm–leg covariation and alluded to it in *Descent of Man*:

> Homologous structures are particularly liable to change together, as we see . . . in the upper and lower extremities. Meckel long ago remarked that when the muscles of the arm depart from their proper type, they almost always imitate those of the leg; and so conversely with the muscles of the legs. [561] (Vol. 1, p. 130)

The existence of a shared limb subroutine is supported by analyses of gene expression in the hand versus foot [1240,2184]. Any tinkering within this algorithm should modify all limbs alike [45,2317,2887]. Walter Gehring (b. 1939), who coined the term "homeobox," tells a cute anecdote about a person with just such a rare mutation [875]:

> It was a hot summer day, and I was on a flight from New York to Seattle to give a talk at the University of Washington. On board I was seated next to a young woman who, judging from her English accent, might have been of Scandinavian origin. As we fastened our seat belts, I noticed that she had unusually short second [index] fingers on both hands. I remembered from lectures in genetics, which I had attended as a young student, that this "short fingeredness" is a heritable trait that was first described as "brachydactylism" in a Norwegian family and is caused by a dominant mutation in a single gene. A defect in this one gene reduces the second bone of the second finger, which means that the normal gene controls the size and shape of just this one bone. . . .
>
> What I did not remember . . . was whether this particular mutation affects only the fingers or whether the corresponding toes are also reduced in size. So I decided to have a look at my co-passenger's toes. This was not easy, however, as there is never enough leg space in tourist class. Fortunately, the young woman was wearing sandals, and I casually dropped my newspaper so that I could inspect her toes while trying to collect the paper from the floor. Indeed, I can assure you that she also had very short second toes. This observation reveals [that] the same gene controls homologous structures.

Intramodule mutations may indeed have helped running quadrupeds whittle their feet concordantly to a 1-digit (*e.g.*, horses [2844]) or 2-digit (*e.g.*, deer [1335]) condition [1159,2060,2724]. In contrast, flying vertebrates (pterosaurs, bats, and birds) had to break this lockstep in order to fashion their forelimbs into wings [868,1894]. The same is true for bipeds like humans where our arms are used for grasping (*vs.* walking) [1103,1370].

Why does our arm differ from our leg?

To make one version of a module different from another (e.g., arm *vs.* leg) requires a selector gene outside the circuitry of the module itself [1566,2727,2762]. In mammals, that role is played by the homeobox gene *Pitx1* (a cousin of the *Pitx2* gene that enforces L–R asymmetry; *cf.* Ch. 2) [1888]. *Pitx1* is expressed and required in the hindlimb but not the forelimb [1641,2343], it transforms a forelimb into a hindlimb when misexpressed there [623,1765], and its absence partly transforms a hindlimb into a forelimb [1477,2540]. A separate *Hox* code of some sort is also involved [1902]. Presumably, pterosaurs used *Pitx1* to prevent their feet from making a huge digit like the one that evolved in their hands.

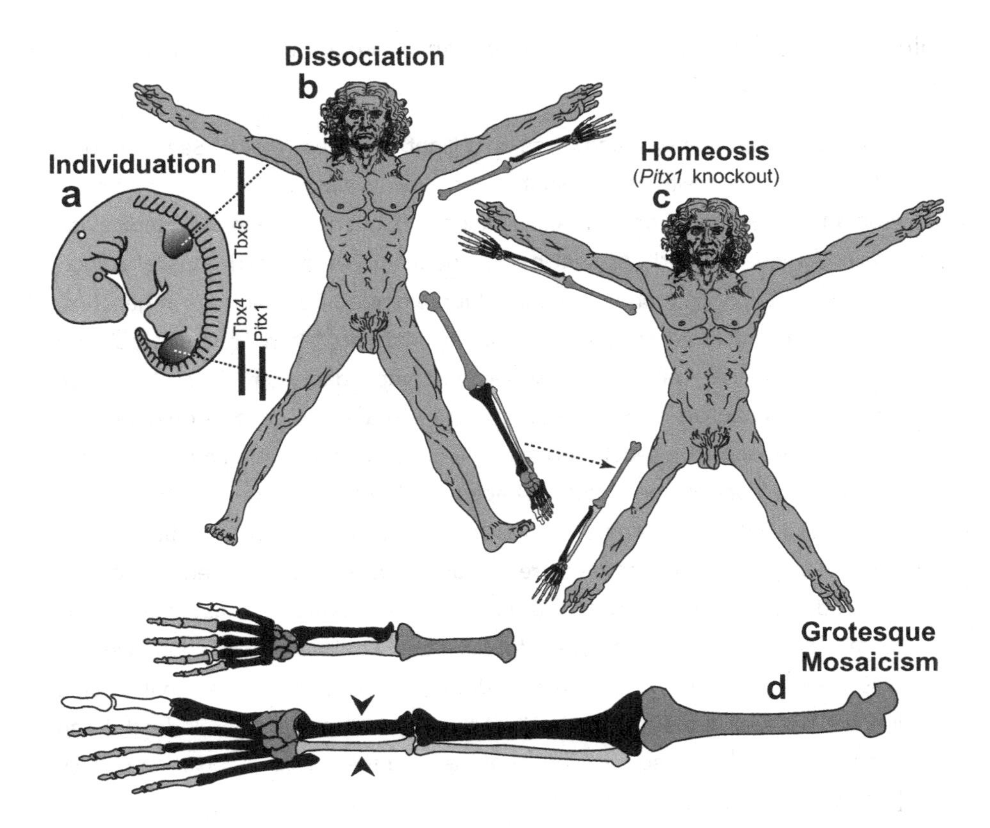

A dramatic case of leg specialization is shown in Figure 4.7d. Frogs have an extra tibia and fibula [1078], which help them jump [715]. These long bones evolved from nodular ankle elements [381,2633], possibly via a *Hox* mutation(s) [2545] that spread through the founder population of the anuran clade [218]. A virtually identical adaptation arose independently in our leaping primate relatives, the tarsier and galago [218,715,873].

Bats exhibit an even more extreme case of limb divergence in terms of size. Their fingers became as long as their entire body to span their umbrella-like forewing, but their toes stayed small [1869]. As with the frog leg, trivial mutations may have been instrumental in this reconfiguration [513]. All it took to enlarge the fingers, apparently, was an excess of BMP2 secretion late in hand development [2328]. Their wing membrane itself may also have arisen just as simply [1372] because interdigital webbing is actually the default condition in tetrapods [2570]. Our own hands would be as webbed as a bat wing (or a duck foot [2591]) were it not for programmed cell death, which carves gaps in the skin between our digits [1970,2924]. One possible factor in the wing–leg disparity of bats has been ruled out: their wings do *not* get a head start in their growth [202].

If it is really so easy to evolve a bat wing, then why hasn't a "Batman" mutant arisen by now in our own species? Webbed digits are found in

Figure 4.7. Development (a, b), transformation (c), and fanciful evolution (d) of the arm and leg.
a. Human embryo at 31 days postfertilization, showing limb buds (shaded protrusions) [1888,2311]. Vertical bars designate genes expressed in each respective limb bud [2568]. In chick embryos, forelimb versus hindlimb identities can be switched by misexpressing *Pitx1* [1567], *Tbx4*, or *Tbx5* [2205,2559]. In mice, *Pitx1* is also effective [623], but *Tbx4* and *Tbx5* are not [1113,1765,1844]—a phyletic difference that is not yet understood [1208,1728,2503]. *Hox* genes along the body axis are also involved in assigning identities (*cf.* Fig. 4.3) [318,1902,2258], although the exact code is unclear [623]. Many other genes are expressed differently in arm versus leg [1245,2382], some of which are undoubtedly downstream effectors.
b. Normal human anatomy. The skeletal similarity of the arm and leg—indeed, of all vertebrate limbs—bolstered Darwin's case for common descent [559]: "What can be more curious than that the hand of a man, formed for grasping, that of a mole for digging, the leg of the horse, the paddle of the porpoise, and the wing of the bat, should all be constructed on the same pattern, and should include the same bones, in the same relative positions?" In other primates, the feet are more handlike in both structure and function [2321].
c. Transformation of leg into arm due to inactivation of *Pitx1* in mice [2540], rendered here in human form. In fact, the observed homeosis is only partial, and no 4-armed humans exist [1638]. However, several families of people who walk on all fours have recently been discovered in Turkey [2564].
d. Human limb bones (arm above; leg below) redrawn to imitate those of the American bullfrog *Rana catesbeiana* [1078]. Arrowheads indicate the extra tibia ("tibiale") and fibula ("fibulare") that characterize all extant frog species [381,2633]. This ability of frogs—and a few primates [715,873]—to convert ankle bones into long bones refutes the cliché [2503] that tetrapod limbs have an immutable (4-zone) bone formula along their length—*viz.*, "1:2:many:≤5," where "≤5" denotes the number of digits. This heretical "magic trick" may not have been as hard as it seems: notwithstanding their later disparity, the *rudiments* of ankle bones are the *same* size as the long bones when they first arise [1622,2526], but they then stop growing in most vertebrates [2844]. Hence, frogs might have lengthened them by merely unleashing their inherent growth potential.

---

certain syndromes [1711,2512], but body-length fingers have never been documented [1427,1638,1939,2582]. The answer to this mystery may come from further studies of bat-wing development at the molecular and morphological levels [428,513,1372,2887].

Could centaurs (or similar hexapods) ever evolve?

One aspect of anatomy that *is* easy to alter is the number of limbs. An extra limb can be elicited by merely exposing the flank of a chick embryo to a dose of FGF proteins [489,1758,1921]. Such appendages can be winglike, leglike, or a patchwork thereof, but they are never a blend that splits the difference [1248,1922,2560]. What this means for humans is that we couldn't easily evolve an "arg" (arm–leg intermediate), despite having all the needed ingredients to make such a limb in our genome [473]. Nevertheless, it should have been trivial to evolve extra legs through the capture of an *FGF* gene by a thoracic transcription factor (see Fig. 4.2) [1577]. Thus, it seems surprising that no centaurs or other multilegged vertebrates ever evolved [275,556,2562,2812]. Perhaps such mutants did arise but could not walk because of a lack of needed locomotory circuitry in their brains [692,1954]. Indeed, whenever extra legs are induced, they have a mirror-image polarity [2409,2543,2565], so an extra pair of (backward) legs would pose a serious "pushmi-pullyu" dilemma for their bearers (a reference to the two-headed beast in Hugh Lofting's Dr. Dolittle stories).

Even if our current status as tetrapods prevents us from becoming hexapods, that does not mean that multilegged vertebrates could not have evolved if history had unfolded differently [959]. When our fish ancestor first "walked" onto land ~375 MYA [27,553], it left behind related species that had more than two pairs of ventral fins (*e.g.*, acanthodians) [1335,2417]. Had one of them crawled onto land instead to become our ancestor, then we might have more than two pairs of appendages today [1059]. Could vertebrates handle an extra pair of arms or legs? Probably easily [556], considering how well insects and spiders manage with three or four pairs of limbs, respectively.

Given that our arms and legs evolved from paired fins, the next obvious question is: Where did paired fins come from? Molecular–genetic evidence (from lampreys [1455], sharks [491,817], and zebrafish [541]) indicates that they are retooled versions of median fins [2]. Median fins, in turn, may have arisen as miniature copies of the body axes as a whole [473,1759,1761]. Asking where the body axes came from takes us back to the Precambrian origin of bilaterians [124,604,1916], when the newly minted *Hox* complex [863,1074] assumed its role as an analog-to-digital converter [386,2010] for body-spanning morphogen gradients [440,884]. We are just starting to learn how those morphogens [17,1386,2792], their gradients [108,2473], and transduction pathways [193,2040] were harnessed as single-celled protozoans became multicellular metazoans [383,395,1828,1879,2293].

Evo-devo researchers are lifting these veils one by one, rewinding the movie of evolution frame by frame as far back as the data permit, and discovering how the sundry parts of our body arose. Darwin would be pleased.

When did we gain the upper hand?

The examples examined in this chapter have shown how members of meristic series evolved as separate compartments through transcription factors whose expression is spatially restricted. There are a few instances in human anatomy where individuality was acquired by the two halves of a single structure, rather than within a series per se. Chapter 2 discussed one such instance: the two halves of our body evolved differences after Pitx2 was restricted (via *nodal*) to our left side.

Another such case involves the two surfaces of our hand (upper *vs.* lower) [1888]. Outgrowth of vertebrate limbs relies on FGF produced along the boundary between the dorsal (D) and ventral (V) surfaces [1666,1888]. The homeobox gene that distinguishes the D versus V domains of our arms and legs (*Lmx1*) [2008,2174] is closely related to the gene that specifies D versus V surfaces in insect wings (*apterous*) [2183], so the underlying genetic circuitry may have evolved in Urbilateria [195,2096,2546].

In 2003, a baby girl in Saudi Arabia was described with "V/V" hands. She had palmar patterns on both the upper and lower sides of her hands, and she had no fingernails except vestigial ones on three fingers [41]. In its extreme manifestation, her case is unique, but the opposite "D/D" phenotype—"circumferential" fingernails and no palms—has been documented in about 20 other patients [40]. Such homeoses are clearly disabling and probably devoid of evolutionary potential, although horses' hooves do resemble D/D fingernails. They might be throwbacks to the primal D/D (?) symmetry that created the fins of our distant fish ancestors [474,2568]. Entire legs with perfect V/V or D/D symmetry are actually quite common in mutant strains of flies [1139,1140]. Not surprisingly, such flies can't walk.

How did evolution sculpt our foot?

Of all the bones in the human body, our fingers and toes are unique insofar as each of them has its own internal plane of symmetry that is separate from the body midline [1188,2489]. All of our other symmetric bones—for example, vertebrae—straddle the midline. This feature may be a legacy of the genetic gadgetry that was co-opted to create digits in our amphibian ancestors [2723,2742].

Our feet differ from those of other primates insofar as our toes are shorter and our big toe is not opposable [1813,2317,2539]. Hominins may have begun to walk bipedally on branches as orangutans still do [1905,2602], but the selective pressures on our feet certainly intensified once we came down from the trees and started walking longer distances on flatter terrain [10,2530]. The biomechanical forces that came into play are well understood [54,2075], and the fossil trail of the foot's history is well documented [1088,2737]. What remains unclear, however, is how the reshaping was implemented developmentally.

A clever explanation was formulated by Louis Bolk in 1926 [232] and popularized by Stephen Jay Gould in his seminal 1977 monograph *Ontogeny and Phylogeny* [956]. Bolk argued that the quirkiness of our big toe is attributable to a pervasive slowdown in our development relative to other primates—a genetic "knob" that could be easily tweaked mutationally. All primate big toes arise in an unrotated posture [2319,2321] and then rotate [2276], except in humans. If toe growth slowed enough in our lineage, then the rotation stage could have been eliminated entirely as a side effect of the available time expiring before rotation could begin.

Any evolutionary change in the rate of development, be it speeding up or slowing down, global or local, is termed "heterochrony" [980,1710]. Gould proposed that many other aspects of our body evolved as by-products (spandrels) because of the integrated way that growth is programmed by our genes (*e.g.*, our skull [1772]). Some of those other features are examined in the light of this hypothesis in later chapters.

If Bolk and Gould are correct, then humans are essentially baby apes who never grew up. How comforting to think that, although our bodies must age, we can all still emulate Peter Pan . . . at least to some extent!

CHAPTER 5

Sexual Dimorphisms

The key lesson from the previous chapter is that no member of a meristic series (finger, tooth, vertebra, etc.) can deviate appreciably in its anatomy until it acquires a distinctive identity. Identities are conferred by the ON/OFF states of selector genes that encode transcription factors.

Once a member(s) of an array *does* come under the jurisdiction of its own selector gene(s), a new "compartment" can form within the genome. If the selector gene later captures a target gene(s) that affects local growth, then this body part can attain a unique size and shape (*cf.* Fig. 4.2).

Why do men and women differ in size and shape?

From the standpoint of our genome, the states of the system that we call Man and Woman are operationally no different from Thumb, Canine, or Sacrum. They're just other compartments [885,1392,1393] without the meristic trappings [151]. In short, men and women differ in anatomy for the same reasons that our thumb differs from our forefinger: (1) they employ different selector genes, and (2) those regulatory genes control different sets of downstream genes that directly affect size and shape.

The sexual equivalent of the "grotesque mosaicism" seen in some meristic patterns (*cf.* Figs. 4.1–4.7) would be a disparity so dramatic that the two sexes look as if they belong to separate species. Gorillas are a good example within the primates [1361,1667], and it is not hard to find similar cases in most animal groups [71,962,1315,2079,2370]. Goldschmidt put it cogently:

> The sexual alternative furnishes a case in which the developmental processes within a species may become so different that the resulting organisms, the two sexes, may exhibit differences of a macroevolutionary order of magnitude. [930] (p. 298)

By comparison, the sexual dimorphisms of *Homo sapiens* are relatively subtle. This chapter surveys those differences and explores how the genetic circuitry of sex determination "captured" them.

One dimorphism of humans that is not seen in our ape relatives concerns our vertebrae. Our backbone's angle relative to Earth's gravitational field changed radically when our hominin ancestors became bipedal ~7 MYA (*cf.* Ch. 7). In response to the new weight distribution, our spine changed from a bridgelike arch to an S-shaped curve [2456]. Pregnant females suffered an added burden from the weight of the fetus in their womb, and a dimorphism eventually evolved that adjusted their center of gravity by lumbar lordosis (bending of the lower back): women have smaller dorsal wedging angles in their L1–L4 vertebrae than do men, and their L3 vertebra in particular has a manifestly different geometry (wedge *vs.* block; Fig. 5.1) [2795].

When did lumbar reshaping evolve in female hominins? The dimorphism must have arisen fairly soon after the onset of bipedalism because it is detectable in the spines of *Australopithecus africanus* [2795].

How was the reshaping achieved genetically? We don't yet know. Presumably, it entailed a new link between the master gene for femaleness and a target gene(s) for vertebral morphogenesis. Whether the link was forged through a novel enhancer site (*cf.* Fig. 4.2) remains to be determined [2164], as does the role of gonadal hormones as intermediate effectors [686,1737,2160,2342].

A more obvious distinction between men and women is the width of our hips [525,2551]. The grace of our bipedal gait (*vs.* the waddling of chimps) was achieved by reducing hip breadth [1579,1581], but the female pelvis can't get too narrow because the baby's head has to traverse its opening (*cf.* Ch. 6) [5,1496, 2320]. These opposing pressures eventually reached a compromise [2081,2626]: the pelvis is proportionally wider in women [525], but just enough so that the baby's head can usually squeeze through [10,502,1807].

The genetic basis for this change remains obscure, although it must have been relatively simple because some pelvic dimorphism is already evident in early hominins [2048,2220,2553]. Separate mutations must have targeted fat deposits to female hips [888]—a site unique among primates [1580]—to accentuate the hip-to-waist ratio that men find attractive [2236,2537].

A correlated mystery is how evolution managed to wire men's brains so that they perceive an hourglass shape as alluring [161]. Can natural selection make *any* shape erotic? A pear? A Volkswagen? If so, how long does such reprogramming require [1811]?

Our most glaring dimorphism, of course, is our adult body size [889,2208]. Males are typically bigger than females in all apes except gibbons [1513,1615], and the same is true for mammals in general [922,1491,2114]. Among animal species overall, however, *females* tend to be bigger [742,2369]. Indeed, in some groups the males are *ridiculously* tiny [1184,1342,1897,2034,2702]—for example, certain species of Darwin's beloved barnacles [2507]. Laugh not, however, male readers, for this fate could have befallen us, too, if we had tread a different ecological path [962].

Various social factors may have fostered sexual selection for unequal size in our lineage (*e.g.*, mating systems, suitor rivalry, or division of labor) [1580, 2045,2381], but none has yet been tested rigorously [72,219,809]. As for the *molecular* causes of size disparity [116,537], the chief suspects are growth hormone [1542] and IGF-1 (Insulin-like Growth Factor 1) [2221]. IGF-1 played a major role in the vast range of body size among dog breeds [2533].

Much more is known about the genetics of dimorphisms in flies than in humans [1424,2830] for the simple reason that experimenting with flies is devoid of ethical restrictions, and it is likely that at least some of what we learn in that model system will eventually apply to us [537]. The traits that have been analyzed most intensely are abdominal pigmentation [1294,1423,2093,2476,2837], wing spot localization [266,939,2094], wing size [1202], and sex comb formation [1425,2567]. In three of the four cases (wing size is the exception), the dimorphisms have been traced to changes in *cis*-regulatory enhancers [537]. However, we still have no idea why male flies have an abdominal muscle (function unknown, lacking in females) that predates the genus ($\leq$62 MYA [103,837,1942]) and may go back before the dawn of dipterans ($\leq$250 MYA [836]).

A major difference between humans and flies is that we have circulating gonadal hormones [221,2160], whereas they do not [2469] (but *cf.* [2351]). Hence, it may have been easier for our forebears to make an organ dimorphic by just expressing androgen or estrogen receptors on its cell surfaces [235,400,2342] instead of having to fiddle with individual target genes to such a great extent [21,65,327,2299].

The puzzle that dimorphic organs pose is how their genes "decide" to turn ON or OFF [21,65,244,327] so that they make an anatomy appropriate to their sex. Looking at the situation from the gene's perspective, the deliberation that governs its decision can be phrased as a first-person imperative:

> **IF** (Input #1) "I am in organ O" {mediated by a region-specific selector gene(s)}
>
> **AND** (Input #2) "I am of sex S" {mediated by a gender-specific selector gene(s)},
>
> **THEN** (Output) "I will turn ON (or OFF)."

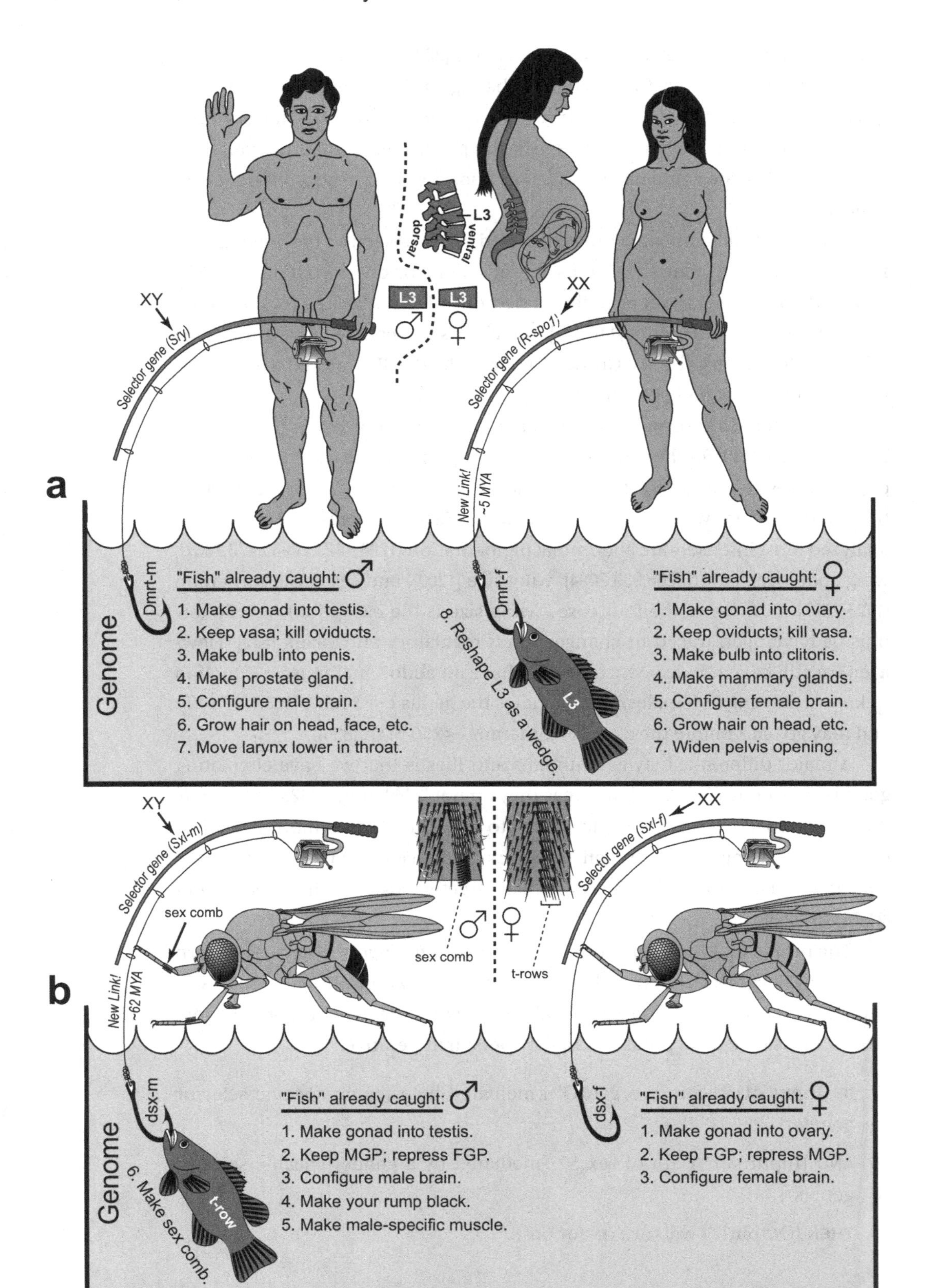

XY
Selector gene (Sry)
dorsal
L3
ventral
L3
L3
♂
♀
XX
Selector gene (R-spo1)
a
New Link!
~5 MYA
Genome
Dmrt-m
"Fish" already caught: ♂
1. Make gonad into testis.
2. Keep vasa; kill oviducts.
3. Make bulb into penis.
4. Make prostate gland.
5. Configure male brain.
6. Grow hair on head, face, etc.
7. Move larynx lower in throat.
Dmrt-f
8. Reshape L3 as a wedge.
L3
"Fish" already caught: ♀
1. Make gonad into ovary.
2. Keep oviducts; kill vasa.
3. Make bulb into clitoris.
4. Make mammary glands.
5. Configure female brain.
6. Grow hair on head, etc.
7. Widen pelvis opening.
XY
Selector gene (Sxl-m)
sex comb
♂
♀
sex comb
t-rows
XX
Selector gene (Sxl-f)
b
New Link!
~62 MYA
Genome
dsx-m
6. Make sex comb.
t-row
"Fish" already caught: ♂
1. Make gonad into testis.
2. Keep MGP; repress FGP.
3. Configure male brain.
4. Make your rump black.
5. Make male-specific muscle.
dsx-f
"Fish" already caught: ♀
1. Make gonad into ovary.
2. Keep FGP; repress MGP.
3. Configure female brain.

Figure 5.1. Genetic basis for sexual dimorphisms in humans versus fruit flies. One example per species is used for illustration, with samples of other sex-limited traits listed in "gene pools" below (*cf.* Fig. 5.2). The fishing theme extends a metaphor begun in Fig. 4.2: fishing poles represent master genes, and the fish symbolize target genes (written here as English imperatives to make specific anatomical traits).

a. Human dimorphisms. Men and women differ in various features, including their lumbar vertebrae, a disparity that evolved after our ancestors became bipedal [2795]. A pregnant woman (side view) bends her back (lordosis) more than a man to bear the weight of the baby. Her L1–L5 vertebrae are magnified alongside, with L3 schematized to contrast its shape (wedge) with that of the male L3 (block). The master gene for maleness (under the XY switch) is *Sry* (*Sex-determining region [on the] Y*), and it has been known for some time [2056,2756], whereas its presumptive counterpart (under the XX switch [221]), *R-spo1* (*R-spondin1*), was only identified recently [2808]. R-spo1 is a diffusible signal that may use Wnt-pathway transcription factors to control its target genes. *R-spo1* is presumably able to reshape L3 because of its having captured a gene(s) that slows growth in the dorsal versus ventral half of the vertebra. That capture must have occurred ~5 MYA (cartoon below where the fish denotes an eighth trait) because it is seen in *Australopithecus africanus* [2795]. The proximate "bait" (at the end of the sex determination pathway but before any target genes) may be *Dmrt*, the human homolog of *dsx* in flies, which, like *dsx*, is spliced differently in males and females (suffixes "-m" and "-f") [2899]. Of the 8 putative *Dmrt* paralogs in humans [1418], only *Dmrt1* (on chromosome 9) seems critical [1957], although much remains to be learned about this circuit [221,270,2333].

b. Fly dimorphisms [327]. Interestingly, the species *D. melanogaster* is named for a dimorphism—namely, the dark (*melano-*) abdomen (*-gaster*) in the male. We've learned a lot about the genetic basis of this trait lately [1294,1423,2476,2837]. MGP and FGP are male and female genital primordia, respectively (*cf.* Fig. 5.2g–i). The "sex comb" (so named because it resembles a hair comb and is present in only one sex) is a row of bristles that may help males grip females during mating [1874], although it could be just an ornament [2055]. The cylindrical tarsal segment where it is located is drawn as a panorama (male *vs.* female) by imaginarily cutting along the dorsal midline and spreading it out flat (proximal above, distal below). Flies are covered with intricate patterns of this sort [1134], but only the sex combs are shown in the whole-body sketches for the sake of clarity. Developmentally, the comb arises as a transverse row (t-row) that rotates ~90 degrees [1138,2567]. Evolutionarily, its bristles became thicker, blunter, darker, and more numerous as it underwent alterations in various lineages [1425]. The taxonomic distribution of sex combs indicates that they originated near the base of the subgenus *Sophophora* (after its divergence from other subgenera but before the splitting of *melanogaster* and *obscura* species groups) [1425] ~62 MYA [103]. Genetically, the sex comb and other dimorphisms depend on a hierarchy of control genes [457,993], beginning with ones that count the number of X chromosomes (*sic, not* the X:A ratio!) [727]. On the basis of that number, the master switch *Sex-lethal* [1576] gets spliced differently in males (*Sxl-m*) versus females (*Sxl-f*) [245] and ultimately controls the gene *doublesex* (*dsx*) [121,526], which also has two isoforms (*dsx-m vs. dsx-f*) [448,1604]. Herein lies the most startling aspect of the human–fly affinity: we use a homolog of the same gene to regulate our dimorphisms (*Dmrt*) [1200,1378,1957]! Thus, the involvement of *dsx* in sex determination must predate the divergence of chordates, arthropods, and nematodes (see text).

An example from *H. sapiens* might be a growth gene that behaves as follows so that the pelvis grows wider in females than in males.

> **IF** (Input #1) "I am in the pelvis"
>
> **AND** (Input #2) "I am female,"
>
> **THEN** (Output) "I will turn ON at higher volume, faster rate, or longer duration."

REFLECTIONS ON FIGURE 5.1

We know a lot more about the genetics of sex determination in flies than in humans, thanks to decades of ingenious research—one of the most engrossing detective stories in the history of science [467]. Nevertheless, there is not a single sex-assignment pathway in any species that makes sense from the standpoint of efficiency [2064,2814]. Every cascade is more complex than it needs to be [2815], especially considering that all they really need is one switch [1172]. Of course, the same accusation could be leveled at lots of gender-neutral signaling pathways as well [1137]. (EGFR comes to mind [1137].) Baroque control systems like these epitomize the layering of evolutionary tinkering [226,683,1265]. They look as if they were wired by a succession of schizophrenic electricians who insisted on "improving" the shambles they found when they arrived. Kooky examples—don't laugh!—include co-opting (1) a catenin to regulate transcription for the Wnt pathway [194,1015] and (2) a kinesin to relay signals for the Hh pathway [749], not to mention all the phosphatases that were recruited to undo what the last bloody kinase did and vice versa [1231,1232]. Indeed, our genome has even more dumb quirks than our body! In a pinch, what matters for evolution is expediency, not efficiency! As for why dimorphisms evolved from an *ecological* perspective, see ref. [2370].

a. One elusive piece of the dimorphism puzzle concerns our brain (item #5 on both lists in the figure). How different are men and women mentally, and to what extent are the disparities hardwired genetically [161]? In mice, surprisingly, the brain's default state appears to be male, not female (as has been assumed for the body as a whole [221]) [2442]. In flies, much progress has been made in dissecting courtship behavior [197,2175,2375,2891] and male combat [660], and those findings might have *some* relevance to us, given that (1) we share *dsx/Dmrt* as a regulator [2273], and (2) we (like flies) express some direct, cell-autonomous (hormone-independent) effects of our XX versus XY constitution within our brains [649] (*cf.* marsupials [2157]), notwithstanding our reliance on gonadal hormones [221]. The virtue of studying fly (*vs.* human) neurobiology is that some clever experiments can be done. For example, when dimorphic neurons were placed under the control of a light trigger, the researchers could turn the male's courtship song ON or OFF with the flip of a switch and thus localize the serenade circuit within the brain [470,673,2892]! See Fig. 6.1 for the source of the human drawings. (Apologies to the artist for doctoring them here!)

b. The 25% difference in body size (female > male) is omitted in this "fly-fishing" diagram [537]. The male-specific muscle in flies can actually be formed from female cells [1385] because it is induced by a neighboring motor neuron whose gender is the deciding factor. The capture of the distalmost t-row by *dsx* evidently occurred via a new link between *dsx* and *Scr* [135]—*viz.*, a *de novo* sex-specific regulation of the *Hox* selector gene *Scr* (*Sex combs reduced*) [1997], which governs the foreleg-bearing body segment [134]. How the enslaved t-row became a comb is actively being researched [1007,2124,2630]. The homology of *dsx* in flies with *Dmrt* in humans is startling. For a survey of similar homologies, consult Tom Brody's *Interactive Fly* Web site. Bristle maps are adapted from ref. [1137].

Presumably, both inputs (site and sex) involve transcription factors that are jockeying for binding sites in the *cis*-regulatory region adjacent to each target gene (*cf.* Fig. 4.2) [1365,1518,2444]. The combinatorial (**AND**) logic would rely on fitting those factors together like jigsaw-puzzle pieces into a configuration [373,1326,1729,2672] that stimulates or represses transcription [826,1428,2517]. Evo-devo is just beginning to scratch the surface of how transcription factors dovetail sterically to implement Boolean logic [240,902,1413,2850] (*e.g.*, [1541]). Solving this riddle should reveal, once and for all, how evolution writes cellular commands in genetic language using protein grammar [572,656,1137,2855].

Why does our species have as many males as females?

Given the randomness of how target genes get recruited in general (*cf.* Fig. 4.2) [897,2835], we might expect men and women to differ in the expression of more than just the genes that overtly affect anatomy. Indeed, microarrays have recently revealed *thousands* of sexual expression differences [713,1251,2186]. Some of the disparities occur in organs that look *mono*morphic (*e.g.*, kidney [2185], liver [2760], and muscle [2874]). Many of these links are probably adaptively neutral [1369,1741,2206] (*cf.* fly genes [300,1062,2084]), but this does not mean that they are clinically negligible. On the contrary, some of them probably contribute to a whole spectrum of sex-linked diseases [139,161,339,1000,1854], and this covert network of haphazard connections might also explain why certain quiescent organs are counterintuitively so prone to cancer (*e.g.*, breast [354] and prostate [2745]) [842,2917].

The target genes being discussed, of course, reside at the *bottom* of the pathway for sex differentiation. Genes at the *top* of the chain of command tend to reside on special chromosomes.

Sex chromosomes determine gender in most animal species [423,808,1174], and ours is no exception. In humans and other placental mammals, eggs fertilized by X-bearing sperm become females (XX), and those fertilized by Y-bearing sperm become males (XY) [762]. Because men make X- and Y-sperm in equal amounts, every fertilization is as random as flipping a coin: "Heads you're male, tails you're female." Our 1:1 ratio of males to females thus stems from segregation symmetry (#X-sperm = #Y-sperm) and viability equality (survival to term of XY *vs.* XX zygotes).

Mutations that distort segregation ("drivers" [403,2080]) or bias viability ("killers" [1205,1206]) are known in flies [542,2572] and humans [326,2176], but their effects tend to be transient [590,775]. Deviations from a 1:1 ratio are typically disfavored by natural selection [1069,1416,2659] for a reason that Darwin explained in *Descent of Man.* To wit, if one sex predominates, then any "cheater" mutations that let individuals leave more offspring of the minority sex (to exploit this resource) will rise in frequency [326], and the ratio will return to equilibrium [1685,2823].

> Let us now take the case of a species producing . . . an excess of one sex—we will say of males—these being superfluous and useless, or nearly useless. Could

> the sexes be equalised through natural selection? We may feel sure, from all characters being variable, that certain pairs would produce a somewhat less excess of males over females than other pairs. The former, supposing the actual number of the offspring to remain constant, would necessarily produce more females, and would therefore be more productive. On the doctrine of chances a greater number of the offspring of the more productive pairs would survive; and these would inherit a tendency to procreate fewer males and more females. Thus a tendency towards the equalisation of the sexes would be brought about. . . . The same train of reasoning is applicable . . . if we assume that females instead of males are produced in excess. [561] (Vol. 1, pp. 316*ff.*)

A single gene on our Y chromosome dictates maleness. Sought since 1959 but not found until 1990, it was named *Sry* (*Sex determining region Y chromosome*) [1122,2056,2756]. It encodes a High-Mobility Group (HMG)-domain transcription factor in the Sox (Sry box) family of DNA-binding proteins [1373,2767]. In the absence of a functional *Sry*, sex is evidently controlled by a master gene for femaleness—possibly *R-spondin1* [2808]—that takes over by default (*cf.* [221, 1956]).

The default nature of our feminine gender is starkly illustrated by a peculiar pair of twins. The twins, born in 1944, were seen at a Paris clinic at age 17, where they were deemed with ≥99% certainty to be monozygotic based on blood-group and skin-graft testing [1515,2653]. The rub was that one was male and the other female! Karyotyping showed the boy to be XY and the girl to be XO with Turner syndrome. (Our Y has so few genes that they are negligible except for *Sry* [2167].) Evidently, an XY embryo lost a Y from a cell(s) after its first mitosis but before it split in two. The XO cells must have cohered to form the girl, while the remainder formed the boy [2469]. Identical twins of opposite gender are as surreal as looking in the mirror and seeing yourself transformed from male to female, or vice versa.

This case was described in a remarkable book by an equally remarkable man. *Genetic Mosaics and Other Essays* (1968) [2469] was based on a series of lectures at Harvard in 1965 (on the 100th anniversary of Mendel's landmark paper) given by Curt Stern (1902–1981) near the end of his career [2612]. Trained as a geneticist in T. H. Morgan's legendary fly lab, Stern taught human genetics at the University of California, Berkeley, and authored what became the leading textbook in that field [1859,2470]. Moreover, like Waddington, he helped to found the field of developmental genetics (which later merged with evolution to form evo-devo), just to name a few of his contributions [1589,1858]. In *Genetic Mosaics*, Stern featured another anomaly at least as eerie as the boy–girl doppelgängers. The following quote is his transcript of a chronicle entry from the town of Piadena, Italy, dated May 26, 1601 (boldface added):

> A weird happening has occurred in the case of a lansquenet [soldier] named Daniel Burghammer. . . . When the same was on the point of going to bed one night he complained to his wife, to whom he had been married by the Church seven years ago, that he had great pains in his belly and felt something stirring therein. An hour thereafter he gave birth to a child, a girl. . . . **He then confessed on the spot that he was half man and half woman**. . . . He also stated that . . . he only slept once with a Spaniard, and he became pregnant therefrom. This, however, he kept a secret unto himself and also from his wife, with whom he had for seven years lived in wedlock, but he had never been able to get her with child. . . . The aforesaid soldier is able to suckle the child with his right breast only and not at all on the left side, where he is a man. He also has the natural organs of a man for passing water. . . . All this has been set down and described by notaries. It is considered in Italy to be a great miracle and is to be recorded in the chronicles. The couple, however, are to be divorced by the clergy.

The hermaphrodite in this case, Daniel Burghammer, might have begun life as an ordinary XY male but then lost the Y chromosome (via nondisjunction [247]) from a cell(s) early in development [410,2469] like the embryo that became the boy–girl twins described earlier, except that Daniel never split in half. This conjecture fails, however, because his XO ovary shouldn't have made fertile eggs (*cf.* Turner syndrome [1809]).

More likely, taking fertility and hormones into account, Daniel may have been totally XX, but one X may have had a recessive mutation for androgen insensitivity expressed mosaically due to X-inactivation (*cf.* McKusick on Gayral *et al.*, 1960; Disorder #313700 [1711]). The rub here, however, is that the patchiness of the X-mosaicism should have reduced the chance of Daniel's having enough androgen-insensitive tissue to form a functional female reproductive tract. Regrettably, the exact etiology is hard to infer from the limited "clinical" information provided.

A similar hermaphrodite was reported in the *New England Journal of Medicine* in 2004 [1333], who appears left–right reversed relative to Daniel. An African-American infant, seen at a clinic in Dallas, Texas, had a penis (with hypospadias) and scrotally enclosed testis (with distal ovarian tissue) on the right, but a hemiuterus, oviduct, and ovary internally on the left. Karyotyping showed the male side to be XY and the female side to be XX. What complicates this case is that the midline divided dark skin on the right from light skin on the left (*cf.* [1083])! The pigmentation boundary suggests a chimera [231] derived from two sperm (one Y, the other X) that fertilized two eggs, which merged to make one composite embryo [2469]—*i.e.*, fraternal twins in a single body. Without any DNA profiling, it is hard to know what really happened.

Were we all once hermaphrodites?

Having testes and ovaries together in one body is a freakish aberration for humans [410,1773,1774], and the same is true for vertebrates in general [378,1788], but it is the *norm* for a third of all animal species excluding insects [262,585,1280]! Darwin was aware that bisexuality had been documented in some primitive chordates (*viz.*, tunicates [1086,1516] but not amphioxus [1471]), and in *Descent of Man*, he conjectured that our ancestors were hermaphrodites at that stage of our evolution [892]. He tacked this speculation onto the end of a lyrical passage in which he listed our former incarnations at successively more remote eras—an exercise that served to (1) distill some lessons of *Descent* and (2) apply *Origin*'s theme of "descent with modification" to our own lineage (boldface added):

> The early progenitors of man were no doubt once covered with hair, both sexes having beards; their ears were pointed and capable of movement; and their bodies were provided with a tail, having the proper muscles. . . . At this or some earlier period, the intestine gave forth a much larger diverticulum or caecum than that now existing. . . .
>
> At a still earlier period the progenitors of man must have been aquatic in their habits. . . . The clefts on the neck in the embryo of man show where the branchiae once existed. . . . These early predecessors of man, thus seen in the dim recesses of time, must have been as lowly organized as the lancelet or amphioxus, or even still more lowly organized. [561] (Vol. 1, pp. 206*ff.*)
>
> There is one other point deserving a fuller notice. It has long been known that in the vertebrate kingdom one sex bears rudiments of various accessory parts, appertaining to the reproductive system, which properly belong to the opposite sex, and it has now been ascertained that at a very early embryonic period both sexes possess true male and female glands. Hence some extremely remote progenitor of **the whole vertebrate kingdom appears to have been hermaphrodite** or androgynous. [561] (Vol. 1, p. 207; *cf.* related passages: Vol. 1, p. 30; Vol. 2, pp. 389*ff.*, and *Origin*, pp. 93*ff.*)

Now that evo-devo has effectively traced all bilaterian phyla back to a common Urbilaterian progenitor, we can ask whether Darwin's speculation about protochordates can be extended to that earlier (pre-chordate) stage as well. Were all bilaterians originally hermaphrodites [2623]? Studies of germ cells in various phyla are consistent with the idea but are inconclusive [736,737], and comparative genomics has not reached a point where such a question can be meaningfully addressed [2814]. Thus, there is no definitive answer yet.

If our distant ancestors were, at some point, hermaphrodites, then this situation begs the question of why (and how) our more recent ancestors split themselves into two separate bodies—a process at least as profound as any creation myth in any religion [349–352,979] (*e.g.*, the Janus-faced, eight-limbed men–women imagined by Aristophanes in Plato's *Symposium* [972,1661]).

Darwin proposed a hypothesis based on the truism that generalists can do many things adequately, while specialists can do a few things superbly. No single individual can be the best possible male *and* female in its reproductive roles [95]. Partitioning the genders would allow a division of labor. He described how the process might work with an imaginary plant species:

> No naturalist doubts the advantage of what has been called the "physiological division of labour;" hence we may believe that it would be advantageous to a plant to produce stamens alone in one flower or on one whole plant, and pistils alone in another flower or on another plant. In plants under culture and placed under new conditions of life, sometimes the male organs and sometimes the female organs become more or less impotent; now if we suppose this to occur in ever so slight a degree under nature, then as pollen is already carried regularly from flower to flower, and as a more complete separation of the sexes of our plant would be advantageous on the principle of the division of labour, individuals with this tendency more and more increased, would be continually favoured or selected, until at last a complete separation of the sexes would be effected. [559] (pp. 93*ff.*)

Ergo, by splitting in half, he/she could allow his/her offspring—now reformed as he and she—to specialize more effectively. Disruptive selection should thus have rewarded any mutations that happened to tip the male–female balance one way or the other (*cf.* Fig. 1.2c.iv).

Indeed, it is thought that *all* sex chromosomes start out in exactly this way [1172,1174,2815]—with random male- or female-biasing alleles that lead (over time) to the conversion of the arbitrary autosome on which they reside [113, 1001,1818] into a bona fide X or Y (or Z or W) sex chromosome [739,1259, 2418]. Our own X and Y appear to have originated ~300 MYA [1467,1967,2070] when they evidently usurped control from a temperature-dependent sex determining system in our reptilian ancestors [1277,2033].

Our earlier loss of bisexuality in favor of two genders (~500 MYA?) may not have been recorded in the fossil record, but it has occurred more recently in other phyla [547]. Nematodes are notorious for shifting back and forth evolutionarily between reproductive styles [262,1856,2039]. In fact, *Caenorhabditis elegans*, which has been a model organism in genetics for the last 35 years [274], looks as if it just couldn't make up its mind: in nature it consists of—are you ready for this?—*males and hermaphrodites* [858,1173,2506]! Imagine *their* creation myths... if they had any! (For American undergraduates, the tragedy of hermaphrodites is not carnal but *social*: they wouldn't be able to join either a fraternity or sorority!)

Why do male fetuses make incipient oviducts?

In the next-to-last quote in the previous section, Darwin mentions the curious fact that male fetuses make the rudiments of female organs and vice versa. Indeed, all mammal embryos build two pairs of tubes linking the gonads to the external genitalia and then destroy one pair, leaving either vasa (males) or oviducts (females). Because the destruction is wasteful, it begs the question of why the process (Fig. 5.2) has been retained.

If Urbilaterians were indeed hermaphroditic, as protochordates appear to have been, then our phylum probably started with bisexual organisms. Hence, it should not be so surprising that our gonad is still bipotential [2809] or that its plumbing manifests bifunctionally redundant pipes when it first appears in the embryo [1395]. The mechanisms may be too entrenched to modify.

Some of the female tubing in mice has recently been shown to employ the same gene(s) for its construction (*Dach1/2*) as comparable tubing in fruit flies (*dachshund*) [583,1358], but the best evidence for a deep conservation of reproductive development all the way back to Urbilateria [1039,2814] comes from a different gene called *doublesex* (*dsx*) in flies [121,526], *mab3* in nematodes [2880], and *Dmrt1* (*Doublesex and mab3-related transcription factor 1*) in vertebrates [1200,1377]. The gene encodes a transcription factor belonging to the zinc-finger class of DNA-binding proteins [725,2911].

Mutations in *doublesex* can transform male and female flies into sterile "hermaphrodites," the reproductive tracts of which have both types of plumbing side by side in jumbled disarray [721]. In normal fly larvae of both sexes, the genital disc contains a pair of rudiments [2270]: the male and female genital primordia (MGP and FGP) [722,948]. Wild-type males repress FGP (leaving MGP to make tubes), whereas wild-type females repress MGP (leaving FGP to make tubes) [733,2269], a strategy similar to that in humans except that the suppression involves prevention of rudiment growth in the first place rather than destruction of rudiments after growth has occurred.

Dmrt1 is expressed differently in males versus females broadly within the vertebrate subphylum [1378,2899]. Deletion of *Dmrt1* in humans can transform XY individuals into hermaphrodites (or females) [1957]. Given that the *dsx* gene of insects regulates its target genes via differently spliced mRNA isoforms (dsx^M and dsx^F) [2240], it is natural to ask whether the same is true for vertebrates [96].

We do not yet know the answer [1200], but human *Dmrt* genes are subject to sex-specific splicing, and the exon structure of the testis isoform does match dsx^M in flies [2899].

Given what has been said so far about master genes, one might expect the most conserved regulator to be at the *top* of the sex determination hierarchy, but *doublesex* is near the *bottom* in both flies and humans. The following are the core components of the fly [457,2064] versus human [2056,2333] pathways as we currently understand them (*Sxl* = *Sex-lethal*; *tra* = *transformer*; M and F superscripts = functional male or female splicing isoforms) [1422]. Omitted (beyond *dsx*) is *fruitless*, which governs fly behavior (via central nervous system circuitry [2872]), rather than morphology [836,2175]. Also omitted is the fly's *Sox9* gene, which functions as it does in humans [618], but possibly only in the gonad. Note that Xs and Ys of humans and flies are not homologous.

Flies: XY (default, no *Sxl*). ... $dsx^M \rightarrow$ male target genes.
XX $\rightarrow$ $Sxl^F \rightarrow tra^F \rightarrow dsx^F \rightarrow$ female target genes.
Humans: XY $\rightarrow$ *Sry* $\rightarrow$ *Sox9* $\rightarrow$ *Dmrt1* $\rightarrow$ male target genes.
XX (default, no *Sry*). ... *R-spondin1* $\rightarrow$ female target genes.

Here then is the dilemma: over the eons it was the genes at the *upper* echelons that changed freely [993], while *dsx/Dmrt1* endured for more than 500 million years near the bottom [448,1411,1753]. For example, usage of *Sxl* is confined to the drosophilid lineage within dipterans [2338,2622], and usage of *Sry* is restricted to the therian lineage (placentals and marsupials) within mammals [2291], whereas *Sox9* is more broadly conserved (among tetrapods [2403] and perhaps beyond [618]). Other mediators that the fly cascade has haphazardly enlisted (not shown) include the JAK/STAT [107,1101,2758,2901], Notch [2012], and proneural [466,2860] signaling pathways.

In terms of the analogy in Figure 5.1, this fluidity of upstream factors forces us to envision the fisherman himself as a fish who could be hooked (from above) by an assortment of lines over time, while keeping all of the fish that he has caught up to now on his own line [2064,2506,2814]. Such "retrograde" retooling of developmental pathways (*vs.* anterograde additions [2815] or intercalary insertions [878]) is not unique to sex determination [1603,1933,2851,2855] (*e.g.*, upstream triggers for the *nodal* circuit [108,1528]; *cf.* Ch. 2), but it is more prevalent there than anywhere else [897,993,2238,2632]. Why?

The answer appears to be that *dsx* enforces its diverse dimorphisms by managing *multiple* target genes in various organ-specific developmental pathways, whereas each of its upstream regulators mainly controls a *single* subordinate gene [2899]. In other words, the cascade is a simple chain down to *dsx*, but then it fans out extensively [84,928]. Extricating such a pleiotropic gene

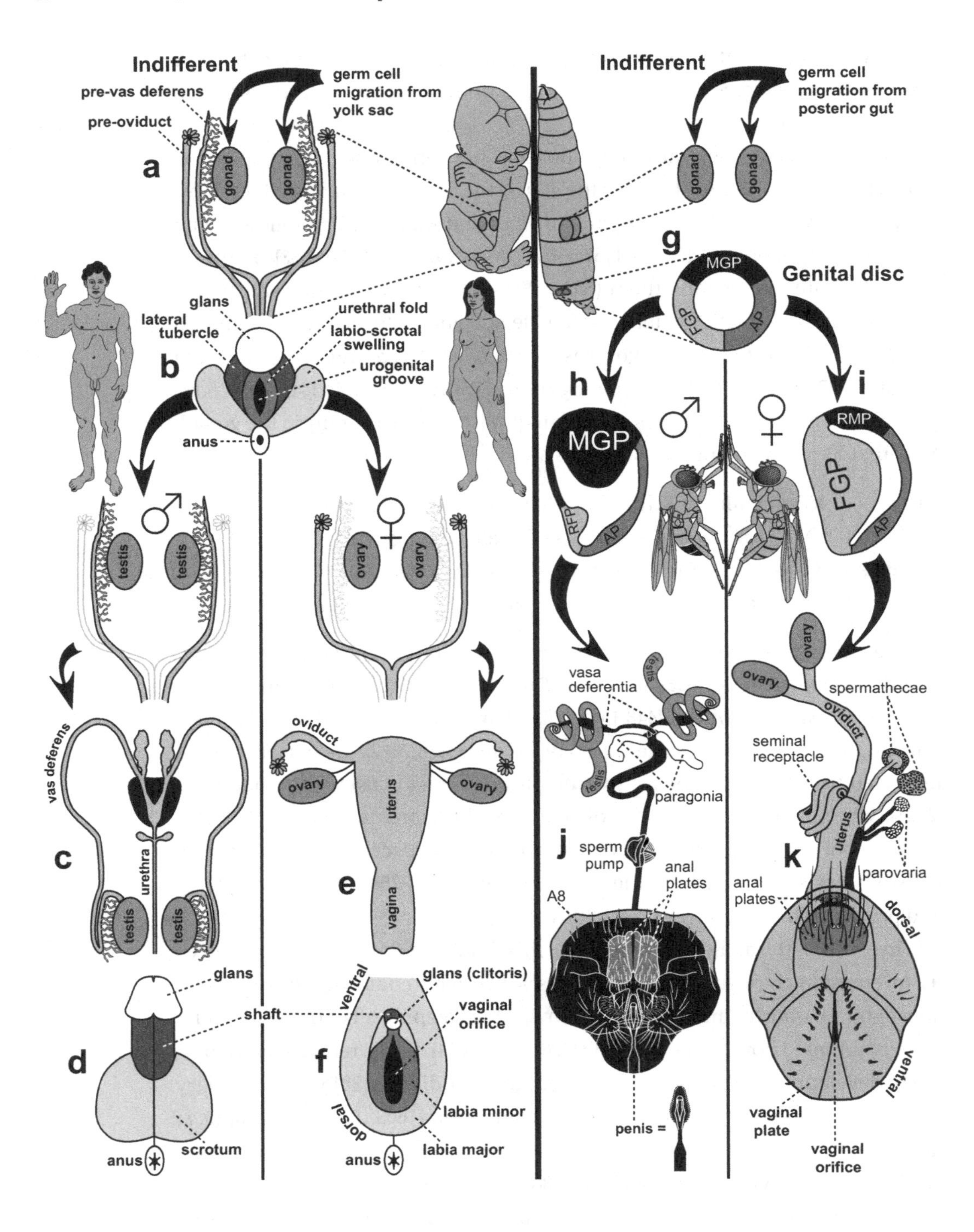

from the cobweb of interactions that it has acquired (by random fishing) and replacing it with a substitute just can't be achieved by piecemeal tinkering. In the extent of its entrenchment, *dsx* resembles the genes of the *Hox* complex, which have likewise insinuated themselves into the inner workings of a host of

Figure 5.2. Genital development in humans (left) versus fruit flies (right). In both species, the incipient genitalia look alike in males and females ("indifferent"), and the same goes for the gonads. Bilaterian gonads typically import their germ cells [923,1393,1453] after the latter have migrated far (using conserved genetic circuitry [2640]), a strange odyssey that may have a simple evo-devo explanation [630,736,737,2623]. Most important, each gender initiates some primordia appropriate for the opposite sex but then either destroys them (humans) or represses them (flies). Anatomy was redrawn from refs. [302,303,1866], and fate maps were derived from refs. [429,2270].

a–f. Development of human genitalia [1866].

a. Internal genitalia in ~10-week embryo (of either sex). Two pairs of tubes flank the gonads. Pre–vasa deferentia (a.k.a. Wolffian ducts) have side branches (mesonephric tubules) medially, whereas pre-oviducts (a.k.a. Müllerian ducts) do not. All tubes converge on the urogenital sinus (not shown), which contains two more rudiments with divergent fates: one forms the prostate (male) or Skene's gland (female) and another becomes Cowper's (male) or Bartholin's gland (female).

b. External genitalia in ~10-week embryo (of either gender). Parts are coded by different shading to mark their homologies in males (**d**) versus females (**f**).

c. Pre-oviducts disintegrate in males, leaving pre–vasa deferentia. Vasa lengthen as the testis descends into the scrotum before birth [1866]. Black organ at the juncture of vasa and urethra is the prostate. Paired glands above are seminal vesicles; paired glands below are bulbourethral (Cowper's) glands.

d. External genitalia in neonate.

e. Pre–vasa deferentia disintegrate in females, leaving pre-oviducts. Ligaments connect the ovaries to the uterus [1866].

f. External genitalia in neonate (urethral opening omitted). Note the homologies: (1) penis ≈ clitoris and (2) scrotum ≈ labia major.

g–k. Development of *D. melanogaster* genitalia [302,303].

g. Fly genitalia come from the genital "disc" [733,948], a midline rudiment that grows during the larval period and differentiates during metamorphosis (like the other 18 imaginal discs [1137]). It has 3 distinct zones: (1) a male genital primordium (MGP), (2) a female genital primordium (FGP), and (3) an anlage for anal plates (AP) [2270]. Each zone comes from a different body segment [429]. Zones are coded by different shading to track their fates in males (**j**) versus females (**k**).

h. In males, growth of the FGP is repressed (RFP), but repression is incomplete because RFP still manages to contribute a thin eighth tergite (A8) to the cuticle (**j**) [2690].

i. In females, growth of the MGP is repressed (RMP), but repression is incomplete because RMP still manages to form parovaria (accessory glands) and a thin strip of the uterine wall (**k**) [2690].

j. The only part of the male made by RFP is a tiny collar (A8) around the genitals proper [429,1359,2690]. Paragonia come from cells recruited from outside the genital disc [28]. Testes change from oval to spiral shape after contact with vasa (*cf.* Fig. 2.1 for chirality) [2467,2468], but the inductive signals have not yet been identified. Because the penis is difficult to discern amid all the clutter, it is rendered alongside as well. Insect genitalia evolve rapidly and are important taxonomically [697,1329,1425,2554,2793], so genital evolution has become a field unto itself [696,698,699,1220]—albeit one that is ill suited for polite discourse in mixed company.

k. The only part of the female made by RMP is a stripe of the uterus and the two parovaria [1359,2690]. In females, anal plates (darker gray AP derivative) are oriented one above the other—with the anus in between—whereas in males (**j**), they are oriented side by side.

---

embryonic pathways to reconfigure the shapes of serially homologous organs [726] (*e.g.*, the revamping of wing anatomy by *Ubx* in crafting a haltere [2474,2763]; *cf.* Ch. 4).

Another mystery about *dsx* that may have been solved recently concerns how it can produce such sharp outcomes—male **OR** female—with so little

REFLECTIONS ON FIGURE 5.2

The dorsal–ventral axis is reversed in these diagrams of humans (**d**, **f**) versus flies (**j**, **k**). The former show humans on their *backs*, and the latter show *standing* flies from the rear.

d, f. Evidence of male–female genital homology is provided most starkly by genital masculinization in female hyenas [802,921,2869], an unfortunate trait that makes giving birth through the enlarged clitoris (≈ penis) [803] as difficult as "pushing a golf ball through a soda straw" [1749]. On the basis of this oddity, Aristotle thought that female hyenas were actually hermaphrodites (*History of Animals*, Book 6, Part 32). A hotly disputed issue in evo-devo (what an understatement!) is whether human female orgasm is merely a spandrel [972,1209,1318,1609,2734]. (Don't try debating this at home with your spouse!) The most exasperating aspect of spandrel debates in general is that they can involve untestable hypotheses [77,330,2105,2424], but meaningful questions can be posed and evaluated with the proper reasoning [2038].

j. Flies may have evolved elongated testes to fit longer sperm [2818], which reduce the chance that eggs will be fertilized by a subsequent suitor's sperm [205,1994]. Indeed, runaway selection in one fly species (*D. bifurca*) has produced the longest sperm of any animal on earth (~6 cm) [2043]! The number of gyres per spiral (N) in *Drosophila* testes is only 2.5 for *D. melanogaster* [2604], but it is 6 for *D. virilis*, ~11 for *D. hydei* [2818], and an incredible ~40 for *bifurca* [2043]. N is correlated with the time available for testis elongation (t) [2467]. For a fixed growth rate "r," evolution could have easily modified N indirectly by changing t, given that $N = rt$ [2781]. Similar logic applies to other spirals (*e.g.*, mollusk shells, mammal horns, human cochlea [510,2133,2498,2596]). Indeed, heterochronic changes may have catalyzed the evolution of many other anatomical shapes as well [1709,1710].

sloppiness (intersexuality) [930,1775]. The genital disc has two signaling pathways that cannot both operate at once: Dpp and Wnt. Only one can "win" the contest in each gender. What *dsx* does is enable one pathway and disable the other [733,2269], effectively tipping this "see-saw," which has no intermediate stable states. Thus, the system consistently yields a digital (binary) output from an analog (upstream) input. The robustness of this toggle switch also helps explain why upstream components of the cascade can be replaced so easily in evolution: *any* genetic (or external) nudge should suffice to bias *dsx* [1382]. The epitome of a continuous analog signal that can still produce discrete outcomes [1832,2261] is *temperature*-dependent sex determination [617,1277,2033,2765].

Mammals, it so happens, use the same trick in their bipotential gonad [272], except that the pathway opposing Wnt is FGF9 instead of Dpp [1382]. Because *dsx* (and *Dmrt1*?) need merely exert leverage at certain key sites within each developing organ, it is mainly those small "organizer" regions that dictate overall organ shape [726,1381]. The subservience of the remaining cells helps explain why XX cells can form sperm and XY cells can form eggs when they find themselves in an alien gonadal environment [15,1252,2159].

This see-saw strategy is reminiscent of an antagonism that exists between the Dpp and Wnt pathways in fly leg development [947,1136]. Indeed, the external genitalia of both flies [734] and humans [657,2870] appear to have evolved from appendage-like precursors [1760] under the "new management" of posterior *Hox* genes [475,1512,2052,2347].

How did we become the only "naked apes"?

Hair is a feature of all mammals [1623,2417]. Its selector gene appears to be *Foxn1* [2299], which encodes a transcription factor of the winged-helix family [1327,1860]. We don't yet know how proto-mammals came to express *Foxn1* in the hair follicle (a novel *cis*-enhancer?) [2299], nor how *Foxn1* captured keratin genes as targets therein (*cf.* Fig. 4.2) [1239,2313], nor, finally, how *Foxn1* acquired an added role in hair pigmentation [144,2771].

Mammals use hair in the same way as birds use feathers—as a layer of insulation to retain heat so that we can maintain constant temperature [2862,2905]. Humans differ from other mammals insofar as we have lost much of our hair, and what remains is useless for insulation, except for the top of our head.

Why did we get so bare? One idea is that our fur stopped being the solution at some point in our past and became the problem [1968]. The inflection point may have come when we began running long distances [267,2915], thus increasing our heat output considerably [1260] (*cf.* [540]). Sweat is only effective if it can reach the air to evaporate [224,1260], and wearing a fur coat limits that access [1793,2212,2794]. As we shed our coat, we would have still been able to keep warm (during infancy, on cold nights, etc.) by tapping our bodily reserves of brown fat to produce heat on demand [359,1500].

How did we get so bare? We don't yet know the genetic changes involved [2009], but we do have a few intriguing clues, which are worth considering briefly.

A number of hereditary syndromes exhibit extra hair [111,154,170,1618,1783, 2787]. The most dramatic is Ambras syndrome [153,2550], in which affected individuals—men *and* women—grow luxuriant hair all over their face, including the forehead and nose [153,765]. Only ~50 cases have been reported since the Middle Ages [153]. The earliest known instance (Petrus Gonsalvus) was documented in a portrait *ca.* 1582 [1524,2134,2787]. A few such people were displayed as curiosities at carnivals in the 1800s [239,1452] under stage names such as the Dog-faced Boy and the Lion-faced Man [678]. Darwin mentioned a famous family with this trait in his 1868 book, *The Variation of Animals and Plants Under Domestication* in which he tried (but failed) to discover the laws of inheritance in general:

> Mr. Crawfurd saw at the Burmese Court a man, thirty years old, with his whole body, except the hands and feet, covered with straight silky hair, which on the shoulders and spine was five inches in length.... This man had a daughter

> who was born with hair within her ears; and the hair soon extended over her body. When Captain Yule visited the Court, he found this girl grown up; and she presented a strange appearance with even her nose densely covered with soft hair.... Of her two children, one, a boy fourteen months old, had hair growing out of his ears, with a beard and moustache. This strange peculiarity has, therefore, been inherited for three generations. [560] (Vol. 2, p. 320)

Ambras Syndrome *appears* atavistic [239,765,2074], but it can't be because apes don't have furry noses [2683]! Other hypertrichosis mutations cause ectopic hair in equally odd places: on the ears [1505,2471], the neck [260], the elbows [171,2058], or on the palms and soles [1262]. The latter two locations also defy an atavistic interpretation because no primate has fur there. Because there is no rhyme or reason to these assorted spots, it's hard to know what to make of them.

If hominin hair was suppressed by a single genetic change, then reversion mutations should have occurred by now to create babies with apelike coverings of hair, and we should know about it. Does the absence of über-furry people refute the Simple-mutation Hypothesis? Not necessarily! A popular conjecture along these lines is Bolk's Fetalization Theory [956]. It holds that we are like fetal apes in many respects—not just in the sparseness of our hair. (Newborn gorillas have hair on the head, but the rest of the coat does not appear until later [232,598].) If hair pattern is an inextricable part of a larger nexus of changes, then it may be too hard mutationally to reverse it without undoing vital aspects of our development. In other words, atavistic fetuses might indeed arise, but they may die in utero before they can make any hair.

Proof that a single mutation *could* have stripped our ancestors of most of their hair has recently come from an unexpected corner of the research world. The genetic basis of hairlessness in the Mexican hairless dog (revered by the Aztecs for 3700 years) has now been ascertained. The trait is due to a frameshift insertion (a 7-base-pair duplication in exon 1) in the *Foxi3* gene [679], which belongs to the same *Forkhead box* family as the *Foxn1* gene that controls hair formation all over the body [1327,2299]. Interestingly, this breed still has plenty of hair on its head (as we do), as well as on its feet and tail. Why this null mutation should leave hair on these sundry body parts remains to be determined.

Why are men hairier than women?

Darwin considered the possibility that the pervasive loss of hair from the hominin body might have evolved as an adaptation for thermoregulation (as argued earlier), but he dismissed the idea, in part because he couldn't figure out why hair should be left on our head in the profuse amount that we make (*cf.* [1260]).

> May we then infer that man became divested of hair from having aboriginally inhabited some tropical land? The fact of the hair being chiefly retained in the male sex on the chest and face, and in both sexes at the junction of all four limbs with the trunk, favours this inference, assuming that the hair was lost before man become erect; for the parts which now retain most hair would then have been most protected from the heat of the sun. The crown of the head, however, offers a curious exception, for at all times it must have been one of the most exposed parts, yet it is thickly clothed with hair. [561] (Vol. 1, p. 149)

Even Aristotle was moved to remark that "no animal has so much hair on the head as man" {PoA:2:14:658b2} [137]. Actually, his assertion is not quite true. Lions rival us.

What is intriguing about lions, of course, is that the mane is only present in the male. Comparably drastic dimorphisms in *primates* evidently led Darwin to his eventual epiphany: many of our traits (including hair pattern) evolved through sexual selection [71].

> Male quadrupeds of many kinds differ from the females in having more hair, or hair of a different character, on certain parts of their faces. . . . In three closely-allied sub-genera of the goat family, the males alone possess beards, sometimes of large size. . . . With some monkeys the beard is confined to the male, as in the Orang, or is much larger in the male than in the female, as in the *Mycetes caraya* and *Pithecia satanas*. So it is with the whiskers of some species of Macacus, and, as we have seen, with the manes of some species of baboons. [561] (Vol. 2, p. 282*ff.*)
>
> In the Baboon family, the adult male of *Cynocephalus hamadryas* differs from the female not only by his immense mane, but slightly in the color of the hair and of the naked callosities. [561] (Vol. 2, p. 291)

Darwin's explanation for hair patterns in men and women, then, is based not on insulation but rather on an entirely separate, secondary function of hair in *ornamentation*! He argued that we use our patchwork of hairy and naked areas in the same way that birds use colored feather patterns [2683]: as decorative markings. Man's beard serves the same role as the peacock's tail [1041,1195], although we are drab compared with the gaudy garb of some other primates. (Boldface is mine. Note his *awe* at the mandrill's quirks.)

> I am inclined to believe . . . that man, or rather primarily woman, became divested of hair for **ornamental** purposes; and according to this belief it is not surprising

> that man should differ so greatly in hairiness from all his lower brethren, for characters gained through sexual selection often differ in closely-related forms to an extraordinary degree. [561] (Vol. 1, p. 149*ff.*)
>
> The faces of many monkeys are **ornamented** with beards, whiskers, or moustaches. The hair on the head grows to a great length in some species of Semnopithecus; and in the Bonnet monkey (*Macacus radiatus*) it radiates from a point on the crown, with a parting down the middle, as in man. It is commonly said that the forehead gives to man his noble and intellectual appearance; but the thick hair on the head of the Bonnet monkey terminates abruptly downwards, and is succeeded by such short and fine hair, or down, that at a little distance the forehead, with the exception of the eyebrows, appears quite naked. It has been erroneously asserted that eyebrows are not present in any monkey. In the species just named the degree of nakedness of the forehead differs in different individuals. [561] (Vol. 1, p. 192)
>
> No other member of the whole class of mammals is coloured in so extraordinary a manner as the adult male mandrill (*Cynocephalus mormon*). The face at this age becomes of a fine blue, with the ridge and tip of the nose of the most brilliant red. According to some authors the face is also marked with whitish stripes, and is shaded in parts with black, but the colors appear to be variable. On the forehead there is a crest of hair, and on the chin a yellow beard. . . . When the animal is excited all the naked parts become much more vividly tinted. Several authors have used the strongest expressions in describing these resplendent colors, which they compare with those of the most brilliant birds. [561] (Vol. 2, p. 292*ff.*)

Once Darwin hit on this insight, other aspects of our hair pattern fell neatly into place. In particular, he now grasped why males of different races have distinct zones of hair growth (*cf.* [650,1371]).

> In individuals belonging to the same race these hairs are highly variable, not only in abundance, but likewise in position: thus the shoulders in some Europeans are quite naked, whilst in others they bear thick tufts of hair. There can be little doubt that the hairs thus scattered over the body are the rudiments of the uniform hairy coat of the lower animals. [561] (Vol. 1, pp. 24*ff.*)

The females of different races simply prefer men with specific hair patterns, but those patterns are not adaptive for any other discernable reason. They are, in short, just frivolous quirks.

> Considering this parallelism there can be little doubt that the same cause, whatever it may be, has acted on mammals and birds; and the result, as far as ornamental characters are concerned, may safely be attributed . . . to the

> long-continued preference of the individuals of one sex for certain individuals of the opposite sex. [561] (Vol. 2, p. 297)

Darwin noticed one last connection that leads us to see our hair pattern in a different light. He sensed some kind of role for these dimorphic features in speciation.

> It is a remarkable fact that the secondary sexual differences between the two sexes of the same species are generally displayed in the very same parts of the organisation in which the different species of the same genus differ from each other. [559] (p. 157)

What role might that be? The hominin family tree was once much bushier that it is today [32,224,964,971,2285], with more than one species often living in the same vicinity [638,2396]. Our unique pattern of hair growth may have evolved as a means of telling the members of our own species apart from sympatric congeners. If those congeners looked confusingly close to us initially, then we would have run the risk of mistakenly mating with them and producing hybrid offspring that were inviable, infertile, or disabled [11,2162,2803]. Under such circumstances, natural selection would have rewarded *any* cue that distinguished the two groups [1021,1481]. Over time, a few features that happened to vary more than others would have diverged—a process termed "character displacement" [167,1157,2461]. Such differences could then have facilitated assortative mating [1391,2330] and intensified reproductive isolation [71,234,1955,2079].

In sum, our hair may convey two distinct, but related, signals [651,1580]: "I'm sexy!" and "I'm a human, not a Neanderthal!" The upshot of Darwin's argument is that our manes and our beards are as purely symbolic as tattoos [469]. It may, therefore, be no accident that humans also exhibit an inordinate fondness for ornamentation of all kinds (tattoos, jewelry, etc.) [2674].

We don't yet know what Neanderthals (or any other hominin) looked like in terms of hair [1063], and, besides, hair pattern may not have been the only (or chief) way that we told one another apart [1355,2046,2791]. (Don't forget their brow ridges [638,2865]!) Generally speaking, premating isolating mechanisms can be auditory (*e.g.*, bird songs), behavioral (*e.g.*, courtship rituals), or olfactory (*e.g.*, pheromones) [71], but in primates, they tend to be visual [2222]—probably because we rely so heavily on our eyes (*vs.* our ears or nose) [59,638,1390,2099].

From an evo-devo perspective, several key questions remain about how the genome controls hair distribution:

1. What selector genes (or *cis*-enhancers?) target hair growth to the sites where it develops (before and after puberty) [2093]? For example, what is

the "area code" of the armpit? Flies use a single genetic locus—the Achaete-Scute Complex (AS-C)—to demarcate the territories where bristles can form throughout the body [1137], but no comparable headquarters has yet been found in our genome [1565]. Do we have one?

2. Do genes regulate hair formation (1) *positively* (with nakedness as a default) through region-specific activation or (2) *negatively* (with full hair cover as a default) through region-specific suppression? Flies dictate bristle sites positively by means of ~8 AS-C *cis*-enhancers [1137], which use "OR" logic, so that it was easy for evolution to add or delete them like plug-in modules. What about humans?
3. What integration, if any, exists between (1) genes dictating hair pattern in males and (2) genes dictating mate choice in the female brain [71,117,1943]? Do they coevolve somehow [46,912] or take steps alternately and independently like a glacially slow cha-cha [1313,1719]?

Why do only men go bald?

Balding is actually not confined to men, but it is much rarer in women [2405,2445]. It is associated with aging, but it cannot be a consequence of the wearing out of hair follicles because it occurs in such well-defined areas only—hence the term "male *pattern* baldness" [1268,2123,2244].

Judging by the number of advertisements for baldness remedies, men *hate* going bald [2404]. If baldness is such a disadvantage in the dating-mating game, then what possible function does it serve? Were Darwin alive today, he might chuckle at this question because it could just be a vestige! Like the appendix and other vestiges (*cf.* Ch. 6), baldness varies greatly from person to person—in age of onset, degree of asymmetry (an obvious detractor from beauty [2165]), rate of progress, etc.—as if it were under reduced selection pressure [561].

If baldness is such a liability now, then what advantage did it offer in the past? The argument put forward above for hair evolution applies equally well here: baldness may have once been as sexy as beards. Early female hominins may have found the bald spot attractive, but at some point their descendants ceased to be turned on by it. Unfortunately (for bald men), the rate at which baldness genes are being purged from our species is apparently much lower than the rate at which mutations are changing the criteria for sex appeal in the female brain [965] (*cf.* [161,2151]).

The only consolation that bald men can have in this regard is that we are not alone. Consider the sad tale of the peacock's tail. Believe it or not, peahens changed their mind at some point in the recent past [2804]. They no longer prefer peacocks with fancier fans [2555]! Extravagant tail feathers were just a passing fad? Apparently so! Another tragedy is the tail of the swordtail fish, which, incidentally, has now been deciphered by evo-devo researchers [708]. For some reason, the females of at least one species (*Xiphophorus birchmanni*) have lost their liking for the male's sword [2846]. O cruel fate! O fickle genes! We are fortune's fools!

Because human balding is testosterone-dependent [2123], it is possible that it was a warning signal in male-male rivalry, instead of—or in addition to—a lure for females. It may have said: "I'm full of testosterone, so get out of my way... or else!"

There is a marvelous treatise on baldness evolution [1750], which is virtually unknown, judging from the fact that it has only been cited twice since it was published in 1931 [2683]. Gerrit Miller was the curator of mammals at the Smithsonian Institution. Having access to a wide variety of primate species allowed him to survey the spectrum of hair patterns in monkeys and apes. Remarkably, he was able to find *all* varieties of the human balding pattern in one primate or another. The stump-tailed macaque, for example, develops a receding hairline after puberty (more strongly in males) [280], and New World uakaris have bald pates that look eerily human [1793]. The existence of such hair patterns in our primate relatives is consistent with the hypothesis that baldness served some sort of adaptive function in our hominin ancestors.

Another revelation from Miller's study concerns hair coloration. He notes that the eyebrows of some monkeys have a different color from their scalp hair, as is also true in humans (*cf.* the "tan points" of Doberman pinschers; Trey Fondon, personal communication). An evo-devotee might infer from this correlation that the same genetic "area codes" that evoke hair growth are also used to evoke hair color. In flies, the master gene for cuticle color (*yellow*) [1852] resides less than 10 kilobases from the command center for bristle patterning (AS-C) [936,1137], but there is no sharing of *cis*-enhancers [881,2839] because of an intervening insulator [934]. Might a comparable system operate in humans? We don't yet know.

Finally, Miller notes that men's beards tend to turn gray before their scalp hair. This temporal modularity chimes with the spatial modularity just described. It affirms the idea that our genome subdivides our skin into territories that operate independently of one other in hair growth (hairy *vs.* bare), hair color (youthful *vs.* gray), plus, perhaps, hair length, density, and texture as well.

Gray hair is typically correlated with old age in our species, but precocious graying is common in certain families, and the same is true for many dog breeds [2307]. Thus, gray hair can't be *caused* by senescence any more than baldness is [2277]. It may play an extant signaling role, as in the silverback gorilla, or it may be a vestige, like baldness.

How genomes control the timing of gene action is one of the biggest outstanding mysteries in evo-devo. The only clue we have so far regarding hair graying is that its time of onset in mice is dependent on the dosage of the *Notch* gene [2314]. In flies, pigmentation must be finely tuned temporally because the *yellow* gene gets turned ON in different body parts in an invariant sequence [1851,2739].

Why do men have nipples?

Mammary glands, like hair, are another novelty of Mammalia (and the basis for our name). The presence of nipples in both sexes cannot, therefore, be interpreted as a legacy from a distant hermaphroditic ancestor (as was argued for male oviducts) because the proximate progenitors of mammals (fish, amphibians, and reptiles) weren't androgynous.

A priori, novelties would be expected to show up more often in *both* sexes than in one sex alone. The reason is simple. It has to do with the Boolean logic discussed earlier in this chapter. If we imagine that mammae arose in the chest, then a single mutational event could theoretically have achieved the following linkage:

> **IF** (Input) "I am in the chest" {mediated by a region-specific selector gene(s)},
>
> **THEN** (Output) "I will turn ON the genes needed to make a mammary gland."

However, at least two separate events would have been required to restrict the formation of such glands to females alone, unless, of course, the novelty arose in an organ that was *already* sexually dimorphic [530].

> **IF** (Input #1) "I am in the chest" {mediated by a region-specific selector gene(s)}
>
> **AND** (Input #2) "I am female,"
>
> **THEN** (Output) "I will turn ON the genes needed to make a mammary gland."

Darwin wondered how easy it would be for artificial selection to convert a trait that starts out being expressed in both sexes into one that is confined to a single sex. On the basis of his intimate knowledge of pigeon breeding, he concluded that it would be quite difficult indeed (*cf.* [560]).

> The equal transmission of characters to both sexes is the commonest form of inheritance, at least with those animals which do not present strongly-marked sexual differences, and indeed with many of these. [561] (Vol. 1, p. 282)
>
> There is one difficult question . . . : namely, whether a character at first developed in both sexes, can be rendered through selection limited in its development to one sex alone. If, for instance, a breeder observed that some of his pigeons (in which species characters are usually transferred in an equal degree to both sexes) varied into pale blue; could he by long-continued selection make a breed, in which the males alone should be of this tint, whilst the females remained unchanged? I will here only say, that this, though perhaps not impossible, would be extremely difficult. [561] (Vol. 1, p. 284)

Mammary glands originated ~310 MYA in a branch of synapsid reptiles [1913], the soft-shelled eggs of which were prone to desiccation [1914]. The initial solution to this dehydration problem, apparently, was for the mother to secrete sweat from apocrine skin glands on her belly during incubation. Later, the oozing fluid was augmented with antibacterial antibiotics [163,927], setting the stage for hatchlings to partake of it as well. Once neonates came to depend on it, the proto-milk was supplemented with nutrients to form true milk [2707]. This plausible scenario is our best guess for how natural selection fostered the conversion of ventral sweat glands into fully functional mammary glands [2801].

Monotremes offer a likely snapshot of this early stage of mammary evolution [2649]. They lack nipples, and their young lick milk from the fur around the gland [69,293,2802]. Nipples arose later (~150 MYA [2232,2807]) in therian mammals [758]. They apparently evolved from hair follicles [1018,1148,1260,1740].

Because mammary glands and nipples were co-opted from gender-neutral structures (sweat glands and hair follicles respectively), Darwin's "Blue Pigeon" argument applies: the existence of nipples in males is most parsimoniously explained in terms of how they first arose. They have presumably persisted because, although useless, they are relatively harmless [2774].

Are male nipples *entirely* useless? Is there no species where the *father* suckles the young? The only putative case is the Malaysian fruit bat *Dyacopterus spadiceus* [651]. In 1994, 10 mature males were reported to have yielded milk upon palpation, but the amount per male (~0.005 ml) was two orders of magnitude less than that obtained from one lactating female (0.35 ml) [801], so even here it is unlikely that the male nipples are adaptively functional [2003].

The worthlessness of extant male nipples begs one last question: were they *ever* useful? Maybe not. In the echidna (a monotreme), the mammae are as well developed in males as females [2649], but there are no reports of

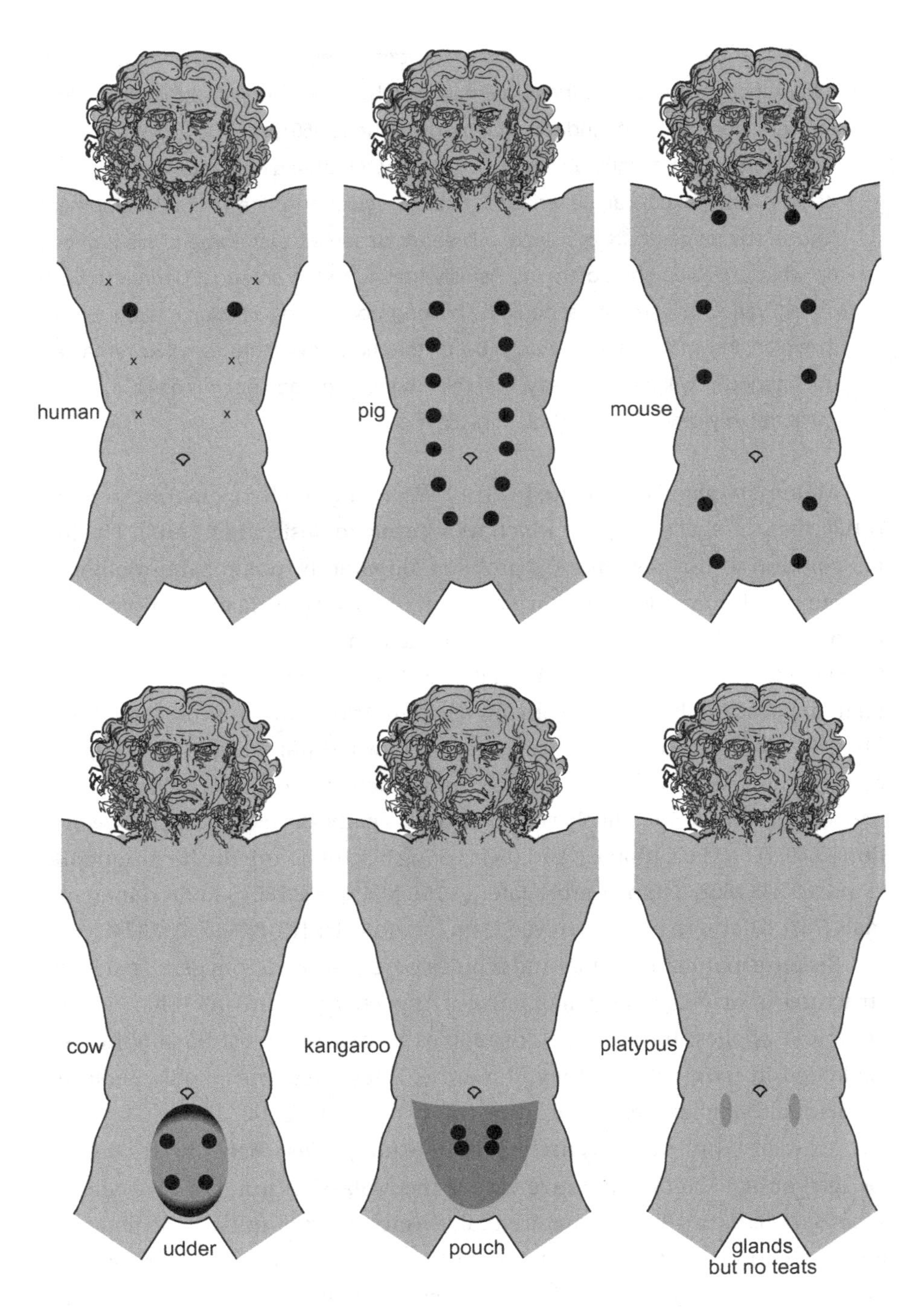

paternal lactation. Sex-limited mutations ("modifiers" [530]) must have arisen in therian clades to yield their anatomical (and functional) dimorphisms. Given how easily a male breast can be hormonally induced to lactate (*i.e.*, gynecomastia [1613]), it is likely that those changes relied on hormonal effectors, but the nature of their control circuitry remains unknown.

Figure 5.3. Mammary gland locations in various mammals. After Turner [2649]. Xs on the torso labeled "human" denote extra nipples in a German soldier (22-year-old male) [2802]. Similar atavisms were cited by Darwin [564], catalogued by Bateson [151], and described by others [187,408,2178,2431]. Most ectopic nipples in humans lie along these same "milk lines" [1023,2683], which stretch from the armpits to the groin [2198,2757]. In one grotesque case, a Thai woman had four big breasts, the extra two growing from her armpits [425]. Monotremes (platypus and echidna) have no nipples [2802]: their young lick the fur over the mammary glands [69,293], and the glands are nearly as well developed in males as in females [2802]. In marsupials, male neonates have fewer mammae than females [2158], or, in the case of Australian marsupials, no mammae at all [2197]. Male mice lack nipples (as do horses [69]) because they regress in response to androgens [687,2757] (but *cf.* [1539]). Other oddities (not shown) include (1) whales, which have two nipples straddling their genital opening [2051,2797]; (2) manatees, which have a nipple in each armpit [188]; and (3) the Virginia opossum, which has one median nipple ringed by 12 others [758]. The world record for teat number is the pouchless didelphia, *Peramya henseli*, which has up to 13 pairs [69].

How did women end up with only two breasts?

Extra nipples are occasionally found in humans, and they typically arise along the same "milk lines" that characterize multiteat species such as dogs and pigs (Fig. 5.3) [69,151,1106]. From this observation, Darwin inferred that humans must be descended from mammals with multiple pairs of mammary glands.

> On the whole, we may well doubt if additional mammae would ever have been developed in both sexes of mankind, had not his early progenitors been provided with more than a single pair. [564] (Vol. 1, p. 37)

Humans (and other simians [2318]) are unusual among mammals in normally having only one pair of mammae [69]. Why so few? Aristotle contemplated this riddle [137] and reasoned that teat number must be functionally related to the number of offspring per litter [652]. He was right about the correlation [313,2649] but wrong about some of the numbers. (Cows and lions have four teats, not two [69,2649].)

> In such animals as produce but few at a birth, whether horned quadrupeds or those with solid hoofs, the breasts are placed in the region of the thighs, and are two in number, while in such as produce litters, or such as are polydactylous, they are either numerous and placed laterally on the belly, as in pigs and dogs, or are only two in number, being set, however in the center of the abdomen, as in the case of the lion. The explanation of this is not that the lion produces few at birth, for sometimes it has more than two cubs at a time, but is to be found in the fact that this animal has no plentiful supply of milk. For, being a flesh-eater, it gets food at but rare intervals, and such nourishment as it obtains is all expended on the growth of its body.

REFLECTIONS ON FIGURE 5.3

What is so surprising about mammary glands from an evo-devo standpoint is how early they develop in the embryo [91], considering that they aren't needed until after puberty [2650]. Aristotle, like Darwin, wondered why males should have nipples {PoA:4:10:688b31*ff*} [137]: "In man there are breasts in the male as well as in the female; but some of the males of other animals are without them. Such, for instance, is the case with horses, some stallions being destitute of these parts, while others that resemble their dams have them." Unlike Darwin, however, he did not venture a guess as to why this is so.

> In the elephant also there are but two breasts which are placed under the pits of the forelimbs. The breasts are not more than two, because this animal has only a single young one at a birth. . . . In such animals as have litters of young, the teats are disposed about the belly; the reason being that more teats are required by those that will have more young to nourish. {PoA:4:10:688a32*ff*}

On average, the number of teats is about twice the litter size across the mammalian spectrum [904,2005,2006,2355]. Litter size is part of the "life-history strategy" of a species, which depends on its ecological niche, resource availability, body size, and other factors [2083,2511]. By contrast, teat number is a feature for which the selective pressures are less clear. Certainly, selection must favor more teats when the litter size increases [155], but to what extent are extra teats a liability when litter size decreases? If we accept Darwin's surmise that our (presimian) forebears had several pairs of mammae, then why did those pairs dwindle to one [1106,2318]? One possible factor in reducing simian litter size (and teat number?) was the difficulty in carrying more than one infant while swinging in the tree canopy. Another was the parental investment needed to raise a large-brained infant [59]. From an evo-devo standpoint, the deeper questions concern the fluidity of linkages in the genome.

1. How evolvable is teat number in different groups of mammals [164,1376, 2751]? Why is it less malleable than litter size [396,652] (*cf.* [2790])?
2. Are these variables completely independent genetically? If so, then how long does it take for one to track the other during evolution?
3. What are the developmental mechanisms for making more teats [1214, 2198]? Fewer teats?

The last question is especially interesting considering that teats constitute a meristic pattern like those surveyed in Chapter 4 (*e.g.*, fingers, teeth, vertebrae). The numbers of teats has changed appreciably throughout evolution (Fig. 5.3), but the spacing along the milk lines has remained relatively uniform [2649]. Is periodicity maintained because the sites arise via lateral inhibition [1724]? Given that a similar device may be used to space embryos apart within the uterus [1199,1988], it is conceivable that these circuits might share genetic components and hence could coevolve—with embryo spacing limiting litter size [296,2877] and teat spacing limiting teat number. We are only now beginning to learn how the milk line is delimited along the dorsal–ventral axis [447] and how mammary buds are patterned along the anterior–posterior axis [1839,2757]. It is worth keeping abreast of this new work... so stay tuned!

CHAPTER 6

Silly, Stupid, and Dangerous Quirks

Anyone who thinks that our body is a marvel of mechanical engineering should get out a phone book and scan the listings of orthodontists, orthopedists, optometrists, and chiropractors—to name just a few specialties. Those doctors are making good livings treating our sundry flaws. Face it: our body has *many* features that could work *much* better. The sad fact is that evolution is no engineer [2657]. It's just a tinkerer [1265,1266].

Worse than that, evolution is a *myopic* tinkerer. It can only solve immediate problems using available variations [226]. If its spur-of-the-moment contrivances turn out to be liabilities at a later date under different circumstances, well, that's just tough luck! Adaptations are typically quite hard to undo. In this way some of our organs have gotten themselves into predicaments akin to people in a bad marriage, who often mutter to themselves, "But it seemed like a good idea at the time!" Indeed, this familiar phrase offers a convenient shorthand for denoting the phenomenon:

> **bislagiatt** /bis-'la-gi-ˌatt/ *adj* [acronym: but it seemed like a good idea at the time!]: suboptimal functioning of a structure due to (1) its being used in a novel context (*e.g.*, a bipedal *vs.* quadrupedal stance) and (2) the inability of evolution to fix it (*e.g.*, due to prohibitive demands of genomic rewiring). Tantamount to a species getting stuck on a low peak in a rugged adaptive landscape [593].

A small sample of our shortcomings is sketched in Figure 6.1, and a larger batch is inventoried in Table 6.1. Other authors have discussed some of them [1103,2384,2824]. A few of the simpler cases have been chosen for closer scrutiny below in the light of relevant data from recent research.

Could choking have been avoided?

There is no better example of the bislagiatt phenomenon than our lungs. The earliest fish had no lungs, nor did they need any [2122]. Their gills extracted

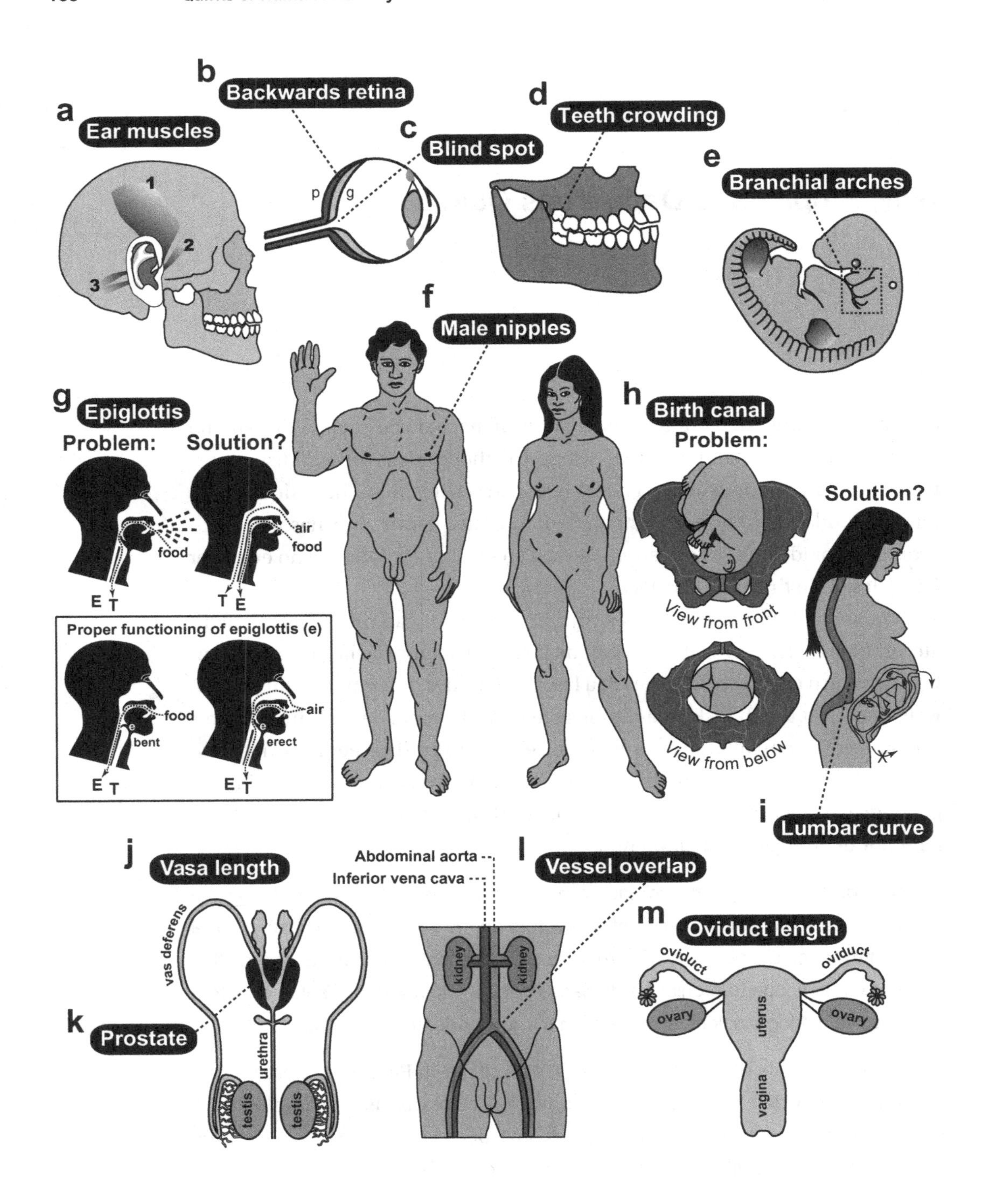

enough oxygen from the water for all of their metabolic needs, and the same is true for ocean-living fish today. However, pond-living fish face a problem that their pelagic cousins do not [1335]. During droughts, smaller ponds start to dry up, the water becomes muddy, and oxygen gets scarce.

Figure 6.1. Rogues' gallery of human anatomical flaws, ranging from the merely silly (a, f, j), to the patently stupid (b, c, d, e, i, k), to the potentially lethal (g, h, l, m). See Table 6.1 for further information.

a. Muscles (dark bands) that permit ear wiggling, a dubious talent that has no apparent function aside from entertaining young children: (1) superior, (2) anterior, and (3) posterior auricularis [1004,1488].

b. Eyeball with optic nerve. The retina is considered to be inside out because its photoreceptors (p) lie behind its ganglion cells (g), an orientation we may have gotten stuck with early in evolution (*cf.* Fig. 6.2). Since then, the photoreceptors have come to rely on the (outer) layer of pigment cells (not shown) [2098], so it may now be impossible to reverse this order. Thickness of retinal layers is exaggerated.

c. Blind spot where ganglion cell axons exit the eye. This suboptimal escape route is due to the inverted geometry shown in **b** and explained in Fig. 6.2. The hole in the retina is sizeable (~9 full moons in the sky), but we don't notice it because (1) the blind spots of the two eyes are in separate parts of the visual field and (2) our brain automatically fills in gaps in our visual field by interpolation [2117].

d. The common crowding of human teeth—especially "wisdom" teeth, which erupt last—is traceable to the evolutionary shortening of our jaw. See text.

e. Human embryo at 31 days postfertilization. Branchial arches (in dashed box) are a vestige from our fish ancestors. They now serve only as precursors for structures which could develop without such clefts (*cf.* Ch. 4).

f. Nipples are useless in human males (*cf.* Ch. 5).

g. Choking (upper left) is caused by malfunctioning of the epiglottis. The epiglottis is a cartilaginous flap (e) that reflexively covers the trachea (T) (lower box) when we swallow so that food or drink goes to the esophagus (E). The silhouette is a sagittal section with cavities as blank areas [612,889,2502,2737]. The line between E and T represents their adjacent lumenal walls [685]. One way to prevent choking (upper right) would be to have separate pipes for breathing and swallowing. In fact, that configuration already exists in babies, who can suckle and nose breathe at the same time: their larynx is so high that the epiglottis can touch the base of the nasal cavity at the midline [1547], with milk flowing around the juncture (see ref. [1469] for a clear diagram). The same is true for other mammals, who rest their epiglottis on the back edge of the soft palate [889,1546,1547]. The problem with having air bypass the mouth entirely, of course, is that it precludes talking [1931], although toothed whales still manage to "sing" to each other via an asymmetric specialization of their nasal passages [1718,2412].

h. Childbirth involves a tight fit between the baby's head and the mother's pelvic opening (*cf.* Fig. 7.1) [525]. Indeed, the passageway is so confining that the head must turn as shown (long axis transverse) to enter the birth canal, but then must rotate 90 degrees (long axis sagittal) to exit and let the shoulders enter [10,2220]. The process would be safer (and less painful) if the path mimicked a cesarean section, with the baby exiting through the navel. Drawn on the baby's head are the fontanels (open sutures) that allow the skull to deform during birth [500,685].

i. Back pain often occurs during pregnancy because of strains in the lumbar region (*cf.* Fig. 5.1). More serious repercussions of the spine's sinusoidal shape include herniated disks [10].

j. The vasa deferentia are tubes that conduct sperm from the testes to the urethra. They are much longer than they need to be in humans because of the circuitous route that they took evolutionarily [2384].

k. Routing a tube like the urethra through a solid organ like the prostate runs the risk of strangulation if the organ overgrows—a condition all too common in older men.

l. In May-Thurner Syndrome, blood clots typically arise in the left—but not right—leg [196,320,1683]. This peculiar ischemic asymmetry is due to a *normal* quirk. The abdominal aorta and the inferior vena cava travel side by side to the groin. Just after they split into the two common iliac arteries and the two common iliac veins (one artery and one vein routed to each leg), the *right* fork of the arteries passes over the *left* fork of the veins. This overlap compresses the left vein between the spine and the overlying artery, resulting in occasional thrombosis. There is no such crossover on the right side.

m. After ovulation, the egg leaves the ovary and enters the flower-shaped opening of the adjacent oviduct. Because fertilization occurs in the oviduct, the embryo can occasionally stick to the oviduct wall before it ever gets to the uterus, resulting in a life-threatening "ectopic" or "tubal" pregnancy [91]. It would have been safer if evolution had kept the length of the oviduct to a bare minimum or done away with it altogether and connected the ovary directly to the uterus.

REFLECTIONS ON FIGURE 6.1

The full-length man and woman are redrawn from a plaque aboard *Pioneer 10*—the first human-made object ever to leave our solar system [2253]. Launched in 1972, this human-sized spacecraft whipped past Jupiter (picking up speed), crossed Pluto's orbit, and began coasting toward Aldebaran, which it should reach in ~2 million years [2125]. The plaque was meant as an interstellar "postcard." If any aliens stumble upon it, they will surely chuckle at our quirks (if they *can* laugh), although turnabout is fair play, and if we were to see their two sexes (or however many they have) *au naturel*, we might chortle right back at them. The standing humans were drawn by Linda Salzman in collaboration with astronomers Carl Sagan (her then-husband) and Frank Drake [573]. Choking diagrams (**g**) are adapted from refs. [1866,2331], and childbirth sketches (**h**) are redrawn from refs. [1579,1786,1866], except for the profile of the gravid woman, which is based on photos of harpist Cheryl Gallagher from her musical compact disc *Pregnant Pause*.

g. As explained in the text, choking became a problem when our lungs arose as a branch of our eating tube. François Jacob ascribed such flaws to the trial-and-error nature of the evolutionary process [1265]: "To make a lung with a piece of esophagus sounds very much like tinkering."

There is also a deeper question here: why must people die when deprived of oxygen? Based on the prevalence of hibernation, estivation, and facultative anaerobiosis among animals [76,1675,2505], the answer is unclear [1171]. Sea turtles, for example, can hold their breath for at least three hours [1600]. Evolution, it would seem, *could* have given us the means to survive episodes of choking, drowning, or suffocation [1316,2229]. Why didn't it? Presumably, the rarity of asphyxiation among primates (by drowning, etc.) reduced the marginal advantage that any salvational mutations might have had to a negligible level.

The greatest irony about oxygen is that it used to be poisonous for living things before ~2 billion years ago [468,774], but eukaryotes evolved ways to detoxify and harness it to our metabolism so that it has become vital for us [12,152,2128]. This abrupt reversal of fortune was as dramatic as any Shakespearean plot.

h. While we are decrying the pain of childbirth, we should also pity the poor kiwi, whose egg comprises 25% of her body weight [915], for she, too, labors mightily [2549]. Some quirk of ratite history (as yet unknown [342,343,2071]) doomed those birds to that lunacy [969]. The depicted solution for humans (*i.e.*, birth through the navel) may seem farfetched, but the spotted hyena has rerouted its birth canal just as drastically: hyena pups must make a ~180-degree turn when they reach the mother's pelvis and then exit through her clitoris (pseudopenis) [803]! The problem with this path is that the diameter of the clitoral meatus is too narrow (~2.2 cm) to pass the pup's head (~6.5 cm), so the pain must be excruciating, which makes one wonder why hyenas laugh at all!

Darwin made an interesting observation about our fontanels. He noted how lucky we are that these hinges were already in place (because of how skull bones grow) before they acquired the function of allowing our skull to deform during the tight squeeze of the birth process. In other words, mammalian sutures were "co-opted" as hominin hinges:

> The sutures in the skulls of young mammals have been advanced as a beautiful adaptation for aiding parturition . . . but as sutures occur in the skulls of young birds and reptiles, which have only to escape from a broken egg, we may infer that this structure has arisen from the laws of growth, and has been taken advantage of in the parturition of the higher animals. [559] (p. 197)

To obtain more oxygen under such dire circumstances, fish have two options that are not mutually exclusive: (1) crawl to a bigger pond or (2) gulp air from the atmosphere. Both tricks are used today—for example, mudskippers do the former [926,1506], whereas lungfish do the latter [946,1225]. Our fish ancestors are thought to have done both at one time or another [1059,1549].

Once a bubble of air is captured, its oxygen can diffuse into the bloodstream wherever it comes to rest along the gut [1634,1934]. Some modern fish keep the bubble in their throat, whereas others store it in their stomach, intestine, or rectum [991,1225].

Hypoxic conditions must have been quite prevalent in ponds during the Devonian period (~400 MYA) because of alternating wet–dry seasons [991,1690]. In the face of prolonged episodes of hypoxia, *any* increase in the area of a fish's bubble-retention zone would have been favored because it would have allowed (1) bigger bubbles, (2) faster diffusion, and (3) greater survival. The fortuitous innovation that came to our fish ancestors' rescue was a ventral outpocketing from their foregut [1690]. This pouch may have arisen as a stunted seventh pair of pharyngeal pouches (*cf.* Ch. 4) [256,2616], although recent data on gene expression in this gut region suggest otherwise [372,430,1384,1754]. Alternatively, it may have been a spurious sac [1949] of the sort so commonly seen in intestinal diverticulosis.

So far so good: the novel throat pouch could henceforth serve as a rudimentary lung to augment the gills when necessary. The difficulties began, however, when a group of these bimodal breathers (leading ultimately to us) later climbed onto land, adapted fully to a terrestrial lifestyle (our reptile phase), and lost their gills entirely. Ever since then, our *only* lifeline to external oxygen has been a breathing tube (the trachea) that is a side branch of our eating tube (the esophagus) [1540]. The danger is obvious: swallow your food the wrong way, and you could choke to death.

This abridged history illustrates a poetic irony of the evolutionary process: the solution to an old problem often leads a new—and sometimes more serious—problem [593,2384,2824]. A partial remedy for the new choking problem did eventually evolve. The epiglottis arose as a hinged flap that reflexively covers the lung duct during swallowing [2737]. Regrettably, it is unreliable [2224,2563], as college students morbidly prove when they get so drunk that they pass out and aspirate their vomit [1969]. The Heimlich maneuver was devised to rescue people who would otherwise die from swallowing "down the wrong pipe" [1131,1132], but choking remains the fourth leading cause of accidental death in the United States [2657]. It has been a bane throughout human history.

TABLE 6.1. FLAWS OF HUMAN ANATOMY[1]	
Quirk	**Why it is suboptimal**
Ankle: ligament asymmetry (*cf.* knee)	**Disabling:** Weakness of lateral (*vs.* medial) ligaments predisposes the ankle to inversion (*vs.* eversion) sprains [10,2331].
Aortic arches: excess (Fig. 4.6a)	**Stupid:** Some arches are made but then destroyed; making unneeded structures is wasteful.
Appendix: risk	**Dangerous:** Susceptible to acute inflammation [1228], which affects ~7% of us during our lifetimes [459].
Blood vessels: routing to and from our legs (Fig. 6.1l)	**Dangerous:** An overlapping artery presses on the left common iliac vein [196] and can cause clots there [320,1683].
Eye: blind spot (Fig. 6.1c & 6.2)	**Stupid:** Puts a hole into the visual image [2098].
Eye: inside-out retina (Fig. 6.1b & 6.2)	**Stupid:** Reduces the resolution of the image (mainly because of blood vessels) [1752].
Eye: shape errors	**Disabling:** Vagaries in eyeball shape can cause focusing errors (near- or far-sightedness) [424,2098].
Hair: patterning	**Silly:** Most of our hair covering is useless, except for our eyebrows, which keep sweat from our eyes.
Heart: pacemaker asymmetry	**Dangerous:** Having a single pacemaker with no backup puts us at mortal risk for heart attacks.
Knee: ligament asymmetry (*cf.* ankle)	**Disabling:** Weakness of medial (*vs.* lateral) ligaments predisposes the knee to injury of the medial (*vs.* lateral) ligament [10,2331,2383].
Lungs: connection to foregut (Fig. 6.1g)	**Dangerous:** May cause choking to death [1132].
Muscles that wiggle our ears (Fig. 6.1a): mere vestiges	**Silly:** useless [1680].
Muscles that cause goose bumps (arrectores pilorum): mere vestiges	**Silly:** useless. Stimulated by sympathetic nervous system in response to cold or fear [1260].
Nerves: long detour of the left recurrent laryngeal (motor) nerve	**Silly:** It arises in front of the aorta, winds around it, and then ascends to the larynx [187,1004].
Nerves: variable routing of the sciatic nerve	**Disabling:** The sciatic nerve can take various "wrong turns" [187] that expose it to pressure from muscles (piriformis syndrome [184,1432]; *cf.* other nerve entrapment syndromes [1698]).
Nipples in men (Fig. 5.3 & 6.1f): mere vestiges	**Silly:** useless; occasionally lethal because of male breast cancer [2774].
Oviduct: length (Fig. 6.1m) [91]	**Dangerous:** If the embryo sticks to the oviduct, a tubal pregnancy can ensue, causing rupture, bleeding, and death [497]; the longer the oviduct, the greater the risk.

How we got stuck with it	Possible solution
Side effect of bipedalism (weight borne by feet *vs.* hands) [10] and heavier body [2212] (greater strain if torqued).	Equalize the strength of lateral *vs.* medial ligaments?
Vestige from our fish ancestors, who used all the arches in this series as precursors for their gill apparatus [1498].	Don't make the ones we don't need?
Vestige from our primate ancestors [446,1756].	Don't make one?
Packing of abdominal organs constrained the routes available?	Route the abdominal aorta *above* the inferior vena cava (*vs.* side by side) and align (*vs.* offset) their forks to dissipate compression?
Bislagiatt[2] legacy from our chordate ancestors (see text) [2278].	Reverse the stacking order of the cell layers in the retina?
Bislagiatt legacy from our chordate ancestors (see text) [2278].	Reverse the stacking order of the cell layers in the retina?
Developmental noise [1065,1402]?	Enable the eyeball (not just the lens!) to adjust its shape homeostatically [1402]?
Vestige from our ape ancestors, who had thick fur that provided insulation [1260].	Don't make hair?
Spandrel of the asymmetric way in which the heart develops [308,1201]?	Make a spare pacemaker in the other atrium? Or make two separate hearts [2824]?
Side effect of bipedalism (weight borne by legs *vs.* arms) [10,1103] and heavier body [2212] (greater strain if torqued).	Equalize the strength of medial *vs.* lateral ligaments?
Bislagiatt legacy from our fish ancestors (see text) [2824].	Join the trachea directly to the nostrils [1931]?
Remnant from our mammal ancestors, who used them to detect the directions of sounds [1488].	Don't make them?
Remnant from our ape ancestors, whose fur was thick enough that erection of hairs would (1) enhance insulation and (2) magnify apparent size as a threat display [1260].	Don't make them?
Legacy of our amphibian ancestors [2384,2731]?	Shorten its path [587]?
Inability of evolution to eliminate minor errors in neuronal pathfinding relative to muscle territories?[3]	Prevent the sciatic nerve from getting offtrack?
Artifact of genetic programming?	Don't make them?
Eutherian viviparity evolved by making embryos stick to the uterus [1610,2725], but they can also stick to the oviduct. (Nonprimates somehow stop such embryos from developing [524].)	Eliminate the oviduct so that embryos only get fertilized in the uterus, or stop tubal embryos from developing somehow [2673]?

(*continued*)

TABLE 6.1 *(continued)*

Quirk	Why it is suboptimal
Pancreas: fusion from two separate rudiments (dorsal and ventral)	**Dangerous:** Mistakes can produce a ring-shaped pancreas that strangles the duodenum [91,685,1436].[4]
Pelvis: tight fit of baby in the birth canal (Fig. 6.1h)	**Dangerous:** Death can result from macrosomia (big baby syndrome) [1146] plus cephalopelvic disproportion [502,1807,2840].
Prostate: encircling of urethra (Fig. 6.1k)	**Stupid:** Can strangle the urethra and block urination if it grows too large (age-related).
Rectum: venous plexus near anal sphincter [50]	**Stupid:** Prolonged sitting can restrict blood flow, clog these vessels, and cause hemorrhoids [2311].
Sebaceous glands: tendency to clog	**Stupid:** High testosterone at puberty induces excess shedding of glandular lining, which can clog pores, causing acne [1260].
Sinuses: tendency to clog with mucus	**Stupid:** Poor drainage can lead to congestion, headaches, and sinus infections.
Spine: curvature, cushioning, and stability (Fig. 6.1i)	**Disabling:** Backaches, herniated disks, etc. [10].
Teeth: crowding (Fig. 6.1d)	**Stupid:** Can cause impacted wisdom teeth [1680].
Testes: descent	**Stupid:** Risks cryptorchidism [1237,1403,2695]; eversion of body wall risks inguinal hernia [459]; superficial location risks injury.
Umbilical cord: length	**Dangerous:** Can kill the baby by tying itself in knots or wrapping around the neck.
Vasa deferentia: length (Fig. 6.1j)	**Silly:** Much longer than necessary [2384].
Yolk sac: vestige	**Dangerous:** Persistence as Meckel's diverticulum of the ileum can lead to a fecal umbilical fistula [91], inflammation, perforation, and bleeding [459].

[1] Quirks are listed alphabetically. Some are illustrated in Figure 6.1. For further examples, see Wiedersheim's 1895 compendium [2802]. Midline (risky) fusions are omitted, as is contralateral (wasteful) neural wiring (*cf.* Ch. 3). Other quirks not listed include tonsils, toenails, inadequate padding of our shin and sole, and vulnerability of a nerve at our elbow ("funny bone"). Moreover, our genome is as infested with vestigial "pseudogenes" as our body is replete with anatomical relics [389].

Equally embarrassing tables could be compiled for our physiological flaws (*e.g.*, menstruation [771], menopause [1962,1963,2354], etc. [2673]) and our pathological flaws [1451], especially our propensity for cancer [1595,1918], which may actually have an evo-devo explanation [842,1843,2243,2917].

Even the fact that we are mortal appears to be a lamentable spandrel [1121,2218,2219,2647,2824], rather than a manifest destiny, as most people assume [672,1722,2732]. Death is not inevitable? Correct! Each of us only dies because natural selection has consistently favored mutations that enhance our ability to reproduce at a young age, even if those same mutations cause us to die after our fertility wanes. As a result of this Faustian trade-off (known as "antagonistic pleiotropy" [1721,2207,2822,2827]), we each carry "time bomb" alleles that will eventually kill us [455]. This mutational constraint may explain why there are no "Methuselah" mutants (life span $\geq$200 years) [785,1932,2523]: they would have to suppress too many adverse genetic side effects *all at once* to live so long.

How we got stuck with it	Possible solution
Legacy of our fish ancestors [2731].	Make a single pancreas *ab initio*?
Bislagiatt legacy from our primate ancestors, exacerbated when hominins evolved big heads (Fig. 7.1) [156,2220].	Make a birth canal that bypasses the pelvis (endogenous equivalent of a cesarean section)?
Legacy of our mammal ancestors [2731].	Attach the prostate to one side of the vas deferens (like the seminal vesicle) [1931]?
Side effect of bipedalism and sedentary lifestyle [1092,2384].	Make better outflow valves in the plexus to relieve pressure [1886]?
Spandrel of high hormone levels (?), although why the face should be more affected than other areas is a mystery [1103].	Make sebaceous glands less hormone-sensitive?
Side effect of bipedalism [1103], which reoriented the angles of the drainage openings.	Fix the plumbing?
Side effect of bipedalism [1103,1720,2456]; upright posture imposes awkward stresses, and strain is exacerbated by pregnancy [2795].	Overhaul it from scratch [1931], or at least make the disks less prone to slippage and herniation?
Disparity in rates of jaw *vs.* tooth size reduction during evolution from long-muzzle apes [1103,2531].	Make smaller teeth, fewer teeth, or a bigger jaw?
Historical constraint: testes (unlike ovaries) cannot function at the core body temperature and hence must move to the surface [2160,2383,2824].	Enable sperm to be made at high temperature [2417,2824] so that testes can stay inside the abdominal cavity?
Legacy of our mammal ancestors [2731].	Make it shorter or prevent fetal gyrations?
Bislagiatt legacy of the caudad route taken by our premammalian ancestors [2383,2824].	Reroute it more directly?
Nutritive remnant from our reptile ancestors, superseded in mammals by the placenta [2731]; retained due to its other (circulatory [2267] and reproductive [2311]) roles (see text).	Don't make it?

[2] "Bislagiatt" ("but it seemed like a good idea at the time") denotes suboptimality of a structure because of its usage in a different context from the one in which it originally evolved. Our ancestors have often coped with challenges by co-opting old structures for new roles, only to find that these "escape routes" became blind alleys.

[3] Many features of our anatomy operate "under the radar" of natural selection (*cf.* Ch. 3), and the exact path that a nerve (or migrating cell) takes to its target is one of them. If routing mistakes occur often enough to give remedial mutations a selective advantage, then evolution might fix the problem. If not, then it never will, and the flaws will endure. Collectively, such defects comprise an irreducible price we must pay for the bislagiatt trap of relying so heavily on stochastic search algorithms to establish interconnections during our development. Insofar as this burden makes us permanently imperfect, it resembles the equally encumbering "mutation load" of our species [1569], although the analogy is weak because the latter has a different genetic basis [1919].

[4] A similar risk affects cephalopods. Their brain wraps around their esophagus like a doughnut [1891], so eating can cause food blockage or neural damage if they bite off more than they can swallow (*cf.* cormorant fishing [1640]). Cephalopods therefore constitute the only known case of natural selection for etiquette. We can imagine a mother squid admonishing her little squidlets, "Now, you all use the table manners that I taught you, or else your brains will explode!"

Aristotle was surprisingly aware of the anatomical flaws that put us at risk for choking. The following passage is from his *Parts of Animals* (*ca.* 350 B.C.E.; boldface added) [137]:

> On the other hand, the windpipe and the so-called larynx are constructed out of a cartilaginous substance. . . . The windpipe lies in front of the esophagus, although this position causes it to be some hindrance to the latter when admitting food. For if a morsel of food, fluid or solid, slips into it by accident, **choking and much distress and violent fits of coughing ensue**. . . . The windpipe then, owing to its position in front of the esophagus, is exposed, as we have said, to annoyance from food. To obviate this, however, nature has contrived the epiglottis. . . . So admirably contrived, however, is the movement both of the epiglottis and of the tongue, that, while the food is being ground in the mouth and passing over the epiglottis, the tongue very rarely gets caught between the teeth and seldom does a particle slip into the windpipe. {PoA:3:3:664a36*ff.*}

Darwin lamented our propensity for choking while extolling the utility of the epiglottis (boldface added).

> Every particle of food and drink which we swallow has to pass over the orifice of the trachea, with **some risk of falling into the lungs,** notwithstanding the beautiful contrivance by which the glottis is closed. [559] (p. 191)

In hindsight, it is easy to see how our fish forebears fell into this bislagiatt trap. They already had a muscular throat pump for imbibing water [2122,2583], which was readily adapted for gulping air [263,1335,1549]. All they needed were a few fortuitous mutations, and voilà, they had a throat sac in which to hold the swallowed air! In contrast, many more mutations would likely have been needed to create (1) a new orifice (*cf.* clinical tracheostomies [602,2541]), (2) a new pump, and (3) a new well-vascularized sac. Indeed, changing the function of an existing structure has often been the path of least resistance in the face of ecological challenges [593,1690]. This phenomenon is termed "preadaptation" or, more inclusively, "exaptation" [77,330,442,973,983].

Humans no longer need the buccal bellows of our fish ancestors to suck air into our lungs because we now have better ways of accomplishing this same end [264,1335,2019]: (1) the rib-cage bellows bequeathed to us by our reptile ancestors [50,1498] and (2) the pleural diaphragm we got from our mammal ancestors [1549,2122]. In theory, therefore, we could have avoided the risk of choking if mutations in our hominin ancestors had detached the trachea from the esophagus and attached it instead to a separate hole(s) such as their existing nostrils (Fig. 6.1g).

Aside from possible genetic difficulties associated with such a reconfiguration [2824], a pivotal problem with this scenario is that it would have kept us from speaking [1931]. Most sounds we make are formed by the same apparatus we use for eating: our lips, teeth, and tongue. Breathing through our nose alone would reduce us to humming.

Babies offer a useful lesson in this regard. They can suckle and breathe (safely) at the same time because their larynx has not yet descended [612,1547]. Their epiglottis is brought into contact with their nasal passage so that air flows directly to the lungs [478] and milk flows around the region of contact [1469]. This proximity restricts their vocal range [2278]. Not surprisingly (given its safety), the larynx of other mammals resembles that of babies more than that of human adults [533,1470,1887], which might explain why chimps can't speak [779,1546] despite their ability to learn English [2073].

Some animals have even gone babies one better [2462]. Toothed whales, for example, evolved laryngeal extensions that let the trachea dovetail directly with their blowhole tube, with food passing safely around the sides [188,2148,2412]. They have thus achieved what was only imagined for hominins, but at the expected cost of forgoing articulate speech [66,2177]. Some whales still manage to sing elaborate songs [612,1179,2233], and dolphins can "talk" using a rich repertoire of clicks, whistles, and squeals [1619].

The opposite trend of *lowering* the larynx is seen in species such as lions [2782] and deer [783,2826], the males of which roar to impress females and outdo rivals [781] (*cf.* trumpeter swans [2798,2913]). Indeed, it is even possible that the *hominin* larynx likewise descended, at least initially, to exaggerate the body size of the suitor, with the ability to speak coming along a serendipitous side benefit [783,1887]. The "Adam's apple" in *Homo sapiens* adult males attests to the courtship role that vocalization must have played in our evolution [889]. We don't yet know how this masculine trait arose in our lineage, but a similar dimorphism has been analyzed in frogs [2608] at both the genetic [400] and hormonal [1647,2609] levels.

Why are our retinas inside out?

Human eyes are beset by a host of optical imperfections [1720,2736], many of which affect our ability to focus near versus far [424,2098,2735]. Some of those flaws can be corrected with eyeglasses. One that cannot concerns the way that cells are layered within our retina. To appreciate the nature of this problem and why it constitutes another bislagiatt trap, it is useful to first consider the relative perfection of another optical component—the lens.

Our lens goes to extreme lengths during its development to make itself transparent by minimizing any variations in the refractive index (RI) of its parts

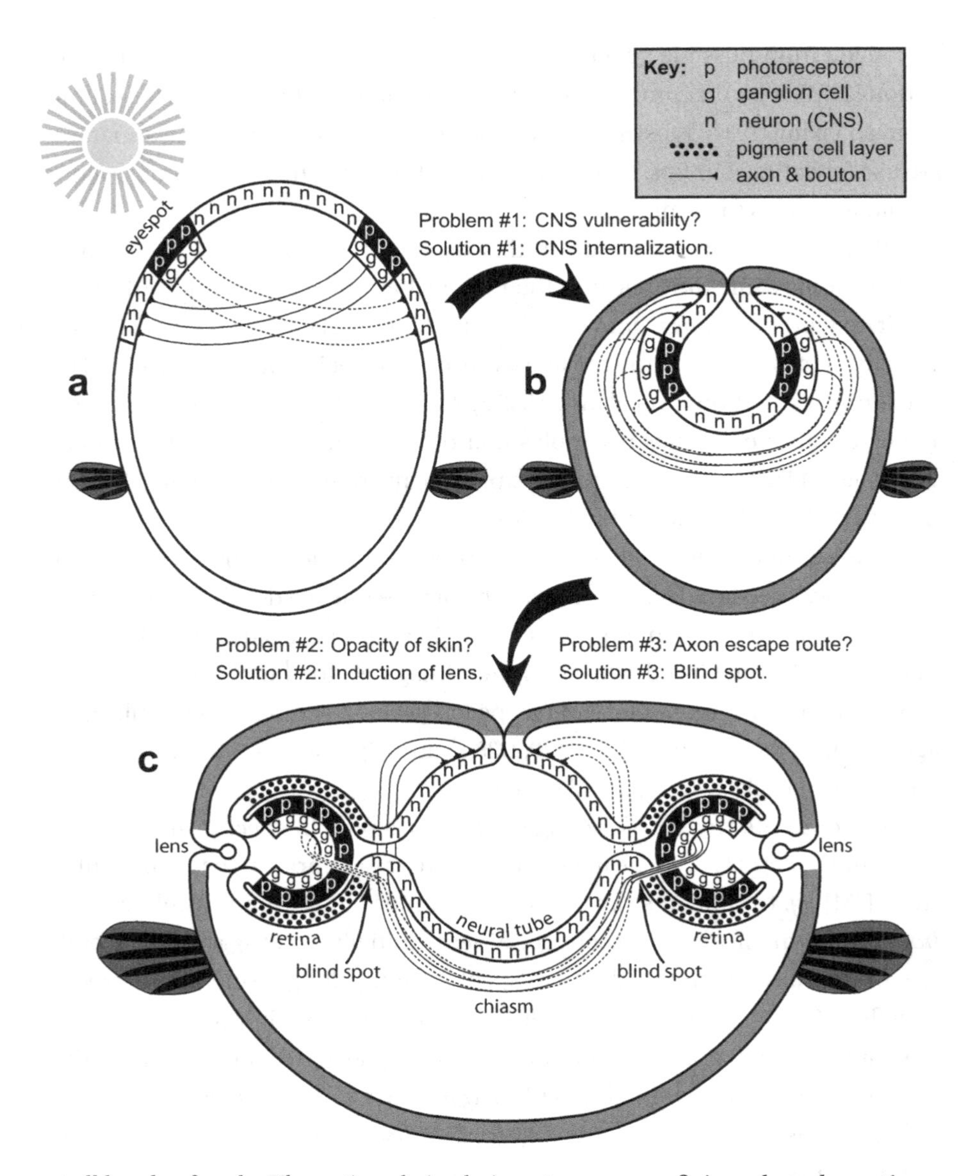

at all levels of scale. The rationale is obvious to camera aficionados: aberrations in the RI of a lens will blur an image, and the greater the deviations, the worse the blur. To convert flesh into glass (or a close approximation thereof), the lens uses a variety of clever tricks [550,908]:

Trick #1: Obliteration of its fluid-filled cavity. The lens arises as an ectodermal placode that invaginates to form a hollow sphere (Fig. 6.2). Cells on the back side then elongate all the way to the front side to occupy the cavity.

Figure 6.2. Hypothetical (bislagiatt) explanation for two flaws of human eyes: our backward retina and our blind spot. This diagram incorporates the inversion scheme of Balfour (1881) [2736], the axonal inferences of Polyak (1957) [2061] and Sarnat and Netsky (1981) [2278], and the opacity conjecture of Walls (1942) [539,2736], all of which are based on the tendency of chordate development to recapitulate its evolution [956]. Cartoons are transverse sections. Fish fins (which evolved later) are added merely to orient the reader (dorsal up, ventral down). Notwithstanding the symbols in the key, photoreceptors (p) and ganglion cells (g) are also neurons (n), albeit part of the peripheral versus central nervous system (CNS). Cell size is greatly exaggerated.

a. Our bilaterian (protochordate?) ancestor is thought to have had (1) a superficial nerve net [1458,1584,2579], (2) eyespots made of photoreceptors and ganglion cells [1461], and (3) criss-crossed wiring where ganglion cells projected axons to motor neurons (subset of n) on the opposite flank [2278]. The advantage of this contralateral wiring was that it allowed each eyespot to turn the body reflexively away from potential harm whenever it detected the shadow of a possible predator [2061,2278]. The problem with having neurons on the surface, of course, is that they could be easily damaged. To put it bluntly, it is not a good idea to "wear your brain on your sleeve"!

b. One solution to this Vulnerability Problem was for the CNS to move inside. The internalization trick that evolved in our chordate ancestors was "neurulation" [1038]—an involution of the dorsal surface to form a tube [1778,2312]—seen as a circle here in cross-section. Our neural tube ultimately forms our brain and spinal cord. During the CNS invagination of our protochordate forebear, the eyespots must have been dragged along and hence turned inside out. Ever since, our clade has, sadly, been saddled with a backward retina. Although this reversal turned out to be a mistake in hindsight, the eyespots were still just shadow detectors (not *image* detectors) at that time, so lower resolution was not yet a problem. This optical system would have worked quite well as long as the skin stayed transparent. However, any skin darkening would have severely reduced the effectiveness of the eyespots, and selective pressure would thereby have compelled (1) the eyespots to move nearer the surface and (2) the skin at the contact site to become transparent [2736].

c. To solve this Opacity Problem (presumably), vertebrates evolved a transparent lens where the prospective retina contacts the surface [1308]. During our development, the retinal outgrowths are initially shaped like balloons [2155] but then collapse, as if burst, into two-layered cups [18]. The outer layer facing the CNS becomes a pigmented epithelium. At some point in our evolutionary history, the lens must have blocked the escape route of ganglion cell axons. To solve this Entrapment Problem, the axons exited through the back wall, hence burdening us with a blind spot in perpetuity. Only a few axons are sketched here to indicate the routing.

Trick #2: Disintegration of its nuclei. Nuclei are destroyed after they have fulfilled their role of providing enough mRNA and proteins for the life of the lens.

Trick #3: Homogenization of its proteins. Lens cells chiefly produce crystallin proteins, the main job of which is to provide as uniform a solute as possible [549,1308]. With such a loose functional constraint, it is not surprising that vertebrate species have conscripted a zoo of arbitrary proteins to serve as crystallins [2028,2029,2833]. Indeed, their variety is one of the best examples ever adduced for the frivolity of evolution's "fishing expeditions" at the genetic level [1281,2631,2814] and its promiscuity at the protein level [26,1420]. Nevertheless, the crystallin menagerie is not *totally* random [1308,2368,2628].

REFLECTIONS ON FIGURE 6.2

These stages suggest how easy it was for evolution to build an eye step by step [591,1478] because any upgrade in imaging would have conferred a significant advantage [1308,2384]. Darwin need not have fretted about the apparent challenge that the eye posed for his gradualistic theory [876]. For a schematic that nicely traces the layering of our retina all the way back to amphioxus, see ref. [59].

a. Surprisingly, all metazoans use virtually the same sensor in their photoreceptors [539,1478]. This gadget first evolved in prokaryotes [2449]. It relies on (1) the vitamin A derivative "retinal" (or a variant [932]) to absorb photons [1019,1723], (2) an "opsin" protein to monitor shape changes in retinal [176,1993,2288], and (3) a signal-relay chain of downstream effectors [787,1466,1971]. The evolutionary riddles posed by this universality are (1) how long did it take for prokaryotic genomes, by random mutation, to stumble upon an opsin-like protein that could cradle a chromophore and transduce its twitching [621,876,2044,2241]?, (2) why was 11-*cis* retinal recruited instead of some other photoactive agent [932,1847,2709]?, and (3) how did it get linked to a particular transduction pathway [763,2521]? Believe it or not, the following creatures actually exist [876,2628]: (1) a jellyfish larva with bona fide photoreceptors but no nervous system to process their output [1896], (2) a fish that uses chlorophyll as a chromophore [675], and (3) a *single-celled* dinoflagellate that sports a humanoid "eye," complete with lens, photosensitive membranes, and a pigment shield [2013]! This odd menagerie prompted one author to propose that the eye evolved before the brain [876]!

b. In arthropods the nervous system arose evolutionarily from a superficial sheet of ectoderm [1779], as did ours [1463]. Strangely, however, arthropods are not all alike in how they solve the CNS Vulnerability Problem. Two classes, chelicerates and myriapods, undergo a variant version of neurulation from the ventral (*vs.* dorsal) surface [2494], but two other classes, insects and crustaceans, develop entirely differently [1778] (*cf.* other phyla [1678]). Instead of ectodermal origami, they "percolate" cells inside one by one to form their nerve cord [1104,1165,2495].

c. Fusions have been omitted to preserve the geometry of (1) neural invagination at the dorsal midline to form a tube and (2) lens invagination on the flank surface to form a sphere. The origami of eye development is different in other phyla [1478], which explains why, for example, the retinas of octopuses and squid are not backward [1732]. Also omitted are (1) the cornea, which is relatively trivial histologically [2103]; (2) the neurons (horizontal, bipolar, and amacrine) that intervene between photoreceptors and ganglion cells [2597,2736]; and (3) the later stages of lens development wherein it abandons its hollow state to simulate a solid glass camera lens [548].

For decades the lens was thought to be induced in a one-step process by the optic cup [1692,2432], but we now know that the optic cup plays more of supporting role along with other actors [670,994,1487,2627]. The optic cups of one unfortunate frog must have taken a wrong turn because its eyes (lenses and all) wound up in the roof of its mouth [2211]: the frog could only see when it yawned!

Inexplicably, our iris is the only muscle in the body that comes from ectoderm instead of mesoderm [191,584]. (How on earth did the ectoderm get the password to unlock the "muscle vault" in the genome?) Thus, when you gaze into your lover's eyes (the iris in particular), you're actually seeing a colorful part of his or her brain [120].

Even stranger is the fact that the iris, which normally develops independently of the lens, can regenerate a lens in newts if the lens is artificially removed [1125,1273,2639]. This ability must be a spandrel of how eye parts are wired in the genome, but we have no clue about how it evolved [344,1056].

One final quirk worth noting is that the iris is *intrinsically* photosensitive—that is, it constricts on its own [2641]. How it does so is unclear, although we do know that it detects light using some sort of cryptochrome rather than an opsin pigment.

Most important, the lens is devoid of blood vessels that could compromise its hard-won transparency (*cf.* [1662]). Lens cells are nourished indirectly by the aqueous humor that bathes them, rather than directly by the bloodstream like most organs of the body [2098].

Given all these efforts by the lens to achieve optical purity, it is shocking that our retina lets blood vessels fan out all over its surface, rather than steering them behind its back where the red cells would be out of the line of fire of incoming photons [1752]. The retina gets around this problem by using a tiny (~1 mm diameter) *avascular* spot—the fovea—for virtually all of its high-acuity needs [324,2287]. (You can prove this constraint for yourself by focusing on any single word and seeing how far afield you can read other words.)

How the problem of vascular interference evolved is unclear, but it may have been linked to another aspect of our retina that also seems absurd. A priori one would think that the photoreceptor cells (rods and cones) would be the foremost stratum in the retina because they could thereby detect photons with the least scatter [588,2824]. Indeed, that is how the eyes of octopuses and squid are built [1478,1732]. However, the rods and cones in vertebrate eyes reside behind several layers of other neurons, the frontmost being the retinal ganglion cells (RGCs) [2736]. Because all visual information collected by the photoreceptors must pass through the RGCs to get to the brain, the RGC axons are left with little choice but to pierce through the retina itself to escape the eyeball. Where they do so becomes our "blind spot" [2098,2118], another optical flaw that mollusks avoid by having their retinas the "right" way 'round.

How did chordate retinas get turned inside out? This odd topology actually makes sense as a bislagiatt consequence of how our eye evolved [90,932]. As shown in Figure 6.2, bilaterian photoreceptors are thought to have arisen as eyespots within a superficial nervous system [2278]. Involution of that nervous system in the chordate lineage (recapitulated in our own development?) would have dragged the eyespots inside and thus inverted them [120]. Any subsequent darkening of the skin would have reduced their access to light unless they could return to the surface and make a "porthole" to see through [2736]. With the induction of a lens, an image-forming eye was formed [1478], but the stratigraphy of its retina (blood vessels included?) was now reversed relative to its original polarity.

The backward layering of our retina, albeit silly, does not prevent us from seeing quite well, and eagles see at least *twice* as well as we do [1311,1665,2163]. How can the fovea achieve such amazing acuity despite this awkward geometry? One of its tricks, explained earlier, is to shun all blood vessels [2761]. Another is its composition [509]: it contains mostly cones [2098], which are specialized for high acuity [2044], as opposed to rods, which are specialized for high sensitivity

[1663]. Moreover, the packing of cones is higher there than anywhere else in the retina because of a narrowing of cone cell diameter [545].

In 1942, the anatomist Gordon Walls published a majestic (now classic) monograph titled *The Vertebrate Eye and Its Adaptive Radiation* [2736]. Until then, most researchers thought that the fovea's key trick was its shape: it is indented because of the pushing aside of all cell layers above the photoreceptors. Perhaps the removal of all of the blurring veils was all that was needed to let the cones see so clearly? Not so, said Walls! Instead, he argued that the pit acts like a magnifying glass. Given an RI difference between retina and vitreous humor, he reasoned that incoming light rays would fan out by refraction when they hit the sloping edges of the pit, and hence they would be detectable at higher resolution because they cover a wider area.

However, it now appears that Walls himself was wrong. A 2008 study showed that sharp vision is retained in the absence of any indentation [1657]! Our best guess is that the fovea attains its acuity by (1) higher cone density and (2) greater pixel fidelity (cones don't merge their outputs onto single RGCs), and (3) longer outer segments [932,2098]. The distinctive shape of the fovea may have evolved merely to put neurons closer to blood vessels [2761].

Evo-devo has recently added an interesting twist to this story. As more homologies have been uncovered between the circuits that make human versus fly eyes [345,768,1435,2031], researchers have raised their estimates of how well our Urbilaterian ancestor could see [763,1278,2628], although their conjectures do come with caveats [2814,2854]. On the basis of the components shared between mice and flies (Table 6.2), our current view of how the Urbilaterian eyespot developed is as follows (gene names are from mice) [1883]:

1. The eye rudiment had two mutually exclusive territories, one of which (*Pax6* ON; *Mitf* OFF) made photoreceptors, whereas the other (*Pax6* OFF; *Mitf* ON) made pigment or played some other non-neural role [1066]. The anlage was also balkanized into areas of *Pax6* and *Pax2* expression [2030,2323].
2. Once the definitive eye region was established by *Pax6*, a wave of *Shh* signaling spread across it like a ripple on a pond [1128,1868], leaving behind a loose array of cells expressing *Math5* [1278], each of which was thereby endowed with the competence to become a photoreceptor.
3. These scattered cells formed a tighter, more neatly tiled (hexagonal) lattice through widespread cell death [258,1025].
4. This initial lattice was used as a scaffolding for assembling the remainder of the retina [1137,2031].
5. Finally, these proneural cells became photoreceptors and grew axons into a neural net (brain) of some sort for processing of the visual information [1635,1865].

TABLE 6.2. HOMOLOGIES OF EYE DEVELOPMENT IN MICE AND FLIES[1]

Gene (mouse ≈ fly)	Protein function	Role in mice	Role in flies
Math5 ≈ *atonal*	Transcription factor (bHLH class)	Enables differentiation of the first retinal neurons (RGCs) [1278,2875]; expression spreads in a wave (*cf.* fish [1353]).	Enables differentiation of the first photoreceptors (R8s) [1279]; expression spreads in a wave [1137].
Mitf ≈ *dMitf*	Transcription factor (bHLH-Zip class)	Specifies RPE identity; LOF transforms RPE into neural retina; opposes *Pax6* to enforce adjacent areas (RPE *vs.* neural retina) [191,1066].	Specifies peripodial membrane (*vs.* eye proper) in eye disc [1066]?; inhibits *Pax6*.
Pax6 ≈ *eyeless*	Transcription factor (paired-homeobox class)	Expressed in optic cup, lens, and iris [2867]; LOF causes small eye [550,751] (*cf.* fish [362]); GOF induces extra eyes in frog [451]; needed for all but one retinal cell type [1658].	Expressed in eye disc [546]; LOF causes eye loss [1441]; GOF induces extra eyes [1051,1331].
Shh ≈ *hh*	Morphogen	Drives a wave of RGC differentiation (*cf.* fish [1868]).	Drives a wave of photoreceptor differentiation [1128].

[1] Genes are listed alphabetically. Their sequence of expression is given in the text. Other homologies between mouse and fly eyes include (1) a regional selector gene besides *Pax6* (*Iroquois*) [435]; (2) two agents of mitotic asymmetry (*numb* [402] and *prospero* [508]); (3) lens crystallins [217]; (4) regulators for opsin expression in photoreceptor subtypes [509]; (5) the strongly conserved "kernel" [572] of *dachshund*, *eyes absent*, and *sine oculis* [1082,1334,2097,2398,2759,2921]; and (6) a gene (*Vsx*) that controls growth and identity of cells that synapse with photoreceptors in both flies (transmedullary neurons) and vertebrates (bipolar cells) [724]. The last homology challenges the conclusion (see text) that pRGCs are remnants of r-type photoreceptors [724]. Much of this circuitry is found in other phyla aside from chordates and arthropods [88,345].

Abbreviations: GOF = gain of function (= ectopic expression); LOF = loss of function; RGC = retinal ganglion cell; pRGC = photosensitive RGC [798]; RPE = retinal pigment epithelium. The eye is only one of many organs where vertebrates and arthropods employ common genetic circuitry. For example, see Chapter 4 for how they use *Hox* genes along the anterior–posterior body axis and Fig. 1.3 for how they use *BMP4* (*dpp*) and *chordin* (*sog*) along the dorsal–ventral body axis. See Tom Brody's *Interactive Fly* Web site for further signaling pathway homologies.

Still unclear is whether the eyespot (1) formed images, (2) had a lens, and (3) was simple or compound. Despite the uncertainty of all of these inferences, one of them (#2 in the list on p. 120) led to a deep and abiding insight.

In 2000, the geneticist Andrew Jarman wrote an enticing review titled "Vertebrates and Insects See Eye to Eye" [1278]. In it, he pondered a paradox emerging from the homology data. The proneural gene *atonal* was already known to specify the initial photoreceptors in fly eyes—the R8 cells [301,1279]. Now new results indicated that *atonal*'s homolog (*Math5*) is essential for the initial neurons of the mouse retina—the RGCs [290,2875]. This relatedness of R8s and RGCs was intriguing because both are patterned in regular arrays [849,1137,1693]. Yet why should such a similarity exist, he wondered, if RGCs aren't photoreceptive? Based on function alone, he had expected homology between R8s and *rods/cones*, not between R8s and *RGCs*!

Two years later, Jarman's Enigma was solved. RGCs, or at least a small subset of them called pRGCs [552,798,2087], turned out to be photoreceptive after all (Figure 6.3) [1115]! An auxiliary sensory pathway of some sort had been suspected because blind mice (or humans [2671]) that lack rods and cones can still adjust their circadian rhythms to light [136,812,1116]. From then on, the more that researchers have looked into how RGCs sense light, the more they have found *functional* parallels (as well as genetic ones) with insect photoreceptors:

1. Our pRGCs use an opsin protein (melanopsin [93]) that is more like the opsin in insect photoreceptors than the rhodopsin in our own rods and cones [1029,1434,2086].
2. Our pRGCs transduce the signal from melanopsin via the type of G-protein cascade that characterizes *insect* photoreceptors [1981], not our own rods and cones [1726,2011,2102].

To sum up, animals in general manifest two kinds of photoreceptors [1884], which differ in the opsin that they use for photon detection and the G-protein that they use for signal transduction [85,90]. They also differ in the way that they make their stacked membranes—a feature that gives them their different names. Insects exhibit the "rhabdomeric" (r) type, whereas our rods and cones exemplify the "ciliary" (c) type [1883]. Until 2002, vertebrates were thought to have only c-type receptors, but we now know that our pRGCs are r-type receptors disguised as ordinary neurons. Given these facts, it would seem reasonable to draw the following conclusions about how animal eyes and eyespots evolved [89,388,763]:

1. Urbilaterians had both r- and c-type receptors.
2. Arthropods kept the r-types and apparently lost the c-types.

3. Chordates kept both types but consigned the r-types to exclusively non-imaging roles (*i.e.*, clock cycle and pupil reflex) while stripping them of the histological hallmarks of photoreceptors (*i.e.*, stacked membranes).

Consistent with this scenario, it turns out that the protochordate amphioxus has both r's and c's, and the r's are just as complex as the c's, although neither is used for image formation [1434,1461]. Indeed, amphioxus has *four* visual organs [59,1463]: two that are r-type (dorsal ocelli and Joseph cells) and two that are c-type (frontal eye and lamellar body, which appear be the antecedents of our lateral and pineal eyes, respectively). Also supportive of this scheme is the fact that both types (or their molecular components) coexist in various protostome phyla as well [89,90,695,1459,2628]. Many questions remain. Chief among them are:

1. Did r- and c-type receptors originate independently [695], or did they diverge from a composite type that used both reception–transduction systems redundantly [85]?
2. What *ecological* conditions led different clades to adopt different systems for image formation [1884]?
3. Should the vertebrate retina be considered backward after all if RGCs functioned as bona fide photoreceptors at one time?

Regardless of how these issues are resolved, the basic point of Jarman's title will endure: vertebrates and insects *do* see eye to eye. Despite the lack of common form at a *macroscopic* level, evo-devo has enabled us to peer more deeply into the structure of both eyes at a *molecular* level, and thus to discover our shared heritage. The proof of our kinship was there all along—under the veils of our retinas—hidden from Aristotle, Linnaeus, and Darwin. Now we've found it at last! A buried treasure, indeed!

Why is childbirth so precarious?

In theory, human birth poses a trivial problem: fit an ovoid head of diameter b (*baby*) through an oval hole of diameter M (*Mom's pelvis*). How hard could it be for evolution to make $M > b$? The answer is: harder than one might think, considering the selection pressures at play during hominin evolution [597,2220,2553]. Selection for greater intelligence tended to maximize b [1496,2320,2515], but pressures for bipedal speed prevented hip width from rising at the same pace [525,1581], thus placing an upper limit on M [5,2112].

This obstetric tug-of-war eventually reached a trade-off where $b > M$, but the skull manages to exit nonetheless thanks to its ability to deform its shape

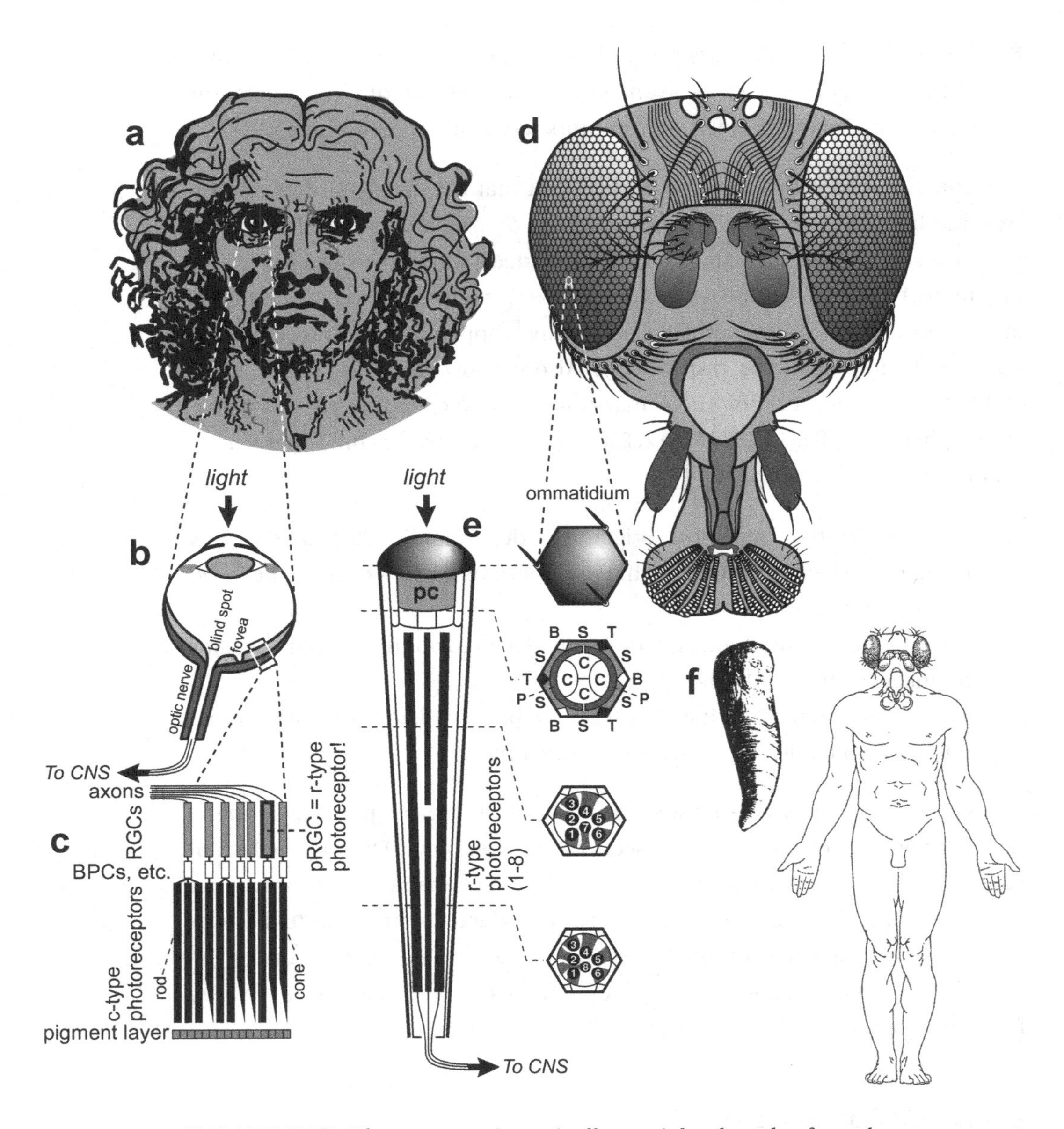

[502,1807,2840]. The squeeze is typically so tight that the fetus has to contort itself like a plumber under a sink to escape [10]. Some babies bend a shoulder so badly during these gyrations that they rip the nerve roots of their brachial plexus away from their spinal cord [58,2385] and emerge with permanent partial paralysis [1853]. Fossils indicate that Neanderthal babies faced an equally difficult gauntlet on their way out of the womb [894,2066].

Even if our genome had somehow been able to optimize *b* relative to *M*, individuals will vary, and half the baby's genes come from the father. Thus, if the father's family carries genes for big *b*, and the mother's has genes for small *M*, then birth could mean death for the mother, the child, or both, as it often did

Figure 6.3. Eye anatomy in humans (a–c) versus flies (d, e). We have a simple eye; flies have a compound one. Human photoreceptors are mainly c-type (ciliary), whereas fly receptors are exclusively r-type (rhabdomeric). Recently (see text) it has been discovered that humans also have an r-type receptor in the guise of what looks like a garden-variety neuron (**c**).

a. It would seem that humans have only two eyes, but a third one (or what's left of it) resides inside our head as the pineal gland (*cf.* Ch. 3).

b. Our eye works like a digital camera. Light is focused by the lens onto a screen (the retina), where it gets pixelated and converted into electrical signals. Those signals, in turn, are processed through various layers of the retina [2788] and sent to the brain via the optic nerve [2042,2098]. Acuity is greatest in the fovea. Axons exit nearby at a spot where we are blind. Abbreviation: CNS = central nervous system.

c. Schematic of retinal stratification (layers not to scale). There are five main types of neurons in the retina [1679] among ~55 types total [1352]. Only three are shown: RGCs (retinal ganglion cells), BPCs (bipolar cells), and photoreceptors. (Horizontal and amacrine cells are omitted.) A small subset (~3%) of RGCs are photosensitive (pRGC) [552,798]. They are r-type receptors, but they lack the stacked membrane system that we associate with bona fide photoreceptors [89]. The pigment layer absorbs photons to prevent backscatter [1478]. Note that multiple rods converge onto single RGCs (the ratio is actually ~20 to 1), whereas every cone typically has its own dedicated RGC [2098]. This difference is one reason our cone-rich fovea has such high acuity.

d. Each of the fly's compound eyes has ~750 simple eyes called "ommatidia" [1137]. Three separate simple eyes (white spots) are visible on top of the head: these "ocelli" mainly detect shadows [1776,1777].

e. The cuticular hexagonal dome serves as a cornea, whereas the pseudocone (pc) acts as a lens. The latter is a gelatinous cylinder secreted by four cone (C) cells (no relation to vertebrate cones). All but two of the eight photoreceptors (R cells) span the length of the ommatidium. As shown by cross-sections, R7 is above R8, an arrangement that enables R7 and R8 (which have orthogonal rhabdomere gratings) to detect polarized light [1090,2634]. Instead of using a pigment backscreen, flies wrap their photoreceptors in pigment tubes [1478] that are assembled like jigsaw puzzles from primary (P), secondary (S), and tertiary (T) pigment cells [1137]. Three of the six vertices of each hexagon are occupied by a bristle (B) instead.

f. Reciprocal human–fly chimeras in a woodcut (1793) by William Blake (1757–1827). This drawing was published by Claudio Stern [2472] from the private collection of Blake biographer Geoffrey Keynes (C. D. Stern, personal communication). It may have been an informal study. Blake did use the chrysalis with a baby's face in the frontispiece (captioned "What is Man!") for his book *Gates of Passion* (1793), but he apparently never published the man with a fly's head [199] (*cf. blakearchive.org*).

---

before the dawn of modern medicine [502]. Analogous disparities arise when big and small dog breeds are hybridized [1951].

Artificial selection for big heads and small hips in English bulldogs has led to an obstetric crisis at least as severe as that which afflicts us (via *natural* selection) [1491,1951,2490]. Cesarean sections are commonly needed to deliver the pups in this particular breed (Trey Fondon, personal communication).

Why are crowded teeth so common?

A similar example of evolution's ineptitude at bivariate geometry is the crowding of teeth. Ideally, our 32 teeth should fit neatly into our jaws—16 above and

REFLECTIONS ON FIGURE 6.3

With the juxtaposition of these two faces we have, in essence, a family portrait. Humans and flies are basically twins separated at birth over half a billion years ago. In many ways, we have come to look overtly different—for example, (1) we have an endoskeleton, whereas they have an exoskeleton; (2) we have four limbs, whereas they have six (not counting their wings); and (3) our eyes are simple, whereas theirs are compound [388,2923]. Yet despite these obvious disparities in gross anatomy, the same genetic gadgetry sculpts each of our respective structures during development [195,392], an abiding legacy of our common ancestry. Our bilaterian ancestor was considerably more primitive than either of these descendants [119,195,886,2663]. The fly portion of the diagram was adapted from ref. [1137].

c. Identities of the various retinal cell types are specified by a hierarchical combinatorial code [1084,1658,1819] that is time-dependent [616,1800,1920]. The regular spacing of the retinal ganglion cell (RGC) tiling pattern is achieved by ~80% apoptotic attrition [1550], homotypic lateral inhibition [705], and small-scale cell rearrangements [2145], without any need for homotypic contact [1568]. Nevertheless, homotypic contacts do play a role in RGC axon guidance [2042]. Instead of an absorptive pigment layer, cats and other nocturnal animals have a reflective tapetum that sends undetected photons back through the photoreceptors a second time [1478].

e. Despite what appears to be a solid-state dioptric apparatus, flies can adjust the amount of light reaching the photoreceptors by means of a myosin-operated "pupil" [2279]. For a primer on how fly eyes work, see refs. [1091,1815,1947]. (*N.B.* for fly aficionados: the upper cross-section is not the minimal "repeat unit" of standard textbook diagrams [2139]; instead, it shows the *entire* ring of cells encircling the photoreceptor core.)

f. William Blake's fascination with the human–fly metaphor was not confined to this woodcut. In *Songs of Innocence and of Experience* (1794), he had a poem called "The Fly": "Little fly, / thy summer's play / my thoughtless hand / has brushed away. // Am not I / a fly like thee? / Or art not thou / a man like me? // For I dance / and drink & sing, / till some blind hand / shall brush my wing. // If thought is life / and strength & breath, / and the want / of thought is death, // then am I / a happy fly, / if I live, / or if I die." Reciprocal human–fly chimeras were also central to the classic 1958 horror/science-fiction movie *The Fly* starring Vincent Price. In the 1986 remake with Jeff Goldblum, a human–fly hybrid was featured instead. (Don't let your kids see either movie!)

16 below. However, for many people (author included), there is just not enough room in the mouth to accommodate all those teeth. Again, we can find the reason by looking to recent trends in hominin evolution [1103,2531]. Jaw and tooth size have both been shrinking lately (relative to body size) [614] as a presumptive consequence of changes in diet [2248,2578], but the former has apparently outstripped the latter in its rate [954,969]. Evidently, these two traits are not integrated in the genome [94,1795], at least not in a functionally useful way [1375,2047]. This independence is probably associated with how their respective morphogens are *spatially* deployed [2349,2416], but it may also be due to the lack of any genetic device for *temporally* coordinating the critical events in jaw and tooth patterning [251].

Is our appendix on the road to oblivion?

The appendix, like the lung, is a cul-de-sac that branches off the gut [2651]. In other primates, it acts as a fermentation chamber [446,1756]. In our hominin ancestors, however, it lost this function as their diets shifted from herbivory toward carnivory, and their habit of cooking made vegetables easier to digest [2660]. Darwin understood that disused organs tend to become vestiges, and he reasoned that vestiges should manifest high variability because they have been freed from any oversight by natural selection [2552].

> We shall have to recur to the general subject of rudimentary and aborted organs; and I will here only add that their variability seems to be owing to their uselessness, and therefore to natural selection having no power to check deviations in their structure. [559] (p. 149)
>
> An organ, when rendered useless, may well be variable, for its variations cannot be checked by natural selection. . . . Rudimentary organs may be compared with the letters in a word, still retained in the spelling, but become useless in the pronunciation, but which serve as a clue in seeking for its derivation. [559] (p. 455)

As per Darwin's diagnostic criterion, our appendix *does* vary a lot. Its length fluctuates over a 6-fold range from person to person: from about 3 to 18 cm [36,2127,2802,2806]. Darwin could not explain *why* vestigial organs deteriorate erratically, but we now know that they do so because of what Sewall Wright called "mutation pressure" [2856]. Mutations occur at random (ergo variability), and they tend to be obstructive (ergo atrophy) [257,2845]. Hence, defects will accumulate unless natural selection can somehow purge them from the population [799,944,945]. By this logic, the appendix should vanish—following ignominiously in the "use-it-or-lose-it" footsteps of snake legs [471,490,2275], whale feet [172], cavefish eyes [807,1285], and all the degenerate organs (and genes [70,1802,2886]) of the despicable parasite world [2914]. Thus, the appendix could become just a historical footnote in the not-too-distant future [1699].

But wait! Might the appendix persist because it plays a role(s) unrelated to fermentation? Purportedly it could be (1) contributing supplementary lymphoid tissue [2063] or (2) providing a reservoir of microbes to reinoculate the gut so as to restore our intestinal flora after pathogen-induced diarrhea [233]. Neither of these reputed duties can be too important nowadays, however, because people who have had their appendix removed don't appear to be compromised immunologically or nutritionally [1228], and people born without an appendix don't seem to be disadvantaged in any obvious way [1767]. Hence, this vermiform vestige really may be devoid of redeeming value and hence doomed to decay.

In all likelihood, therefore, our appendix finds itself in the same evolutionary purgatory as kiwi wings [915] and manatee fingernails [559,1052]. Both of those outmoded structures are on their way to outright elimination [626], but they are not there yet [2802], and in the meantime they look ridiculous. The larger lesson is that *Homo sapiens*, like most species, is just one still frame in an epic movie of anatomical ebb and flow that is playing itself out at a glacially slow pace on a geological timescale [268]. Revisit Earth in a few million years, and we might look quite different... if we're still here!

From an evo-devo standpoint, the most curious thing about the appendix is how easily it can be removed entirely by a simple genetic manipulation. In normal mice, all *Hoxd* genes, except *Hoxd12* and *Hoxd13*, are expressed where the small and large intestines meet. If *Hoxd12* is artificially turned ON at this juncture, then the appendix fails to grow out [2895]. Mutations of this sort might be responsible for the aforementioned cases of appendicular agenesis in humans [1767].

Why do we make a yolk sac if we have no yolk?

Humans make several *external* culs-de-sac when we are embryos. They are called "extra-embryonic membranes" [2311]. The most famous is the amnion, the diaphanous aquarium where we each spent nine months *in utero*. Another is the yolk sac, a bag that evolved in our reptile ancestors to contain yolk [24,2731]. Anyone who has ever fried an egg knows that yolk is a yellow nutrient liquid. Unlike reptiles and birds, however, we don't need to provision our eggs with any such larder because we feed our babies continually inside the womb through an umbilical cord and placenta [2232].

Thus, the yolk sac fell out of fashion at the dawn of placental mammals ~150 MYA [271,2200,2232,2807]. Hence, it has had much longer to disintegrate than our appendix, which may have stopped being useful (1) ~5 MYA when we became omnivores and no longer needed a fermentation chamber or (2) ~2 MYA when hunting replaced scavenging of rotting carcasses, and we no longer needed a bacterial reservoir to remedy severe bouts of diarrhea [1522,2485]. Nevertheless, we still make a yolk sac, albeit a smaller one proportionally than reptiles. Why? A useful metric here is that dormant *genes* suffer enough disabling mutations after ~10 MYA that they can't be reawakened [1608,1659], but the genes (or *cis*-enhancers) that construct a yolk sac have endured for ~15 times longer than that!

The reason for the yolk sac's longevity is that, unlike the appendix, it made itself indispensable by adopting new roles as (1) a source of blood stem cells [2267] and (2) a rest stop on the migration path of primordial germ cells [2311,2834]. This phenomenon of outmoded organs "saving themselves" from evolutionary annihilation by finding other jobs in the nick of time is familiar to evo-devotees [1056] because it occurs so often at the genetic level (following gene duplication) [794,2631]. Other examples are not hard to find throughout our body [2384,2802]:

1. Our tongue evolved from some leftover gill muscles (of our fish ancestors) when our amphibian ancestors started breathing air [1256].
2. Two of our three middle ear bones evolved from redundant jaw fragments (of our reptile ancestors) when our mammal ancestors reconfigured their chewing apparatus [59,758,1670,1690,2168].
3. Our notochord used to be a hydroskeleton in pre-vertebrates, but it surrendered its supporting role when the backbone evolved [833]. Ironically, the notochord rescued itself by becoming the sole inducer for the nervous system around which the backbone is built [908,2130]! It is now an *integral* part of the phylotypic stage through which all vertebrate embryos must funnel [101,2683] (but *cf.* [201]).

The best example of co-option, though, is right in front of your eyes. Our hands were put to good use for tool manipulation fairly quickly after we became bipedal ~7 MYA [1929,2161]. Had they been left idle instead, we might have suffered the same humiliating fate as the mighty *Tyrannosaurus rex*, whose muscular arms withered to tiny twigs after it became bipedal, probably because it continued to eat its prey with its muzzle alone (as dogs do) [2278,2456].

Why can't we regrow arms or legs?

Salamanders regenerate their limbs completely after amputation [864]. We obviously do not. How wonderful it would be if we had their talent! It's not that we can't regenerate *any* lost parts [949,2538]. Our liver, for instance, has remarkable powers of regeneration [55,2577]. Why not our arms or legs [847]? After all, amphibians were our ancestors. Have mammals forgotten the recipe [1072]? Has the mechanism atrophied (like our appendix) from disuse?

Some of the factors that may be preventing regeneration in mammals (*vs.* salamanders) are (1) an inability of our muscle cells to reenter S-phase when prodded by the thrombin pathway [285,1910], (2) an inability of our neurons to regrow their axons fast enough to support blastema initiation [434,1449], and

(3) a silencing of our *Hoxc6* gene at the completion of limb development [284]. The greatest impediment, though, appears to be the tendency of our skin to scar [1829], rather than to heal with a rejuvenated wound epidermis [2492].

Herein lies the greatest medical relevance of the entire evo-devo field [285,2268,2492]. Researchers are actively exploring every available means of coaxing mammalian appendages to regrow [206,2566,2881]. Someday we will crack the cryptic passwords that salamanders use to reboot their limb-growth software. When that day comes—and it may come sooner rather than later—amputees will be able to discard their prosthetics and grow back their own arms or legs [1187,1829].

Nor is limb regeneration the only riddle that might eventually yield to the withering glare the evo-devo searchlight [291,1185]. Efforts are also under way to decipher how planaria regenerate their central nervous system (CNS) [404] in the hopes of enabling our spinal cord to do the same [160,421,436,494]. If we could solve the mystery of human CNS regeneration, then paraplegics might someday abandon their wheelchairs and walk again [937,2098].

As far-fetched as such cures may have sounded a decade ago, they have already left the world of science fiction and entered the realm of science [1419,2189,2603]. For the next generation of researchers, the clinic looms just over the horizon [105,1013,1243,2491,2909], and the first rays of hope are lighting their way [1527]. Our keenest students are already charting their careers to seek those goals. No greater tribute could be paid to Darwin than to use the tools he gave us to make these dreams come true!

CHAPTER 7

Mind and Brain

As you are reading this sentence, your brain is (1) importing a stream of pixels from your retinas, (2) detecting features of these squiggles to perceive them as letters, (3) combining the letters into words, (4) finding the meaning of each word in the mental dictionary you memorized in childhood, (5) assembling the phrase fragments into sentences by hardwired rules of syntax, and finally—and here we cross from the subconscious to the semiconscious—(6) checking the validity of each statement vis-à-vis your mental encyclopedia of accrued knowledge [179,2271,2648,2693]. We take reading for granted. We shouldn't. It is a remarkable feat of information processing [2298,2842].

Our brain is the most complex machine on earth. It is a "great raveled knot" [1003] that sets us apart from all other animals [638,2575] by enabling us to imagine what *could* be, not just to see things as they are [594,2037]. Despite centuries of probing its properties and decades of tracing its wiring, no one has ever managed to crack its code. Many have died trying, including titans such as Francis Crick [534,2393,2480], who had earlier, with James Watson, solved the jigsaw puzzle of DNA. That puzzle was child's play, however, compared with the Gordian knot of the human mind. How the mind works is the last great mystery in biology [1686,2356]. If only we knew the answer to that riddle, then we might be able to figure out how it evolved [213,253].

Our mind is the one quirk that's always been our greatest source of pride [2466], but are we deluding ourselves to believe that the way we think is *qualitatively* unique on planet Earth [1014,1247,2078,2866]? That conclusion seems a bit premature given how little we understand about the nature of intelligence (*cf.* autistic savants [913,2423,2624,2625,2667]) or the circuitry that underlies it [47,1930,2037,2138].

How did humans acquire intelligence?

Despite the exasperating opacity of our mental machinery, one thing is abundantly clear: our brain's abilities *far* exceed what it needs to ensure survival. Max Delbrück put it well:

> If [we imagine] that mind arose from mindless matter by a Darwinian evolutionary process of natural selection favoring caveman's reproductive success, then how did this process give rise to a mind capable of elaborating the most profound insights into mathematics, the structure of matter, and the nature of life itself, which were scarcely needed in the cave? [624]

The brain must therefore have spandrels [959,973]—that is, incidental features which happened to evolve in connection with adaptive traits that were selected directly [982] (*cf.* Ch. 4). The challenge is to determine what those spandrels are [77,770,1689,2824]. One is certainly our ability to read [2651]: writing is too recent (~5000 years old [1909]) to have *caused* our cognitive evolution, so it must have emerged as a fortuitous side *effect.* Another is probably our love of music [127,2037]. The issue of whether language is a spandrel has been vigorously debated [1096,1118,2036,2038] and is still unresolved [776,781,782,1926]. What about intelligence itself?

In *Ontogeny and Phylogeny,* Stephen Jay Gould proposed that our intelligence is a lucky by-product of (1) our larger brain, which is, in turn, a by-product of (2) our clearly extended period of brain growth relative to that of chimps (*cf.* [612,643,981,2166,2694]).

> The evolution of consciousness can scarcely be matched as a momentous event in the history of life; yet I doubt that its efficient cause required much more than a heterochronic extension of fetal growth rates and patterns of cell proliferation. [956] (p. 409)

How on earth could extra growth, by itself, lead to a new level of consciousness? Studies of brain evolution have uncovered three main rules that cause the brain to get more complex as it gets bigger. These developmental constraints have been nicely delineated by the evo-devotee Georg Striedter (with apologies to him for tinkering with his phrasing) [2515,2516]:

1. ***Later becomes larger:*** The last parts of the brain to develop in the ancestor tend to automatically expand the most in the descendant. This principle explains why our prefrontal cortex has become disproportionately larger relative to chimps (Fig. 7.1) [612,769,770,2181].

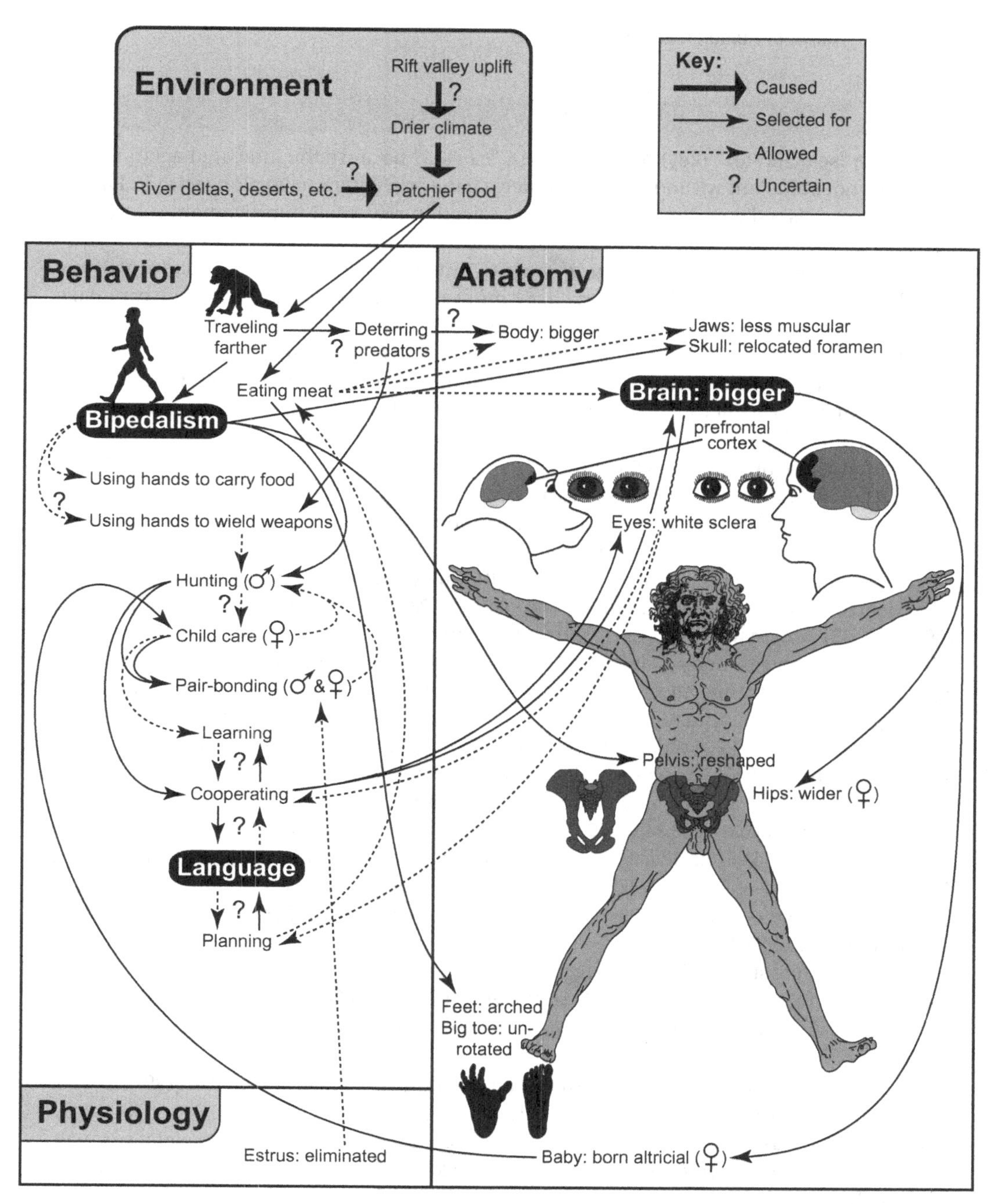

Figure 7.1. Selective forces that may have led to some important quirks that distinguish us from other apes—notably, bipedalism, language, and big brains. Question marks denote the most speculative links. Divergence of hominins from other apes apparently began when a change in climate nudged a population from an arboreal to a terrestrial lifestyle [2144]. Positive feedback loops may have driven the depicted outcomes [1708,2753]. Comparisons with chimps include our feet [1813], pelvis [786,1866], eyes, and brain [128,259,385]. Key for brain shading: black = prefrontal cortex [253,611]; dark gray = rest of cerebrum; light gray = cerebellum. The enlargement of our prefrontal cortex is important because it is the seat of rational decision making [249,2259] and voluntary action [115,2173,2515] (≈ free will? [2413]). Its expansion may have been driven by the fibroblast growth factor FGF17 [449], although the extent of that increase is disputed [611,788,2181,2309], as is the degree to which it arose due to allometry alone [332].

REFLECTIONS ON FIGURE 7.1

The phrase "selected for" (key) does not mean "caused" because the indicated agent is necessary but not sufficient without aptly conducive genetic changes to go along with it. Rather, it means "established a context where natural selection rewarded improvements along this line" [2753].

As for the genetics, how much *neural* rewiring was needed to convert a quadruped to a biped? Maybe not much [385,2730]: (1) a single mutation in an axon-guidance gene can make a mouse hop like a rabbit [1257] (*cf.* kangaroo rats [1351]), and (2) a human syndrome (due to one allele?) was found in Turkey where affected adults revert to walking on all fours [2564]!

Genes that might have fostered the enlargement of our brain include *ASPM*, the nonsense alleles of which cause microcephaly [2907]; *BF-1*, which reduces forebrain size in mice when it is knocked out [59]; plus at least a dozen others [411,776,910,1927,2015].

The "Man the Hunter" scenario, although still viable [1120], has been roundly criticized [204,316,1119,2455], in part because (1) carnivory can involve scavenging instead of hunting [2457], and (2) *females* are the ones who hunt in bonobo society [609]. Bonobos (*Pan paniscus*) are as close to us as chimps (*Pan troglodytes*) [608].

References to the depicted links include: Rift valley uplift →...patchier food [224,515]; river deltas, deserts, etc.→patchier food [224,628,2849]; patchier food→...bipedalism [786,1522,1578] and carnivory (as a part of omnivory) [1522,1756,2456]; traveling farther (overland)→...bigger body [145,2456,2737]; bipedalism → freed hands [42,607,938], relocated foramen magnum [10,2278], reshaped pelvis [1579], arched foot [1366], and unrotated big toe [956]; manipulation → hunting [2286,2753]; carnivory → bigger body [31,1756], bigger brain [59,789,1522], and lighter jaws [2014]; bigger brain ↔ cooperating (*i.e.*, society and culture) [255,2017]; bigger brain → wider pelvis [525,1579,2753] and altricial childbirth [2456,2679,2737]; altricial childbirth → greater infant dependency [956,1578,2220,2334]; carnivory–hunting cycle [2353,2455,2485] → cooperating [1716,2170,2457]; cooperating → white sclera [1398,1410,2515,2613]; division of labor (hunting *vs.* child care [341]) ↔ pair-bonding [204,418]; child care → learning [1476]; learning ↔ cooperating [1476]; cooperating ↔ language [689,2461]; language ↔ planning [253,1476]; bigger brain → planning [253]; and loss of estrus → pair-bonding [316,1578,2002,2463].

Other links (not shown) include bigger body → bigger brain [770] (via allometry [2181]); bipedalism → raising our head so high off the ground that our sense of smell waned in value [389,701,2278] → loss of olfactory genes [1882]; hunting with weapons → less need for shearing teeth → reduction in canine size [145,2753]; carnivory → changes in gut anatomy and enzyme allocation [2310]; and monogamy (pair-bonding) → larger brain size [2294]. Use of fire by hominins dates back 0.8 MYA or more [1186].

For rebuttals of aspects of this scheme, see refs. [228,406,628,2286,2576]. For skeletal comparisons (*Homo vs. Australopithecus vs. Pan*), see refs. [267,1522,2915]. For a picture of chimp feet that look hauntingly like human hands, see ref. [10]. Knuckle-walking chimp silhouette was adapted from ref. [1476].

Finally, it is important to dispel a widespread misinterpretation of diagrams like this. Yes, evolution entails problem solving (*e.g.*, the problem of patchier food), but it is neither progressive nor teleological, and it is certainly not Lamarckian. Species do not solve ecological problems, although such wording is sometimes used as a convenient shorthand [2779]. Mutations and recombination *do*, but only in the same way that you might eventually find the door in a dark room by feeling your way along the wall.

2. ***Larger becomes more subdivided:*** If morphogens have finite ranges, which certainly seems to be the case [2542,2920], then the number of subregions may automatically increase as the surface area expands [1907,1908,2078, 2141,2337]. A richer landscape of diverse modules might thereby emerge [1322,1324,1442,1444,2284].
3. ***Larger becomes more interconnected:*** As different parts of the brain enlarge, their surplus of extra axons may automatically be displaced to new targets [610,612]. This trend may explain why we can speak but chimps can't [2515,2516]: our neocortical axons have evidently invaded regions of the medulla where they gain direct (*conscious*) control over motor neurons of the face, jaws, tongue, and larynx [2358–2360,2515] (*cf.* [1046]). If this argument is valid, then chimps would still be unable to speak even if their larynx descended as low as ours [779,1546].

In *How the Mind Works*, cognitive scientist Steven Pinker rejected Gould's argument and rebutted all other spandrel-based explanations for the origin of human intelligence [2037] (p. 187*ff.*; boldface added):

> Human ingenuity has been explained away as a by-product of blood vessels in the skull that radiate heat, as a runaway courtship device like the peacock's tail, as a stretching of chimpanzee childhood, and as an escape hatch that saved the species from the evolutionary dead end of bearing fewer and fewer offspring. Even in theories that acknowledge that intelligence itself was selected for, **the causes are badly underpowered in comparison with the effects.**

Instead, Pinker asserted that our mental talents evolved as bona fide adaptations in their own right, and he advocated using reverse engineering to deduce their adaptive roles [2169]. He ruled out a number of putative possibilities, which he dismissed as Just-So stories:

> In various stories the full human mind sprang into existence to solve narrow problems like chipping tools out of stone, cracking open nuts and bones, throwing rocks at animals, keeping track of toddlers, following herds to scavenge their dead, and maintaining social bonds in a large group. There are grains of truth in these accounts, but they lack the leverage of good reverse-engineering. Natural selection for success in solving a particular problem tends to fashion an idiot savant like the dead-reckoning ants and stargazing birds. We need to know what the more general kinds of intelligence found in our species are good for.

Ultimately, he endorsed the notion that intelligence evolved as an adaptation for *hunting*, which requires a multitude of mental talents, including anticipation, simulation, and communication (boldface added) [2615]:

> The only theory that has risen to this challenge comes from John Tooby and the anthropologist Irven DeVore. . . . Humans, Tooby and DeVore suggest, entered the **"cognitive niche."** Remember the definition of intelligence . . . : using knowledge of how things work to attain goals in the face of obstacles. By learning which manipulations achieve which goals, **humans have mastered the art of the surprise attack.** . . . Humans analyze the world using intuitive theories of objects, forces, paths, places, manners, states, substances, hidden biochemical essences, and, for other animals and people, beliefs and desires. **People compose . . . plans by mentally playing out combinatorial interactions among these laws in their mind's eye.** . . . Humans have the unfair advantage of attacking in this lifetime organisms that can beef up their defenses only in subsequent ones. Many species cannot evolve defenses rapidly enough, even over evolutionary time, to defend themselves against humans.

The clashing views of Gould and Pinker lie at opposite poles of a wide theoretical spectrum. The theories along that spectrum can only be evaluated by examining the *ecological* factors that contributed to our psychological awakening.

Why did humans acquire intelligence?

What drove our split from other ape species and ultimately led to our idiosyncratic intellect? Most of the hypotheses that have been proposed rely on other quirks aside from intelligence as stepping stones that led to it inadvertently (Table 7.1), and most of them invoke climate change as the starting point [2144]. A composite flow chart that interweaves the threads of the chief arguments is presented in Figure 7.1. Similar diagrams abound (*e.g.*, [789,1221,2278]), but there is no consensus about which, if any, of them is correct [365,2458].

Implicit in Darwin's theory is the notion that "each species has proceeded from a single birthplace" [559] (p. 357). Ours was Africa. Why there? Why then? According to the "East Side Story" [224,515,516], our odyssey began with the uplift of Africa's rift valley, which left an abiding rain shadow over the eastern plateau (our home then) ~8 MYA. As the climate became drier, the forests dispersed into smaller fragments, and the intervals between the remnants increased [2849]. Those apes who found themselves in this predicament were impelled to travel farther across open land in pursuit of food.

On such terrain, walking on two legs (or running [267]) is more efficient than ambling about on all fours [1522,2451]. If modern apes are any guide, then we can safely assume that the apes of that period already walked upright for

TABLE 7.1. ADAPTIVE QUIRKS OF HUMAN ANATOMY[1]

Quirk	Evolutionary explanation
Brain: huge size	Adaptation for thought, language, etc. [612]?
Eyes: white sclera (exposed area around the iris) *vs.* dark sclera of *all* other primates [1786]	Adaptation that aided cooperation [325,2613] by letting others discern your gaze angle [1398,1410], intention [158,1935], and emotional state [19]?[2] Even babies no older than 6 mos. are attuned to gaze shifts [824,2336].
Face: variable features [1544] and a knack for detecting such differences [305,523,2598]	Adaptation for identification or group bonding? Is facial *variability* itself a selectable trait [11,1141,2903]? Can the genome *target* variability to certain compartments [367,1180,2230]? These are all outstanding issues in evo-devo [366,1064,1989].
Foot: arched sole and unrotated big toe	Adaptations for bipedalism [10,2737].
Hair: loss of hair from most of the body	Adaptation for thermoregulation [1260]? Or capricious quirk of sexual selection [561]? Eyebrows may have been retained to keep sweat from our eyes [1371] and to accentuate facial expressions [1103].
Larynx: descended ("Adam's apple" in males) [1546]	Adaptation that enhanced our ability to speak [779,1183,2607]? Because of the deeper descent of the larynx, men's voices are an octave lower than women's, which women may have found sexy [889,890].
Lips: darker than rest of the face, everted, and plumper	Adaptation that facilitated cooperation in social groups [325] by enhancing our ability to discern facial expressions [1514]? May also have aided articulation (phonemes) and parsing (lip-reading) of speech. Kissing is probably just a fortuitous by-product [2737,2738]. O happy spandrel!
Pelvis: changed shape from "spoon" to "bowl" [1534]	Trade-off between selective forces for walking (narrower = better) and parturition (wider = better) [10].
Sexual traits: larger size of adult penis and breasts [650]	Adaptations for pair bonding that evolved through sexual selection [360,651,2381,2737]? Erotic role of breasts (exposed by lack of fur) makes sense as an indicator of fecundity [68].[3]

[1] This table contrasts with Table 6.1 (useless or detrimental quirks) insofar as these features appear to have been *directly* selected for their utility in hominins. Only a few of our salient quirks relative to other simian primates are listed. For broader surveys, see refs. [788,1706,1786,2212]. For more on the sequelae of bipedalism, see ref. [2286]. Our prominent chin and nose are discussed in refs. [956,982,2824]. Sexual behavior is covered in ref. [651].

[2] Darwin sketched the value of cooperation (although not a white sclera per se) in *Descent* [561] (Vol. 1; p. 162): "When two tribes of primeval man, living in the same country, came into competition, if the one tribe included . . . a greater number of courageous, sympathetic, and faithful members, who were always ready to warn each other of danger, to aid and defend each other, this tribe would without doubt succeed best and conquer the other."

[3] This may be a good guess as to *why*, but the abiding mystery is *how* men's brains got wired to be aroused by a pair of pendulous globes of fatty tissue with little pink knobs on them.

brief errands. Under the newly arid conditions, natural selection would have rewarded any mutations that enhanced this ability to permit foraging on longer treks [50]. As we became more bipedal, the force vectors acting on our skeleton would have shifted [1534], and selection would have favored any structural adjustments that made striding more graceful (Fig. 7.1) [10,2286]. Among the adaptive responses that emerged as a result were a shifted foramen magnum [2278], a reshaped pelvis [1579], and an unrotated big toe [1366].

This scenario exemplifies an important principle of animal evolution: changes in behavior typically precede—and foster—changes in anatomy [1016, 1476,2437,2456], especially in the face of ecological challenges [77,1690,2790]. Most animals have large repertoires of behaviors, some of which they are more adept at executing than others. In desperate circumstances, of course, even a clumsy response is better than no response at all. Over the long term, natural selection will reward any anatomical changes that make the behaviors more effective [1493]. If the structural alterations facilitate further improvements in behavior, then positive feedback loops can emerge [1690]. Once such a process of "runaway selection" begins, the pace of evolution accelerates [600,1708,1719,2753], just as when predators and prey are locked in a mortal race to outrun one another [587].

One possible instance of this phenomenon is the superhuman vision seen in tribes of sea gypsies in Southeast Asia [918]. Children of those tribes have been diving for food from the sea floor for thousands of years (= behavior), and their underwater acuity (= anatomy) is *twice* that of Europeans. This sharper acuity has been traced to an astounding ability on the part of these youths to (1) constrict their pupils and (2) alter the focal lengths of their lenses. However, until adequate controls are done (*e.g.*, testing tribal members raised elsewhere), we can't rule out a nonevolutionary explanation: it is equally possible that the children's eyes are changing as they grow in response to the optical demands of their diving habit [2789]. Phenotypic plasticity is common in animal development and may be instrumental in facilitating evolution when ecological pressures persist over many generations [2790].

Darwin was intrigued by the transformative power of the behavior–anatomy feedback loop. In *Origin of Species*, he mused about how a progressive ratcheting of reinforcing interactions might have spurred the evolution of some especially peculiar species (*cf.* bats [210]). Following are a few passages from *Origin* to illustrate his train of thought.

On flying squirrels: "Let the climate and vegetation change . . . and all analogy would lead us to believe that some at least of the squirrels would decrease in numbers or become exterminated, unless they also became modified and improved in structure in a corresponding manner. Therefore, I can see no difficulty, more especially under changing conditions of life, in the continued preservation of individuals with fuller and fuller flank-membranes, each modification being useful, each being propagated, until by the accumulated effects of this process of natural selection, a perfect so-called flying squirrel was produced." [559] (pp. 180–181)

On whales: "In North America the black bear was seen by Hearne swimming for hours with widely open mouth, thus catching, like a whale, insects in the water. Even in so extreme a case as this, if the supply of insects were constant, and if better adapted competitors did not already exist in the country, I can see no difficulty in a race of bears being rendered, by natural selection, more and more aquatic in their structure and habits, with larger and larger mouths, till a creature was produced as monstrous as a whale." [559] (p. 184)

On penguin-like diving birds: "The acutest observer by examining the dead body of the water-ouzel would never have suspected its sub-aquatic habits; yet this anomalous member of the strictly terrestrial thrush family wholly subsists by diving—grasping the stones with its feet and using its wings under water." [559] (p. 185)

He extended this basic argument to the evolution of hominin bipedalism in *Descent of Man*. Here, too, he focused on the versatility of locomotory behavior as a starting point for anatomical change (boldface added):

On hominins: "I can see no reason why it should not have been advantageous to the progenitors of man to have become more and more erect or bipedal. They would thus have been **better able to have defended themselves with stones or clubs, or to have attacked their prey, or otherwise obtained food.** . . . Thus the gorilla runs with a sidelong shambling gait, but more commonly progresses by resting on its bent hands. The long-armed apes occasionally use their arms like crutches, swinging their bodies forward between them, and some kinds of Hylobates [gibbons], without having been taught, can walk or run upright with tolerable quickness; yet they move awkwardly, and much less securely than man. We see, in short, with existing monkeys various gradations between a form of progression strictly like that of a quadruped and that of a biped or man." [561] (Vol. 1, pp. 142*ff.*)

As Darwin surmised, the adoption of bipedalism freed our hands for other uses [938], including (eventually) the grasping of weapons for hunting [224]. Meat added a rich new source of nutrition in the face of diminishing returns from the sparser fruits of the shrinking forests [1522]. The trend toward omnivory is well documented in the fossil skulls of our hominin ancestors [386,2014]: their heavy facial muscles (needed for chewing leaves) began to dwindle after bipedalism arose [11,224,1522]. According to the "Man the Hunter" scenario [1508], it was hunting that set the stage for the evolution of bigger brains [1120,2455]. Meat supplied the metabolic fuel for brain growth and brain maintenance, while cooperative hunting behavior established a selective environment that rewarded even the slightest increments in our ability to plan, socialize, and communicate with others [255].

Is intelligence attributable to brain size?

The dependency of mental acuity on brain size was broached earlier, but it bears more scrutiny now that the trajectory of hominin evolution has been sketched as a backdrop for further discussion. A second pass through the topic is thus warranted with that history in mind.

Gould's "bigger is better" postulate for intelligence [900,956,2251] makes sense to anyone familiar with computers and their need for memory capacity [2636] and processing capability [788,2254]. Sophisticated software (consciousness? [336,903,909,2327]) *does* require sufficient hardware (circuitry [1364,1596, 1789]) to run successfully [1671,2037,2864]. However, if *absolute* brain size were all that mattered, then men would be smarter than women [57,161,2515], pygmies would be stupid [2348], and sperm whales would put us all to shame because their brains are six times bigger [1885,2228,2253,2278]. Because none of these conclusions is valid, the premise itself must be suspect [1196].

The definitive treatise on brainpower vis-à-vis brain size is Jerison's 1977 book *Evolution of the Brain and Intelligence* [1295]. In it, he argued that brain size *relative to body size* provides a more meaningful metric of brainpower than brain size alone [745,1223,1296]. By that standard, men and women approach parity [744], and sperm whales dive well below our level [59,2228].

Interestingly, Jerison's formula implies that dolphins should rival us in wits [59,624,1653], and they apparently do [157,1149,1651,1885]. Like us, dolphins recognize themselves in a mirror [2152], use tools [1446], comprehend abstract concepts [1649], teach skills to their offspring [1446], communicate using a symbolic language [1652], and obey grammatical rules when they engage in conversation [1804]. If dolphins ever evolve opposable thumbs, Heaven help us [14,80]!

Despite its intuitive appeal, the brain:body ratio actually makes *less* sense cybernetically than does absolute brain size [461,1197,1325,2309]. Think about it. The brain's most basic duty is to maneuver the body (toward food, away from danger, etc.) in response to sensory input [1440], and body size should have little relevance to the central processor's ability to achieve this goal. Consider a banal example. The brain of one man can steer an 18-wheeler (= motor output) as easily as a Volkswagen, and the only modality that should scale with body size (= sensory input) is touch [770], assuming that receptor density stays constant [624,1445]. These quibbles may sound naïve, but Darwin himself had similar misgivings (boldface added) [1986]:

> No one, I presume, doubts that the large size of the brain in man, relatively to his body, in comparison with that of the gorilla or orang, is closely connected with his higher mental powers. . . . On the other hand, **no one supposes that the intellect of any two animals or of any two men can be accurately gauged by the cubic contents of their skulls.** It is certain that there may be extraordinary mental activity with an extremely small absolute mass of nervous matter: thus the wonderfully diversified instincts, mental powers, and affections of ants are generally known, yet their cerebral ganglia are not so large as the quarter of a small pin's head. Under this latter point of view, the brain of an ant is one of the most marvellous atoms of matter in the world, perhaps more marvellous than the brain of man. [561] (Vol. 1, p. 145)

Notwithstanding these musings, Nature does display a consistent trend in brain:body ratios. Brain size increases at about the $^2/_3$ power of body size among vertebrates [1295], although there is appreciable scatter, and mammals exhibit an exponent closer to $^3/_4$ [59,1108]. Jerison advocated using the extent to which a species departs from the regression line of its affiliated group—in our case, primates—as an estimate of its mental potential [624]. He called it "EQ"—the encephalization quotient [612,1109].

Hominins began veering away from our chimp cousins (brain volume: ~400 cc) with our greater reliance on bipedalism ~7 MYA [168,169,895,2847], but our EQ remained marginal (*Australopithecus spp.*: ~5 MYA; ~450 cc) [1383,2610] until ~2 MYA (*Homo habilis*: ~650 cc) [743,1476] when it began to rise fairly steadily [612,638,1706,2309] up to our current level (*H. sapiens*: ~0.2 MYA; ~1300 cc) [654,1222,1706,2749,2796]. Overall, our brain tripled in size, while the chimp brain stayed the same [385,484,1324,2182].

During the phase of greatest brain expansion (2 MYA to now), our body size remained virtually constant [612]. Evidently, brain:body allometry is just a *trend* [1341], not an inviolable *constraint* [1197,2309,2514]. Otherwise, our EQ could never have increased to the extent that it did [33,259]. The trend might reflect a sort of default state [770] that prevails in the absence of selective forces for (or against) particular brain functions [1442–1445].

Did we ascend from brute to aesthete suddenly or gradually?

When did we cross the Rubicon to a human level of self-awareness [1360,2309]? How quickly did it happen [1534]? Is the concept of a *discrete* brute–aesthete threshold even justifiable [1197,1594]? No one knows [332,1295,2515].

In a lecture titled "Termites and Telescopes," [1811] the physicist-*cum*-polymath Philip Morrison reasoned that a mere threefold difference in our working memory (*cf.* [315,505,2865]) could have made a *huge* difference in our language and thinking ability (*cf.* [374,2309,2636]). Trained chimps, he noted, can concatenate $\sim$3 symbols at a time [743,2073], for a total of $\sim 3^3$ possible "sentences," whereas we routinely use sentences of $\sim$10 words [518], yielding $\sim 10^{10}$ possibilities. The totals are actually smaller, because permutation factorials (3! and 10!) should be used instead of combinations, but the basic point remains: our increase in processing ability has risen *exponentially* as a function of short-term memory capacity, not arithmetically!

Of course, the mere *correlation* between brain size and cognition does not prove *causation* [386,1650]. The idea that size alone was the critical factor in pushing us over the sentience threshold could be tested by a simple experiment, but one which is so utterly unethical that it should never be done: to create (by genetic engineering) a 1000-pound chimp, whose brain should attain 1300 cc (based on extrapolating the primate trend line of brain:body size) [612], and see if this Frankenchimp Monster is as smart as we are [2456].

Would such an über-chimp perforce become the first *Pan sapiens*? Probably not! Our neocortex is not just a scaled-up version of the primate prototype [2181,2290,2515]: the layout of our cortical areas has been revamped in a mosaic manner [146,2670]. Among other neurological quirks, we have a larger-than-expected prefrontal cortex but a smaller-than-expected visual cortex (based on the primate allometric regression) [770,1198,1534,2181].

Our neocortical distortions presumably occurred by heterogeneous growth in a balkanized patchwork of (1) regional growth-factor morphogens [147, 1907,1908,2237], (2) selector genes [73,1633,2121], and (3) transcription factors [236,1393,1533,2509] (*cf.* the cerebellum [2397]). It probably did *not* occur in as straightforward a way as Gould initially imagined [956]—that is, by merely extending the period of brain growth [654,1717].

Unfortunately, we can't hope to make much headway in dissecting human brain evolution without first deciphering these regional changes within the brain at the genetic and developmental levels [1908]. To do that we must turn to evo-devo [910,1899].

Alas, having foresworn the best tools of experimental genetics for the ethical reasons just stated, evo-devotees have had to resort to approaches that are correspondingly weaker [638]. Not surprisingly, their contributions have been disappointingly meager [1697]. Their chief approach has been to (1) find inherited disorders for the talent of interest (intelligence in general or language in particular) [723], (2) identify the defective genes responsible [411], (3) study the molecular nature of their mutational lesions [225], and (4) use comparative genomics to assess how they have changed vis-à-vis other primates [63,2394,2743]. About half a dozen candidate genes have been analyzed in this way [411,776].

One example should suffice to illustrate the preliminary status of evo-devo's forays into brain research. The most promising gene uncovered thus far is *FoxP2*. It encodes a transcription factor in the same *Forkhead-box* family [1468] as the selector gene for hair formation (*Foxn1* [2299]; *cf.* Ch. 5) [1327, 2654,2655]. Mutant alleles impair vocalization and behave as dominants [717] (indicating that gene dosage matters [2680]), and comparative genomics offered evidence that the gene has recently undergone the sort of selective sweep that we'd expect for a hominin-specific adaptation [1437]. Might this gene be the Rosetta Stone for language evolution? After a flurry of initial excitement [385,387] and intriguing findings in songbirds [1042,1043], the exuberance faded after it was found that (1) mice use *FoxP2* more broadly for learning motor skills [2586] and (2) bats show extreme variations in *FoxP2* that are correlated with their mode of echolocation [1538]. The emerging consensus is that *FoxP2* helps enable sensorimotor coordination in general, not just language articulation in particular. It is not so magical after all.

Clearly, we have more questions than answers about the origin of human thought, as well as about most of the other topics covered in this book [910, 1224,2015]. There is much to learn and much to think about.

Epilogue

Our anatomy is a palimpsest, and so is our genome. Both of them harbor layer upon layer of residual traces of ourselves when we looked quite different. The Quest that *Quirks* set for itself was to peel away those veils to reveal how we acquired the body we have now. During the process, we have uncovered abundant relics of our former incarnations as reptiles, amphibians, fish, protochordates, and, ultimately, as Urbilaterians.

By focusing on our Urbilaterian ancestors in particular, we found a closer kinship to flies and their ilk than anyone had ever guessed. Indeed, the reality of that affinity has shattered our prejudices and transported us beyond the outermost reaches of science fiction. Who would ever have thought that fact could be *so much* stranger than fiction!

As we have wandered through this Wonderland where up seems down and down seems up, we have encountered a menagerie of other species that we *thought* we already knew. We used to just yawn and mutter, "Oh, yes, I know them," but now we see how strange they look through the lenses of evo-devo. Our earthly zoo has been transformed from a dusty old museum into an amusement park full of funhouse mirrors that show us *ourselves* as we could have looked if the aeolian breezes of natural selection had blown us along a different course through Morphospace.

Darwin's odyssey aboard the *Beagle* changed his life and led him to write a book that changed ours. On the 150th anniversary of *Origin*, *Quirks* reminds us of what he gave us. For evo-devotees, his most lasting legacy, ironically, may be more spiritual than intellectual. Yes, he gave us a seaworthy ship for our explorations, but he also filled our sails with an inexhaustible supply of insatiable curiosity. Ask the right questions, he showed us, and you will someday find the answers.

APPENDIX

Quirks of Human Behavior

Listed here are a few of the most curious behaviors of our species, along with commentaries and references. Some of the reflexes occur in nonhuman species as well (as noted; see Fig. A.1). Several of the emotions were treated by Darwin in his 1872 book *Expression of the Emotions in Man and Animals* [562]. To the extent that those feelings are innate, they must be encoded genetically (*cf.* [2050,2120]), and to that same extent they must have been subjected to natural selection. It's unclear how many of them arose as by-products of other traits [1093].

CRYING Infants of many mammal and bird species use extravagant displays to beg for food, express distress, and inform the mother of their health [831]. In these respects, the crying by human babies is akin to the bleating of lambs and the squealing of piglets [59,2426]. Sobbing in human adults is more enigmatic [1601]. We are the only primate to emit tears while crying [132]. Indeed, we are the only *animal* to do so [2737].

We do not yet know how our emotions got linked to our lacrimal glands [1476]. The same question can be asked about (1) our chin, which, in some people, quivers before crying, and (2) our throat, which gets a "lump" [2737]. The latter link is also interesting because it usually precludes speaking [612]. Crying resembles laughing insofar as it is partly contagious [612] (*cf.* [1935]).

DANCING Why do people in every culture like to dance [707]? The reason must be rooted somehow in our love of music [294], but is it the melody or just the rhythm? Conceivably, our ability to mentally record the sequence of dance steps or musical notes may even have offered a preadaptation for language [648]. See Musicophilia.

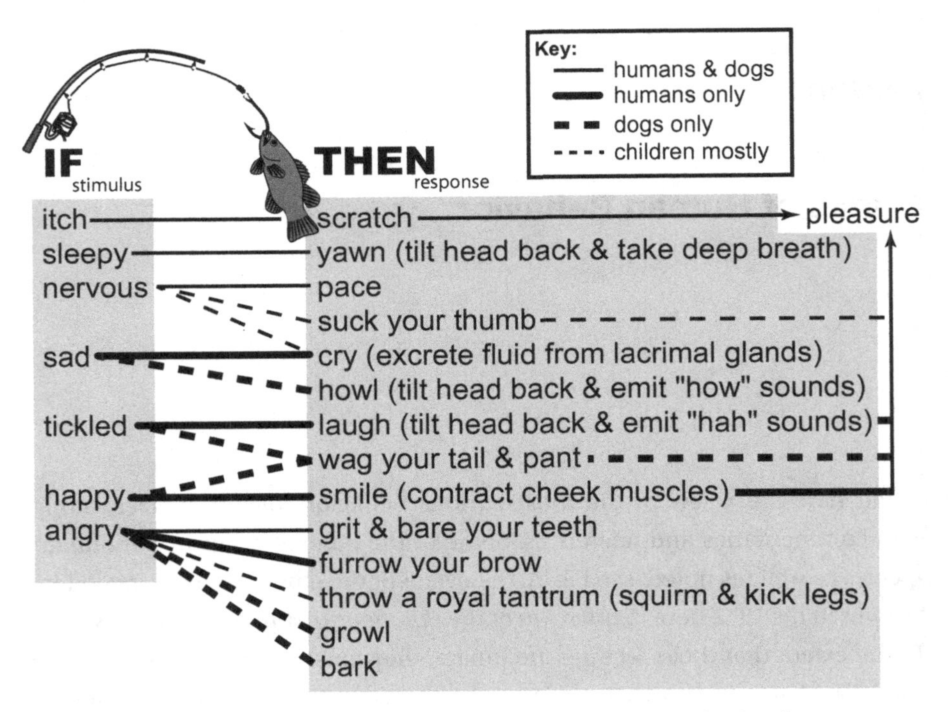

Figure A.1. Whimsical sample of hardwired behaviors in humans versus dogs. The *involuntary* nature of these actions (*e.g.*, laughing, crying, and yawning) is underscored by their retention despite loss of *voluntary* facial control in Anterior Operculum Syndrome [198,454,1494]. As implied by the fishing icon, many of the responses (right) seem to have been *arbitrarily* linked (neurally and genetically) to emotions (left), because they make little sense functionally [325,2050,2120]. For example, why should humans and dogs both be ticklish on our bellies [1097]? The only links that do seem reasonable are the baring of teeth and barking in response to anger, which serve as threat displays.

DREAMING Throughout human history the dream world has furnished a rich reservoir of raw material for psychoanalysts, dramatists, shamans, and charlatans [1170,1572]. For neuroscientists, the most fascinating aspect of dreams is how free they are from the constraints of waking thought [1880]. Ideally, we should all be given, at birth, an "operating manual" concerning dreams that would contain the following sort of disclaimer:

> There are well-formed sensory impressions, though sometimes quite bizarre. A series of actions seems to occur, often with an improbable juxtaposition of people, places, and times. Strong emotions may be experienced.
>
> And the dreamer seems to accept, uncritically, what's happening in the dream—memory recall doesn't work very well in sleep. Making new memories is even more impaired. If you wait as much as five minutes after the end of the rapid eye movement sleep to awaken sleepers, they won't remember much of the dream. Dreaming is mostly visual, with some auditory and tactile sensations

REFLECTIONS ON FIGURE A.1

The aim of this figure is the same as that of Jonathan Swift's *Gulliver's Travels*: to reveal the lunacy of behaviors we take for granted as mundane. By contrasting humans with dogs, we can see the symmetry of the absurdity: they look as silly to us as we must look to them. If they could laugh (and gossip), they might be snickering behind our backs.

This "switchboard" is based on Darwin's *Expression of the Emotions in Man and Animals* [562], on his *Variation of Animals and Plants Under Domestication* [560], on reflex physiology [1037,1270,1490], and on child psychology [314,2150,2894]. The schematic is not universal because there are so many deviations from the norm. For example, some people break out in hives when they get angry (not shown), and behavior varies tremendously among dog breeds [1312]. Dogs are an ideal foil for exposing our foibles because we are so familiar with them, and they share so many of our social inclinations [1742].

Tail-wagging exhibits bizarre asymmetries [2104,2357]. Darwin was struck by the vestigial nature of a few dog habits, which will be comically familiar to any dog lover:

> Dogs, when they wish to go to sleep on a carpet or other hard surface, generally turn round and round and scratch the ground with their fore-paws in a senseless manner, as if they intended to trample down the grass and scoop out a hollow, as no doubt their wild parents did, when they lived on open grassy plains or in the woods.... Dogs after voiding their excrement often make with all four feet a few scratches backwards, even on a bare stone pavement, as if for the purpose of covering up their excrement with earth, in nearly the same manner as do cats. [562] (pp. 42*ff*.)

Darwin was also fascinated with the novel quality of dog barking because it has no obvious precedent among wild wolves.

> The habit of barking, however, which is almost universal with domesticated dogs, forms an exception, as it does not characterize a single natural species of the family. [562] (p. 27)

No one knows how barking evolved, except to observe that it arose during domestication [1489,1742,1806].

How odd that by tugging on one "strand" of the genomic "spiderweb" (domestication), we should elicit such a surprising change in a radically different part of the web (barking)! An even more stunning instance of network integration was seen during the recent domestication of foxes [338,1093,2635]: a suite of *anatomical* traits (floppy ears, short tails, etc.) emerged as a result of the artificial selection for *behavior*! Such counterintuitive results make one wonder how many of our own quirks might of arisen likewise as side effects of selection for adaptive traits (*cf.* Fig. 7.1).

Stimulus–response links may be easier to revamp than we might think. For example, a recessive mutation puts Doberman pinschers to sleep instantly when they're excited [2386], and a missense mutation paralyzes goats instantly when they're startled [159]. In humans, this condition, known as "hyperekplexia" [1110], can be fatal due to its associated apnea. Analogous mutations might have led to the tactic of "playing dead" employed by possums and other animals. Another bizarre link, between a spasmotic trigger and the uttering of curse words [1456], afflicts nearly 1% of us [1263] in the form of Tourette's Syndrome [1820].

Other mutations that drastically change behavior include (1) *miffy*, which makes mice hop like rabbits [1257], (2) $Hoxb8^{null}$, which makes mice groom themselves and their cagemates to the point of bleeding [1006,1010], and (3) a myostatin mutation that enhances the racing ability of whippets and turns them into übermuscular "superdogs" [1510,1816]. In the fly world, the weirdest example may be the ability to turn normal males into homosexuals by overexpressing a pigmentation gene [2908].

REFLECTIONS ON FIGURE A.1 (*continued*)

By genetically linking a perception, response, or action to the pleasure centers of the brain [1158,1503,2715], natural selection can radically reconfigure the reward-seeking behavior of a species [178,319,1070,1643]. The epitome is orgasm [1488,2068], which links ecstasy to intercourse and compels us to reproduce [1417,2308,2428]. A more mundane example is the itch–scratch–pleasure chain shown here [75,1242,1747,2528], which sadly goes terribly awry in some people [901].

Such "aah that feels good!" reinforcement helps explain (1) why cognition is so often overruled by emotion, despite our supposedly exalted intellect [253,667,1328,1643] and (2) why so many people get trapped in self-destructive spirals of obsession, compulsion, and addiction [415,901,1005]. (Just visit a casino in Las Vegas!) Subtler polymorphisms in these links are likely responsible for the inexplicable idiosyncrasies of tastes and temperaments [1127,2677]—for example, the "search images" that guide mate choice (*e.g.*, What does she see in *him*?) [1827,2165,2236,2537,2791], although some of that imagery may be imprinted during infancy [1363].

A related question concerns the degree to which fear is innately linked to the perception of certain shapes or movements. For example, primates appear to have a hardwired fear of snakes, which makes sense given our arboreal origins [1250], but it begs the deeper question of how evolution was able to program our brains thusly (*cf.* [2274]). What other old circuits still steer our behavior [1412,1555,2091,2254]? No matter how many there may be, we cannot justifiably claim that we're just pawns of our genes [1543,1835,2169].

> as well as a sense of movement—but seldom does anyone report pain, or smells or tastes.
>
> . . . Furthermore, real dreams are like delusions; you have very little insight, and while the dreams are occurring, you tend to think it's all real. [346] (pp. 25*ff.*)

Dreams may be a wild version of the tamer "simulation software" that we routinely use for daily thought [594]. That same software can also sometimes mislead us into *waking* delusions (*cf.* [1168,1169]).

Using rapid eye movements (REMs) as a sign of dreaming [2459], it appears that most mammals dream [1877], with the extent ranging from none for dolphins [1602,2388] up to 8 hours per day for the platypus [2392] at the base of the mammal clade [1347,2390,2391]. Sleep per se is pervasive throughout the animal kingdom [460,2389,2918] and more ancient—stretching back at least to Urbilaterians [56,1087,2888].

Aside from its restorative physiological benefits [364,1739], sleep fosters memory consolidation [1261,2483] through memory playback [735,2126]. However, the playback is only weakly linked to dreaming [1748], and it does not rely on the REM (dream) state [1660,2387]. In what may be humanity's oddest quirk, some people enact this playback by sleepwalking [1677,2156].

LAUGHING Laughter is not limited to humans: monkeys emote as we do [562,1736], chimps chuckle when tickled [2088], and even rats enjoy a good belly laugh now and then [317,1632,1983,1984]. Nevertheless, we laugh more than any other species on earth [2089].

In young primates, laughter is a signal for social bonding during play behavior [887,1982,2037,2737]. In rats it acts as a "come hither" signal [2841]. In humans, it is typically associated with surprise and relief that an expected difficulty did not materialize [2116], and it is a potent signal of an affable emotional state [1935].

People who are born deaf laugh normally [1629], so it can't just be a learned response. The trigger may be in the supplementary motor area where stimulation evokes laughter [821], whereas joy itself is mediated by the mesolimbic reward centers [1781].

As for why laughing is so contagious [1664] (especially in children and adolescents [419,2894]), it may use the same receiver–generator circuit as yawning, albeit in an auditory (*vs.* visual) mode [887,2088] (*cf.* mirror neurons [663, 2190,2191]). Darwin thought that humor "tickles" the mind [562], a conjecture for which there is some empirical support [820]. He also wondered about the paradoxical fact that excessive laughter can actually lead to tears (crying?).

Laughing can still occur in the absence of any voluntary control of the facial muscles [198,454,1494]. The same is true for crying and yawning.

MUSICOPHILIA This term, the title of a book by Oliver Sacks [2249], denotes people's love for music [1532]. No one knows why we like music so much [1314, 2769,2900], nor why it can move us to tears [1271], nor why tastes vary so greatly from person to person and culture to culture [1320,2037].

Monkeys manifest no preference for music whatsoever [1475,1701], so it is unclear how our affinity for it arose [867,1700,1744]. It could just be a by-product of how our brains are wired for language [127,211,2037], although Darwin argued the converse—that language evolved from music [292,780].

Some people exhibit an atonality just as profound as that of monkeys. The etiology of this "amusia" is unknown [2481]. Nor do we know why those of us who are musically inclined vary so widely in musical aptitude [2289]. Indeed, we are only now beginning to dissect how our brain processes music [126,129,507, 2450]. In this regard, consider the astonishing musical aptitudes of people with Williams Syndrome [1337], who are otherwise profoundly mentally retarded [1520,2249].

Our ability to sing reveals a range of vocal control that far exceeds what we need to be able to speak [612]. Is that range just a fortuitous spandrel?

PLAYING Play behavior is widespread among juvenile mammals [295,741,1476], and placental mammals engage in it more than do marsupials [337]. Play offers a variety of adaptive advantages [173], including the chance to practice responding to unexpected events [2439], but it could just be a spandrel [2824].

Children often play pretend games in which one thing symbolizes another [707,1227]. The symbolic and social aspects of play suggest that it might even have facilitated language evolution [638,1233] (*cf.* [688]). The persistence of play in adult humans may be a by-product of our pervasive retention of juvenile traits (a.k.a. "paedomorphism") [59,956]. The universal popularity of comedy attests to this persistence.

READING How our brain processes the squiggles on a page such as this into meaningful concepts remains a riddle [2842]. Written language is only ~5000 years old [1909], so it must have emerged as a side effect of our intellect [2651], rather than being a driving force (*cf.* Ch. 7). Dyslexia, which impairs reading [2842], is a developmental disorder [838] that may stem from the misrouting of axons across the midline of the brain (*cf.* Ch. 3) [1080].

SUICIDE The most enigmatic of all human behaviors from the standpoint of evolution is suicide. It would only make sense if it enhanced the survival chances of one's offspring or some other close relative [445], but it patently does not in most cases [2646].

Suicide is probably a nonadaptive side effect of self-awareness (*cf.* Hamlet's soliloquy) [740,1901,2587,2916], which deepens during adolescence [2329]. It has been linked to the same serotonin circuitry as depression [279,1032,1439].

A disorder that seems like *slow* suicide is anorexia nervosa [1030]. Is anorexia a dysgenic side effect of our fragile self-image? What other animal starves itself in the presence of plentiful food?

TICKLISHNESS Aristotle noted that "when men are tickled they are quickly set a-laughing" {PoA:3:10:673a3} [137]. Dogs and rats are ticklish on their bellies [317,1632,1984]. In humans the armpit is the most sensitive trigger [1097]. Normal people can't tickle themselves [464], but schizophrenics can [216]. See Laughing.

YAWNING Yawning appears to be universal among vertebrates [2090]. It is associated with fatigue but also occurs upon waking [74], and it is often accompanied (enigmatically) by stretching. Like laughing, yawning is contagious [2090], even among ostriches [2281], where preening and dust-bathing are also contagious [1972]. The value of yawning as a social signal, if any, is unclear [1935].

References

1. Aamodt, S. and Wang, S. (2008). "Welcome to Your Brain: Why You Lose Your Car Keys but Never Forget How to Drive and Other Puzzles of Everyday Life." Bloomsbury, New York.
2. Abe, G., Ide, H., and Tamura, K. (2007). Function of FGF signaling in the developmental process of the median fin fold in zebrafish. *Dev. Biol.* **304**, 355–366.
3. Abel, R.M., Robinson, M., Gibbons, P., and Parikh, D.H. (2004). Cleft sternum: case report and literature review. *Pediatr. Pulmonol.* **37**, 375–377.
4. Aberg, T., Wang, X.P., Kim, J.H., Yamashiro, T., Bei, M., Rice, R., Ryoo, H.M., and Thesleff, I. (2004). Runx2 mediates FGF signaling from epithelium to mesenchyme during tooth morphogenesis. *Dev. Biol.* **270**, 76–93.
5. Abitol, M.M. (1987). Obstetrics and posture in pelvic anatomy. *J. Human Evol.* **16**, 243–255.
6. Aboitiz, F. and Montiel, J. (2003). One hundred million years of interhemispheric communication: the history of the corpus callosum. *Braz. J. Med. Biol. Res.* **36**, 409–420.
7. Abu-Issa, R. and Kirby, M.L. (2007). Heart field: from mesoderm to heart tube. *Annu. Rev. Cell Dev. Biol.* **23**, 45–68.
8. Abzhanov, A., Protas, M., Grant, B.R., Grant, P.R., and Tabin, C.J. (2004). Bmp4 and morphological variation of beaks in Darwin's finches. *Science* **305**, 1462–1465.
9. Achenbach, J. (2002). All thumbs: the cell phone and evolution. *Nat. Geogr.* **202** #6, 1.
10. Ackerman, J. (2006). The downside of upright. *Nat. Geogr.* **210** #1, 126–145. [See also (2008) *BMJ* **337**, a2469.]
11. Ackermann, R.R. (2007). Craniofacial variation and developmental divergence in primate and human evolution. *In* "Tinkering: The Microevolution of Development," *Novartis Found. Symp. 284* (G. Bock and J. Goode, eds.). Wiley, Chichester, U.K., pp. 262–279.
12. Acquisti, C., Kleffe, J., and Collins, S. (2007). Oxygen content of transmembrane proteins over macroevolutionary time scales. *Nature* **445**, 47–52.
13. Ádám, G., Perrimon, N., and Noselli, S. (2003). The retinoic-like juvenile hormone controls the looping of left-right asymmetric organs in *Drosophila. Development* **130**, 2397–2406.
14. Adams, D. (1979). "The Hitchhiker's Guide to the Galaxy." Harmony Books, New York.
15. Adams, I.R. and McLaren, A. (2002). Sexually dimorphic development of mouse primordial germ cells: switching from oogenesis to spermatogenesis. *Development* **129**, 1155–1164. [See also (2007) *Development* **134**, 3401–3411.]
16. Adams, L.A. and Eddy, S. (1949). "Comparative Anatomy: An Introduction to the Vertebrates." John Wiley & Sons, New York.
17. Adamska, M., Matus, D.Q., Adamski, M., Green, K., Rokhsar, D.S., Martindale, M.Q., and Degnan, B.M. (2007). The evolutionary origin of hedgehog proteins. *Curr. Biol.* **17**, R836–R837.
18. Adler, R. and Canto-Soler, M.V. (2007). Molecular mechanisms of optic vesicle development: complexities, ambiguities and controversies. *Dev. Biol.* **305**, 1–13.
19. Adolphs, R. (2008). Fear, faces, and the human amygdala. *Curr. Opin. Neurobiol.* **18**, 166–172.
20. Affolter, M., Bellusci, S., Itoh, N., Shilo, B., Thiery, J.-P., and Werb, Z. (2003). Tube or not tube: remodeling epithelial tissues by branching morphogenesis. *Dev. Cell* **4**, 11–18.
21. Affolter, M. and Mann, R. (2001). Legs, eyes, or wings—selectors and signals make the difference. *Science* **292**, 1080–1081.
22. Affolter, M., Slattery, M., and Mann, R.S. (2008). A lexicon for homeodomain-DNA recognition. *Cell* **133**, 1133–1135.
23. Afonin, B., Ho, M., Gustin, J.K., Meloty-Kapella, C., and Domingo, C.R. (2006). Cell behaviors associated with somite segmentation and rotation in *Xenopus laevis. Dev. Dynamics* **235**, 3268–3279.

24. Ager, E., Suzuki, S., Pask, A., Shaw, G., Ishino, F., and Renfree, M.B. (2007). Insulin is imprinted in the placenta of the marsupial, *Macropus eugenii*. *Dev. Biol.* **309**, 317–328.
25. Aguinaldo, A.M.A., Turbeville, J.M., Linford, L.S., Rivera, M.C., Garey, J.R., Raff, R.A., and Lake, J.A. (1997). Evidence for a clade of nematodes, arthropods and other moulting animals. *Nature* **387**, 489–493.
26. Aharoni, A., Gaidukov, L., Khersonsky, O., Gould, S.M., Roodveldt, C., and Tawfik, D.S. (2005). The 'evolvability' of promiscuous protein functions. *Nature Gen.* **37**, 73–76.
27. Ahlberg, P.E. and Clack, J.A. (2006). A firm step from water to land. *Nature* **440**, 747–749.
28. Ahmad, S.M. and Baker, B.S. (2001). Sex-specific deployment of FGF-signalling in *Drosophila* recruits mesodermal cells into the male genital imaginal disc. *Proc. 42nd Ann. Drosophila Res. Conf.* **Abstracts Vol., a49.**
29. Ahmed, S., Liu, C.-C., and Nawshad, A. (2007). Mechanisms of palatal epithelial seam disintegration by transforming growth factor (TGF) β3. *Dev. Biol.* **309**, 193–207.
30. Ahn, D. and Ho, R.K. (2008). Tri-phasic expression of posterior *Hox* genes during development of pectoral fins in zebrafish: implications for the evolution of vertebrate paired appendages. *Dev. Biol.* **322**, 220–233.
31. Aiello, L.C. (1992). Body size and energy requirements. *In* "The Cambridge Encyclopedia of Human Evolution" (S. Bunney, S. Jones, R. Martin, and D. Pilbeam, eds.). Cambridge Univ. Pr., New York, pp. 41–45.
32. Aiello, L.C. and Andrews, P. (2006). The australopithecines in review. *In* "The Human Evolution Source Book," 2nd ed. *Advances in Human Evolution Series* (R.L. Ciochon and J.G. Fleagle, eds.). Pearson Prentice Hall: Upper Saddle River, N. J., pp. 76–89.
33. Aiello, L.C., Bates, N., and Joffe, T. (2001). In defense of the Expensive Tissue Hypothesis. *In* "Evolutionary Anatomy of the Primate Cerebral Cortex" (D. Falk and K.R. Gibson, eds.). Cambridge Univ. Pr., New York, pp. 57–78.
34. Aiello, V.D. and Xavier-Neto, J. (2006). Full intrauterine development is compatible with cardia bifida in humans. *Pediatr. Cardiol.* **27**, 393–394.
35. Ainsworth, C. (2007). Tails of the unexpected. *Nature* **448**, 638–641.
36. Ajmani, M.L. and Ajmani, K. (1983). The position, length and arterial supply of vermiform appendix. *Anat. Anz.* **153**, 369–374.
37. Akam, M. (1998). *Hox* genes: From master genes to micromanagers. *Curr. Biol.* **8**, R676–R678.
38. Akam, M. (1998). *Hox* genes, homeosis and the evolution of segment identity: no need for hopeless monsters. *Int. J. Dev. Biol.* **42**, 445–451.
39. Akiyama-Oda, Y. and Oda, H. (2006). Axis specification in the spider embryo: *dpp* is required for radial-to-axial symmetry transformation and *sog* for ventral patterning. *Development* **133**, 2347–2357.
40. Al Qattan, M.M. (2004). On the emerging evidence of a new category of duplication in the human hand: the dorsoventral duplication. *Plast. Reconstr. Surg.* **114**, 1233–1237.
41. Al-Qattan, M.M. (2003). Congenital duplication of the palm in a patient with multiple anomalies. *J. Hand Surg. (Br.)* **28B**, 276–279.
42. Alba, D.M., Moyà-Solà, S., and Köhler, M. (2003). Morphological affinities of the *Australopithecus afarensis* hand on the basis of manual proportions and relative thumb length. *J. Human Evol.* **44**, 225–254.
43. Alberch, P. (1980). Ontogenesis and morphological diversification. *Amer. Zool.* **20**, 653–667.
44. Alberch, P. (1982). Developmental constraints in evolutionary processes. *In* "Evolution and Development" (J.T. Bonner, ed.). Springer-Verlag, Berlin, pp. 313–332.
45. Alberch, P., Gould, S.J., Oster, G.F., and Wake, D.B. (1979). Size and shape in ontogeny and phylogeny. *Paleobiol.* **5**, 296–317.
46. Albert, A.Y.K. and Otto, S.P. (2005). Sexual selection can resolve sex-linked sexual antagonism. *Science* **310**, 119–121.
47. Albright, T.D., Jessell, T.M., Kandel, E.R., and Posner, M.I. (2000). Neural science: a century of progress and the mysteries that remain. *Cell* **100 (Suppl.)**, S1–S55.
48. Alexander, R.M. (1982). "Locomotion of Animals." Blackie, London.
49. Alexander, R.M. (1985). Body support, scaling, and allometry. *In* "Functional Vertebrate Morphology" (M. Hildebrand, D.M. Bramble, K.F. Liem, and D.B. Wake, eds.). Harvard Univ. Pr., Cambridge, Mass., pp. 26–37.
50. Alexander, R.M. (1992). "The Human Machine." Columbia Univ. Pr., New York.
51. Alexander, R.M. (1993). Joints and muscles of hands and paws. *In* "Hands of Primates" (H. Preuschoft and D.J. Chivers, eds.). Springer-Verlag, New York, pp. 199–205.
52. Alexander, R.M. (1994). "Bones: The Unity of Form and Function." MacMillan, New York.
53. Alexander, R.M. (1998). Finding purpose in life. *Science* **281**, 927.
54. Alexander, R.M. (2004). Bipedal animals, and their differences from humans. *J. Anat.* **204**, 321–330.
55. Alison, M.R. (2002). Liver regeneration with reference to stem cells. *Sems. Cell Dev. Biol.* **13**, 385–387.
56. Allada, R. and Siegel, J.M. (2008). Unearthing the phylogenetic roots of sleep. *Curr. Biol.* **18**, R670–R679.
57. Allen, J.S., Bruss, J., and Damasio, H. (2004). The structure of the human brain. *Am. Sci.* **92**, 246–253.

58. Allen, R.H. (2007). On the mechanical aspects of shoulder dystocia and birth injury. *Clin. Obstet. Gynecol.* **50**, 607–623.
59. Allman, J.M. (1999). "Evolving Brains." Sci. Am. Lib., New York.
60. Alonso, C.R. (2008). The molecular biology underlying developmental evolution. *In* "Evolving Pathways: Key Themes in Evolutionary Developmental Biology" (A. Minelli and G. Fusco, eds.). Cambridge Univ. Pr., New York, pp. 80–99.
61. Alonso, C.R. and Wilkins, A.S. (2005). The molecular elements that underlie developmental evolution. *Nature Rev. Genet.* **6**, 709–715.
62. Alvarez, W., Claeys, P., and Kieffer, S.W. (1995). Emplacement of Cretaceous-Tertiary boundary shocked quartz from Chicxulub crater. *Science* **269**, 930–935.
63. Amadio, J.P. and Walsh, C.A. (2006). Brain evolution and uniqueness in the human genome. *Cell* **126**, 1033–1035.
64. Amemiya, C.T. and Wagner, G.P. (2006). Animal evolution: When did the "Hox system" arise? *Curr. Biol.* **16**, R546–R548.
65. An, W. and Wensink, P.C. (1995). Integrating sex- and tissue-specific regulation within a single *Drosophila* enhancer. *Genes Dev.* **9**, 256–266.
66. Andersen, H.T., ed. (1969). "The Biology of Marine Mammals." Acad. Pr., New York.
67. Anderson, J., Lavoie, J., Merrill, K., King, R.A., and Summers, C.G. (2004). Efficacy of spectacles in persons with albinism. *J. AAPOS* **8**, 515–520.
68. Anderson, P. (1983). The reproductive role of the human breast. *Curr. Anthrop.* **24**, 25–45.
69. Anderson, R.R. (1978). Embryonic and fetal development of the mammary apparatus. *In* "Lactation: A Comprehensive Treatise," **Vol. 4** (B.L. Larson, ed.). Acad. Pr., New York, pp. 3–40.
70. Andersson, J.O. and Andersson, S.G.E. (1999). Insights into the evolutionary process of genome degradation. *Curr. Opin. Gen. Dev.* **9**, 664–671.
71. Andersson, M. (1994). "Sexual Selection." Princeton Univ. Pr., Princeton, N. J.
72. Andersson, M. and Wallander, J. (2004). Relative size in the mating game. *Nature* **431**, 139–141.
73. Andoniadou, C.L., Signore, M., Sajedi, E., Gaston-Massuet, C., Kelberman, D., Burns, A.J., Itasaki, N., Dattani, M., and Martinez-Barbera, J.P. (2007). Lack of the murine homeobox gene *Hesx1* leads to a posterior transformation of the anterior forebrain. *Development* **134**, 1499–1508.
74. Andrews, M.A.W. (2002). Why do we yawn when we are tired? And why does it seem to be contagious? *Sci. Am.* **287** #5, 99.
75. Andrews, M.A.W. (2007). How do itches come about, and why does it feel good to scratch them? *Sci. Am.* **296** #6, 104.
76. Andrews, M.T. (2007). Advances in molecular biology of hibernation in mammals. *BioEssays* **29**, 431–440.
77. Andrews, P.W., Gangestad, S.W., and Matthews, D. (2002). Adaptationism—how to carry out an exaptationist program. *Behav. Brain Sci.* **25**, 489–553.
78. Angelini, D.R. and Kaufman, T.C. (2005). Comparative developmental genetics and the evolution of arthropod body plans. *Annu. Rev. Genet.* **39**, 95–119.
79. Annett, M. (2002). "Handedness and Brain Asymmetry: The Right Shift Theory." Taylor & Francis, New York.
80. Anon. (2000). Dolphins evolve opposable thumbs. "Oh, shit," says humanity. *www.theonion.com* #36:30, Aug. 30, 2000.
81. Anon. (2007). The biologists strike back. *Nature* **448**, 18–21.
82. Aoyama, H. and Asamoto, K. (2000). The developmental fate of the rostral/caudal half of a somite for vertebra and rib formation: experimental confirmation of the resegmentation theory using chick-quail chimeras. *Mechs. Dev.* **99**, 71–82.
83. Applequist, J. (1987). Optical activity: Biot's bequest. *Am. Sci.* **75**, 58–68.
84. Arbeitman, M.N., Fleming, A.A., Siegal, M.L., Null, B.H., and Baker, B.S. (2004). A genomic analysis of *Drosophila* somatic sexual differentiation and its regulation. *Development* **131**, 2007–2021.
85. Arendt, D. (2003). Evolution of eyes and photoreceptor cell types. *Int. J. Dev. Biol.* **47**, 563–571.
86. Arendt, D. and Nübler-Jung, K. (1997). Dorsal or ventral: similarities in fate maps and gastrulation patterns in annelids, arthropods and chordates. *Mechs. Dev.* **61**, 7–21.
87. Arendt, D. and Nübler-Jung, K. (1999). Rearranging gastrulation in the name of yolk: evolution of gastrulation in yolk-rich amniote eggs. *Mechs. Dev.* **81**, 3–22.
88. Arendt, D., Tessmar, K., de Campos-Baptista, M.-I.M., Dorresteijn, A., and Wittbrodt, J. (2002). Development of pigment-cup eyes in the polychaete *Platynereis dumerilii* and evolutionary conservation of larval eyes in Bilateria. *Development* **129**, 1143–1154.
89. Arendt, D., Tessmar-Raible, K., Snyman, H., Dorresteijn, A.W., and Wittbrodt, J. (2004). Ciliary photoreceptors with a vertebrate-type opsin in an invertebrate brain. *Science* **306**, 869–871.
90. Arendt, D. and Wittbrodt, J. (2001). Reconstructing the eyes of Urbilateria. *Phil. Trans. Roy. Soc. Lond. B* **356**, 1545–1563.
91. Arey, L.B. (1946). "Developmental Anatomy: A Textbook and Laboratory Manual of Embryology," 5th ed. W. B. Saunders, Philadelphia.
92. Argue, D., Donlon, D., Groves, C., and Wright, R. (2006). *Homo floresiensis*: Microcephalic, pygmoid, *Australopithecus*, or *Homo*? *J. Human Evol.* **51**, 360–374.

93. Arnheiter, H. (1998). Eyes viewed from the skin. *Nature* **391**, 632–633.
94. Arnold, S.J. (2005). The ultimate causes of phenotypic integration: lost in translation. *Evolution* **59**, 2059–2061.
95. Arnqvist, G. and Rowe, L. (2005). "Sexual Conflict." Princeton Univ. Pr., Princeton, N.J.
96. Artamonova, I.I. and Gelfand, M.S. (2007). Comparative genomics and evolution of alternative splicing: the pessimists' science. *Chem. Rev.* **107**, 3407–3430.
97. Arthur, W. (2000). Intraspecific variation in developmental characters: the origin of evolutionary novelties. *Amer. Zool.* **40**, 811–818.
98. Arthur, W. (2002). The emerging conceptual framework of evolutionary developmental biology. *Nature* **415**, 757–764.
99. Arthur, W. (2004). "Biased Embryos and Evolution." Cambridge Univ. Pr., New York.
100. Arthur, W. (2006). D'Arcy Thompson and the theory of transformations. *Nature Rev. Genet.* **7**, 401–406.
101. Arthur, W. (2008). Conflicting hypotheses on the nature of mega-evolution. *In* "Evolving Pathways: Key Themes in Evolutionary Developmental Biology" (A. Minelli and G. Fusco, eds.). Cambridge Univ. Pr.: New York, pp. 50–61.
102. Asami, T., Cowie, R.H., and Ohbayashi, K. (1998). Evolution of mirror images by sexually asymmetric mating behavior in hermaphroditic snails. *Am. Nat.* **152**, 225–236.
103. Ashburner, M. (1989). "*Drosophila*: A Laboratory Handbook." CSH Pr., Cold Spring Harbor, N.Y.
104. Ashe, H.L. and Briscoe, J. (2006). The interpretation of morphogen gradients. *Development* **133**, 385–394.
105. Atala, A., Lanza, R., Thomson, J., and Nerem, R. (2007). "Principles of Regenerative Medicine." Acad. Pr., New York.
106. Auerbach, B.M. and Ruff, C.B. (2006). Limb bone bilateral asymmetry: variability and commonality among modern humans. *J. Human Evol.* **50**, 203–218.
107. Avila, F.W. and Erickson, J.W. (2007). *Drosophila* JAK/STAT pathway reveals distinct initiation and reinforcement steps in early transcription of *Sxl*. *Curr. Biol.* **17**, 643–648.
108. Aw, S., Adams, D.S., Qiu, D., and Levin, M. (2008). H,K-ATPase protein localization and Kir4.1 function reveal concordance of three axes during early determination of left-right asymmetry. *Mechs. Dev.* **125**, 353–372. [See also (2009) *Development* **136**, 355–366.]
109. Ayala, F.J. (2007). Darwin's greatest discovery: Design without designer. *PNAS* **104 Suppl. 1**, 8567–8573.
110. Aylsworth, A.S. (2001). Clinical aspects of defects in the determination of laterality. *Am. J. Med. Genet.* **101**, 345–355.
111. Azziz, R., Carmina, E., and Sawaya, M.E. (2000). Idiopathic hirsutism. *Endocr. Rev.* **21**, 347–362.
112. Babcock, L.E. (1993). The right and the sinister. *Nat. Hist.* **102** #7, 32–39.
113. Bachtrog, D. (2006). A dynamic view of sex chromosome evolution. *Curr. Opin. Gen. Dev.* **16**, 578–585.
114. Bada, J.L. (1995). Origins of homochirality. *Nature* **374**, 594–595.
115. Badre, D. (2008). Cognitive control, hierarchy, and the rostro-caudal organization of the frontal lobes. *Trends Cogn. Sci.* **12**, 193–200.
116. Badyaev, A.V. (2002). Growing apart: an ontogenetic perspective on the evolution of sexual size dimorphism. *Trends Ecol. Evol.* **17**, 369–378.
117. Badyaev, A.V. (2004). Developmental perspective on the evolution of sexual ornaments. *Evol. Ecol. Res.* **6**, 1–17.
118. Baer, C.F., Miyamoto, M.M., and Denver, D.R. (2007). Mutation rate variation in multicellular eukaryotes: causes and consequences. *Nature Rev. Genet.* **8**, 619–631. [See also (2008) *Genetics* **180**, 1501–1509.]
119. Baguñà, J., Martinez, P., Paps, J., and Riutort, M. (2008). Unravelling body plan and axial evolution in the Bilateria with molecular phylogenetic markers. *In* "Evolving Pathways: Key Themes in Evolutionary Developmental Biology" (A. Minelli and G. Fusco, eds.). Cambridge Univ. Pr., New York, pp. 217–238.
120. Bainbridge, D. (2008). "Beyond the Zonules of Zinn: A Fantastic Journey through Your Brain." Harvard Univ. Pr., Cambridge, Mass.
121. Baker, B.S. and Wolfner, M.F. (1988). A molecular analysis of *doublesex*, a bifunctional gene that controls both male and female sexual differentiation in *Drosophila melanogaster*. *Genes Dev.* **2**, 477–489.
122. Baker, C.V.H. and Bronner-Fraser, M. (1997). The origins of the neural crest. Part II: an evolutionary perspective. *Mechs. Dev.* **69**, 13–29.
123. Baker, R.E., Schnell, S., and Maini, P.K. (2006). A clock and wavefront mechanism for somite formation. *Dev. Biol.* **293**, 116–126.
124. Ball, E.E., de Jong, D.M., Schierwater, B., Shinzato, C., Hayward, D.C., and Miller, D.J. (2007). Implications of cnidarian gene expression patterns for the origins of bilaterality—is the glass half full or half empty? *Integr. Comp. Biol.* **47**, 701–711.
125. Ball, P. (1999). "The Self-made Tapestry: Pattern Formation in Nature." Oxford Univ. Pr., New York.
126. Ball, P. (2008). Facing the music. *Nature* **453**, 160–162.
127. Balter, M. (2004). Seeking the key to music. *Science* **306**, 1120–1122.
128. Balter, M. (2007). Brain evolution studies go micro. *Science* **315**, 1208–1211.
129. Balter, M. (2007). Study of music and the mind hits a high note in Montreal. *Science* **315**, 758–759.
130. Baltzinger, M., Ori, M., Pasqualetti, M., Nardi, I., and Rijli, F.M. (2005). *Hoxa2* knockdown in *Xenopus* results in hyoid to mandibular homeosis. *Dev. Dynamics* **234**, 858–867.

131. Bandyopadhyay, A., Tsuji, K., Cox, K., Harfe, B.D., Rosen, V., and Tabin, C.J. (2006). Genetic analysis of the roles of BMP2, BMP4, and BMP7 in limb patterning and skeletogenesis. *PLoS Genet.* **2** #12, 2116–2130 (e216).

132. Bard, K.A. (2003). Are humans the only primates that cry? *Sci. Am.* **289** #5, 107.

133. Barkai, N. and Shilo, B.-Z. (2007). Variability and robustness in biomolecular systems. *Molec. Cell* **28**, 755–760.

134. Barmina, O., Gonzalo, M., McIntyre, L.M., and Kopp, A. (2005). Sex- and segment-specific modulation of gene expression profiles in *Drosophila*. *Dev. Biol.* **288**, 528–544.

135. Barmina, O. and Kopp, A. (2007). Sex-specific expression of a HOX gene associated with rapid morphological evolution. *Dev. Biol.* **311**, 277–286.

136. Barnard, A.R., Hattar, S., Hankins, M.W., and Lucas, R.J. (2006). Melanopsin regulates visual processing in the mouse retina. *Curr. Biol.* **16**, 389–395.

137. Barnes, J., ed. (1984). "The Complete Works of Aristotle: The Revised Oxford Translation." Vol. 1. Princeton Univ. Pr., Princeton, N.J.

138. Barolo, S. and Posakony, J.W. (2002). Three habits of highly effective signaling pathways: principles of transcriptional control by developmental cell signaling. *Genes Dev.* **16**, 1167–1181.

139. Baron-Cohen, S., Knickmeyer, R.C., and Belmonte, M.K. (2005). Sex differences in the brain: implications for explaining autism. *Science* **310**, 819–823.

140. Barr, M., Jr. and Cohen, M.M., Jr. (1999). Holoprosencephaly survival and performance. *Am. J. Med. Genet. (Semin. Med. Genet.)* **89**, 116–120.

141. Barrallo-Gimeno, A. and Nieto, M.A. (2008). Riding the right wave: would the real neural crest please stand up? *Evol. Dev.* **10**, 509–510.

142. Barreiro, L.B., Laval, G., Quach, H., Patin, E., and Quintana-Murci, L. (2008). Natural selection has driven population differentiation in modern humans. *Nature Genet.* **40**, 340–345.

143. Barry, A. (1951). The aortic arch derivatives in the human adult. *Anat. Record* **111**, 221–238.

144. Barsh, G. and Cotsarelis, G. (2007). How hair gets its pigment. *Cell* **130**, 779–781.

145. Bartholomew, G.A., Jr. and Birdsell, J.B. (1953). Ecology and the protohominids. *Amer. Anthrop.* **55**, 481–498.

146. Barton, R.A. and Harvey, P.H. (2000). Mosaic evolution of brain structure in mammals. *Nature* **405**, 1055–1058.

147. Basson, M.A., Echevarria, D., Ahn, C.P., Sudarov, A., Joyner, A.L., Mason, I.J., Martinez, S., and Martin, G. (2008). Specific regions within the embryonic midbrain and cerebellum require different levels of FGF signaling during development. *Development* **135**, 889–898.

148. Bastida, M.F. and Ros, M.A. (2008). How do we get a perfect complement of digits? *Curr. Opin. Gen. Dev.* **18**, 374–380.

149. Bates, D., Taylor, G.I., Minichiello, J., Farlie, P., Cichowitz, A., Watson, N., Klagsbrun, M., Mamluk, R., and Newgreen, D.F. (2003). Neurovascular congruence results from a shared patterning mechanism that utilizes Semaphorin3A and Neuropilin-1. *Dev. Biol.* **255**, 77–98.

150. Bateson, G. (1972). "Steps to an Ecology of Mind." Ballantine Books, New York.

151. Bateson, W. (1894). "Materials for the Study of Variation Treated with Especial Regard to Discontinuity in the Origin of Species." MacMillan, London.

152. Baudouin-Cornu, P. and Thomas, D. (2007). Oxygen at life's boundaries. *Nature* **445**, 35–36.

153. Baumeister, F.A.M., Egger, J., Schildhauer, M.T., and Stengel-Rutkowski, S. (1993). Ambras syndrome: delineation of a unique hypertrichosis universalis congenita and association with a balanced pericentric inversion (8) (p11.2; q22). *Clin. Genet.* **44**, 121–128.

154. Baumeister, F.A.M., Schwarz, H.P., and Stengel-Rutkowski, S. (1995). Childhood hypertrichosis: diagnosis and management. *Archiv. Dis. Child.* **72**, 457–459.

155. Bautista, A., Mendoza-Degante, M., Coureaud, G., Martínez-Gómez, M., and Hudson, R. (2005). Scramble competition in newborn domestic rabbits for an unusually restricted milk supply. *Anim. Behav.* **70**, 1011–1021.

156. Beadle, G. and Beadle, M. (1966). "The Language of Life: An Introduction to the Science of Genetics." Doubleday, Garden City, N.Y.

157. Bearzi, M. and Stanford, C.B. (2008). "Beautiful Minds: The Parallel Lives of Great Apes and Dolphins." Harvard Univ. Pr., Cambridge, Mass.

158. Becchio, C., Bertone, C., and Castiello, U. (2008). How the gaze of others influences object processing. *Trends Cogn. Sci.* **12**, 254–258.

159. Beck, C.L., Fahlke, C., and George, A.L., Jr. (1996). Molecular basis for decreased muscle chloride conductance in the myotonic goat. *Proc. Natl. Acad. Sci. USA* **93**, 11248–11252.

160. Becker, C.G. and Becker, T., eds. (2007). "Model Organisms in Spinal Cord Regeneration." Wiley-VCH Verlag, Weinheim, Germany.

161. Becker, J.B., Berkley, K.J., Geary, N., Hampson, E., Herman, J.P., and Young, E.A., eds. (2008). "Sex Differences in the Brain: From Genes to Behavior." Oxford Univ. Pr., New York.

162. Becker, L., Poreda, R.J., Basu, A.R., Pope, K.O., Harrison, T.M., Nicholson, C., and Iasky, R. (2004). Bedout: a possible end-Permian impact crater offshore of northwestern Australia. *Science* **304**, 1469–1476.

163. Becker, M.I., De Ioannes, A.E., Leon, C., and Ebensperger, L.A. (2007). Females of the communally breeding rodent, *Octodon degus*, transfer antibodies to their offspring during pregnancy and lactation. *J. Reprod. Immunol.* **74**, 68–77.

164. Beckman, J., Banks, S.C., Sunnucks, P., Lill, A., and Taylor, A.C. (2007). Phylogeography and environmental correlates of a cap on reproduction: teat number in a small marsupial, *Antechinus agilis. Molec. Ecol.* **16**, 1069–1083.
165. Becskei, A., Séraphin, B., and Serrano, L. (2001). Positive feedback in eukaryotic gene networks: cell differentiation by graded to binary response conversion. *EMBO J.* **20**, 2528–2535.
166. Beddington, R.S.P. and Robertson, E.J. (1999). Axis development and early asymmetry in mammals. *Cell* **96**, 195–209.
167. Begon, M., Townsend, C.R., and Harper, J.L. (2006). "Ecology: From Individuals to Ecosystems," 4th ed. Blackwell, Malden, Mass.
168. Begun, D.R. (2003). Planet of the apes. *Sci. Am.* **289** #2, 74–83.
169. Begun, D.R. (2004). The earliest hominins—Is less more? *Science* **303**, 1478–1480. [See also (2008) *Science* **319**, 1662–1665.]
170. Beighton, P. (1970). Congenital hypertrichosis lanuginosa. *Arch. Derm.* **101**, 669–672.
171. Beighton, P. (1970). Familial hypertrichosis cubiti: hairy elbows syndrome. *J. Med. Genet.* **7**, 158–160.
172. Bejder, L. and Hall, B.K. (2002). Limbs in whales and limblessness in other vertebrates: mechanisms of evolutionary and developmental transformation and loss. *Evol. Dev.* **4**, 445–458.
173. Bekoff, M. and Byers, J.A. (1998). "Animal Play: Evolutionary, Comparative and Ecological Perspectives." Cambridge Univ. Pr., New York.
174. Bell, M.A., Ellis, K.E., and Sirotkin, H.I. (2007). Pelvic skeleton reduction and *Pitx1* expression in threespine stickleback populations. *In* "Tinkering: The Microevolution of Development," *Novartis Found. Symp. 284* (G. Bock and J. Goode, eds.). Wiley: Chichester, U.K., pp. 225–244.
175. Bell, M.A., Khalef, V., and Travis, M.P. (2007). Directional asymmetry of pelvic vestiges in threespine stickleback. *J. Exp. Zool. (Mol. Dev. Evol.)* **308B**, 189–199.
176. Bellingham, J. and Foster, R.G. (2002). Opsins and mammalian photoentrainment. *Cell Tissue Res.* **309**, 57–71.
177. Bellone, R.R., Brooks, S.A., Sandmeyer, L., Murphy, B.A., Forsyth, G., Archer, S., Bailey, E., and Grahn, B. (2008). Differential gene expression of *TRPM1*, the potential cause of congenital stationary night blindness and coat spotting patterns (*LP*) in the appaloosa horse (*Equus caballus*). *Genetics* **179**, 1861–1870.
178. Belova, M.A., Paton, J.J., Morrison, S.E., and Salzman, C.D. (2007). Expectation modulates neural responses to pleasant and aversive stimuli in primate amygdala. *Neuron* **55**, 970–984.
179. Ben-Shachar, M., Dougherty, R.F., and Wandell, B.A. (2007). White matter pathways in reading. *Curr. Opin. Neurobiol.* **17**, 258–270.
180. Benkman, C.W. and Lindholm, A.K. (1991). The advantages and evolution of a morphological novelty. *Nature* **349**, 519–520.
181. Bennett, M.R. and Hasty, J. (2008). Genome rewired. *Nature* **452**, 824–825.
182. Benson, M.T., Dalen, K., Mancuso, A.A., Kerr, H.H., Cacciarelli, A.A., and Mafee, M.F. (1992). Congenital-anomalies of the branchial apparatus—embryology and pathological anatomy. *Radiographics* **12**, 943–960.
183. Bentley, R. (1969). "Molecular Asymmetry in Biology." Acad. Pr., New York.
184. Benzon, H.T., Katz, J.A., Benzon, H.A., and Iqbal, M.S. (2003). Piriformis syndrome. *Anesthesiology* **98**, 1442–1448.
185. Berg, J., Willmann, S., and Lässig, M. (2004). Adaptive evolution of transcription factor binding sites. *BMC Evol. Biol.* **4**, e42.
186. Berger, M.F., Badis, G., Gehrke, A.R., Talukder, S., Philippakis, A.A., Peña-Castillo, L., Alleyne, T.M., Mnaimneh, S., Botvinnik, O.B., Chan, E.T., Khalid, F., Zhang, W., Newburger, D., Jaeger, S.A., Morris, Q.D., Bulyk, M.L., and Hughes, T.R. (2008). Variation in homeodomain DNA binding revealed by high-resolution analysis of sequence preferences. *Cell* **133**, 1266–1276.
187. Bergman, R.A., Afifi, A.K., and Miyauchi, R. (2009). "Illustrated Encyclopedia of Human Anatomic Variation." Available at *www.anatomyatlases.org*.
188. Berta, A., Sumich, J.L., and Kovacs, K.M. (2006). "Marine Mammals: Evolutionary Biology," 2nd ed. Elsevier, New York.
189. Beverdam, A., Merlo, G.R., Paleari, L., Mantero, S., Genova, F., Barbieri, O., Janvier, P., and Levi, G. (2002). Jaw transformation with gain of symmetry after *Dlx5/Dlx6* inactivation: mirror of the past? *Genesis* **34**, 221–227.
190. Bhalla, U.S. and Iyengar, R. (1999). Emergent properties of networks of biological signaling pathways. *Science* **283**, 381–387.
191. Bharti, K., Nguyen, M.-T.T., Skuntz, S., Bertuzzi, S., and Arnheiter, H. (2006). The other pigment cell: specification and development of the pigmented epithelium of the vertebrate eye. *Pigment Cell Res.* **19**, 380–394.
192. Bhat, K.M. (2005). Slit-Roundabout signaling neutralizes Netrin-Frazzled-mediated attractant cue to specify the lateral positioning of longitudinal axon pathways. *Genetics* **170**, 149–159.
193. Bhattacharyya, R.P., Reményi, A., Yeh, B.J., and Lim, W.A. (2006). Domains, motifs, and scaffolds: the role of modular interactions in the evolution and wiring of cell signaling circuits. *Annu. Rev. Biochem.* **75**, 655–680.
194. Bienz, M. (2005). β-catenin: a pivot between cell adhesion and Wnt signaling. *Curr. Biol.* **15**, R64–R67.
195. Bier, E. and McGinnis, W. (2004). Model organisms in the study of development and disease.

In "Inborn Errors of Development: The Molecular Basis of Clinical Disorders of Morphogenesis," Oxford *Monographs on Medical Genetics, No. 49* (C.J. Epstein, R.P. Erickson, and A. Wynshaw-Boris, eds.). Oxford Univ. Pr.: New York, pp. 25–45.

196. Billakanty, S., Burket, M.W., and Grubb, B.P. (2006). May-Thurner syndrome: a vascular abnormality encountered during electrophysiologic study. *Pacing Clin. Electrophysiol.* **29**, 1310–1311.

197. Billeter, J.-C., Rideout, E.J., Dornan, A.J., and Goodwin, S.F. (2006). Control of male sexual behavior in *Drosophila* by the sex determination pathway. *Curr. Biol.* **16**, R766–R776.

198. Billeth, R., Jörgler, E., and Baumhackl, U. (2000). Bilaterales anteriores Operkulumsyndrom. *Nervenarzt.* **71**, 651–654.

199. Bindman, D. (1978). "The Complete Graphic Works of William Blake." Putnam, New York.

200. Bininda-Emonds, O.R.P., Cardillo, M., Jones, K.E., MacPhee, R.D.E., Beck, R.M.D., Grenyer, R., Price, S.A., Bos, R.A., Gittleman, J.L., and Purvis, A. (2007). The delayed rise of present-day mammals. *Nature* **446**, 507–512.

201. Bininda-Emonds, O.R.P., Jeffery, J.E., and Richardson, M.K. (2003). Inverting the hourglass: quantitative evidence against the phylotypic stage in vertebrate development. *Proc. Roy. Soc. Lond. B* **270**, 341–346.

202. Bininda-Emonds, O.R.P., Jeffery, J.E., Sánchez-Villagra, M.R., Hanken, J., Colbert, M., Pieau, C., Selwood, L., ten Cate, C., Raynaud, A., Osabutey, C.K., and Richardson, M.K. (2007). Forelimb-hindlimb developmental timing changes across tetrapod phylogeny. *BMC Evol. Biol.* **7**, 182 (7 pp.).

203. Binkley, S. (1976). Comparative biochemistry of the pineal glands of birds and mammals. *Amer. Zool.* **16**, 57–65.

204. Bird, R. (1999). Cooperation and conflict: the behavioral ecology of the sexual division of labor. *Evol. Anthrop.* **8** #2, 65–75.

205. Birkhead, T.R. and Pizzari, T. (2002). Postcopulatory sexual selection. *Nature Rev. Genet.* **3**, 262–273.

206. Birnbaum, K.D. and Sánchez Alvarado, A. (2008). Slicing across kingdoms: regeneration in plants and animals. *Cell* **132**, 697–710.

207. Bisazza, A., Cantalupo, C., Robins, A., Rogers, L.J., and Vallortigara, G. (1996). Right-pawedness in toads. *Nature* **379**, 408.

208. Bisgrove, B.W., Essner, J.J., and Yost, H.J. (2000). Multiple pathways in the midline regulate concordant brain, heart and gut left-right asymmetry. *Development* **127**, 3567–3579.

209. Bisgrove, B.W. and Yost, H.J. (2006). The roles of cilia in developmental disorders and disease. *Development* **133**, 4131–4143.

210. Bishop, K.L. (2008). The evolution of flight in bats: narrowing the field of plausible hypotheses. *Q. Rev. Biol.* **83**, 153–169.

211. Bispham, J. (2006). "Music" means nothing if we don't know what it means. *J. Human Evol.* **50**, 587–593.

212. Black, D.L. and Zipursky, S.L. (2008). To cross or not to cross: alternately spliced forms of the Robo3 receptor regulate discrete steps in axonal midline crossing. *Neuron* **58**, 297–298.

213. Black, I.B. (1998). Genes, brain, and mind: the evolution of cognition. *Neuron* **20**, 1073–1080.

214. Black, S.D. and Gerhart, J.C. (1986). High frequency twinning in *Xenopus* eggs centrifuged before first cleavage. *Dev. Biol.* **116**, 228–240.

215. Blader, P. and Strähle, U. (1998). Ethanol impairs migration of the prechordal plate in the zebrafish embryo. *Dev. Biol.* **201**, 185–201.

216. Blagrove, M., Blakemore, S.J., and Thayer, B.R. (2006). The ability to self-tickle following Rapid Eye Movement sleep dreaming. *Conscious Cogn.* **15**, 285–294.

217. Blanco, J., Girard, F., Kamachi, Y., Kondoh, H., and Gehring, W. (2005). Functional analysis of the chicken *δ1-crystallin* enhancer activity in *Drosophila* reveals remarkable evolutionary conservation between chicken and fly. *Development* **132**, 1895–1905.

218. Blanco, M.J., Misof, B.Y., and Wagner, G.P. (1998). Heterochronic differences of *Hoxa-11* expression in *Xenopus* fore- and hind limb development: Evidence for lower limb identity of the anural ankle bones. *Dev. Genes Evol.* **208**, 175–187.

219. Blankenhorn, W.U. (2000). The evolution of body size: what keeps organisms small? *Q. Rev. Biol.* **75**, 385–407.

220. Blaustein, A.R. and Johnson, P.T.J. (2003). Explaining frog deformities. *Sci. Am.* **288** #2, 60–65.

221. Blecher, S.R. and Erickson, R.P. (2007). Genetics of sexual development: a new paradigm. *Am. J. Med. Genet. A* **143A**, 3054–3068.

222. Bloch, J.I. and Boyer, D.M. (2002). Grasping primate origins. *Science* **298**, 1606–1610.

223. Blum, M., Steinbeisser, H., Campione, M., and Schweickert, A. (1999). Vertebrate left-right asymmetry: old studies and new insights. *Cell. Mol. Biol.* **45**, 505–516.

224. Boaz, N.T. (1997). "*Eco Homo*: How the Human Being Emerged from the Cataclysmic History of the Earth." Basic Books, New York.

225. Bock, G. and Goode, J., eds. (2007). "Cortical Development: Genes and Genetic Abnormalities." Wiley, Hoboken, N.J.

226. Bock, G. and Goode, J., eds. (2007). "Tinkering: The Microevolution of Development." Wiley, Chichester, U.K.

227. Bock, W.J. (1965). The role of adaptive mechanisms in the origin of higher levels of organization. *Syst. Zool.* **14**, 272–287.

228. Boesch, C. (1999). A theory that's hard to digest. *Nature* **399**, 653.

229. Bok, J., Dolson, D.K., Hill, P., Rüther, U., Epstein, D.J., and Wu, D.K. (2007). Opposing gradients of Gli repressor and activators mediate Shh signaling along the dorsoventral axis of the inner ear. *Development* **134**, 1713–1722.
230. Boklage, C.E. (1981). On the timing of monozygotic twinning events. *In* "Twin Research 3: Twin Biology and Multiple Pregnancy." Alan R. Liss: New York, pp. 155–165.
231. Boklage, C.E. (2006). Embryogenesis of chimeras, twins and anterior midline asymmetries. *Hum. Reprod.* **21**, 579–591.
232. Bolk, L. (1926). "Das Problem der Menschwerdung." Gustav Fischer, Jena.
233. Bollinger, R.R., Barbas, A.S., Bush, E.L., Lin, S.S., and Parker, W. (2007). Biofilms in the large bowel suggest an apparent function of the human vermiform appendix. *J. Theor. Biol.* **249**, 826–831.
234. Bolnick, D.I. and Doebeli, M. (2003). Sexual dimorphism and adaptive speciation: two sides of the same ecological coin. *Evolution* **57**, 2433–2449.
235. Bolton, E.C., So, A.Y., Chaivorapol, C., Haqq, C.M., Li, H., and Yamamoto, K.R. (2007). Cell- and gene-specific regulation of primary target genes by the androgen receptor. *Genes Dev.* **21**, 2005–2017.
236. Boncinelli, E., Mallamaci, A., and Muzio, L. (2000). Genetic control of regional identity in the developing vertebrate forebrain. *In* "Evolutionary Developmental Biology of the Cerebral Cortex," *Novartis Found. Symp. 228* (G.R. Bock and G. Cardew, eds.). Wiley, New York, pp. 53–66.
237. Bond, J.E. and Opell, B.D. (1998). Testing adaptive radiation and key innovation hypotheses in spiders. *Evolution* **52**, 403–414.
238. Bondeson, J. and Allen, E. (1989). Craniopagus parasiticus. Everard Home's two-headed boy of Bengal and some other cases. *Surg. Neurol.* **31**, 426–434.
239. Bondeson, J. and Miles, A.E.W. (1996). The hairy family of Burma: a four generation pedigree of congenital hypertrichosis lanuginosa. *J. Roy. Soc. Med.* **89**, 403–408.
240. Bonifer, C. (2000). Developmental regulation of eukaryotic gene loci. *Trends Genet.* **16**, 310–315.
241. Bonn, S. and Furlong, E.E.M. (2008). *cis*-Regulatory networks during development: a view of *Drosophila*. *Curr. Opin. Gen. Dev.* **18**, 513–520.
242. Bonner, J.T. (2006). Matters of size. *Nat. Hist.* **115**, 54–59.
243. Boorman, C.J. and Shimeld, S.M. (2002). The evolution of left-right asymmetry in chordates. *BioEssays* **24**, 1004–1011.
244. Bopp, D. (2001). Merging sex and position. *BioEssays* **23**, 304–306.
245. Bopp, D., Calhoun, G., Horabin, J.I., Samuels, M., and Schedl, P. (1996). Sex-specific control of *Sex-lethal* is a conserved mechanism for sex determination in the genus *Drosophila*. *Development* **122**, 971–982.
246. Borneman, A.R., Gianoulis, T.A., Zhang, Z.D., Yu, H., Rozowsky, J., Seringhaus, M.R., Wang, L.Y., Gerstein, M., and Snyder, M. (2007). Divergence of transcription factor binding sites across related yeast species. *Science* **317**, 815–819.
247. Boschi, M., Belloni, M., and Robbins, L.G. (2006). Genetic evidence that nonhomologous disjunction and meiotic drive are properties of wild-type *Drosophila melanogaster* male meiosis. *Genetics* **172**, 305–316.
248. Bottke, W.F., Vokrouhlicky, D., and Nesvorny, D. (2007). An asteroid breakup 160 MYR ago as the probable source of the K/T impactor. *Nature* **449**, 48–53.
249. Botvinick, M.M. (2008). Hierarchical models of behavior and prefrontal function. *Trends Cogn. Sci.* **12**, 201–208.
250. Bouchard, B.L. and Wilson, T.G. (1987). Effects of sublethal doses of methoprene on reproduction and longevity of *Drosophila melanogaster* (Diptera: Drosophilidae). *J. Econ. Entomol.* **80**, 317–321.
251. Boughner, J.C. and Hallgrímsson, B. (2008). Biological spacetime and the temporal integration of functional modules: a case study of dento-gnathic developmental timing. *Dev. Dynamics* **237**, 1–17.
252. Bourlat, S.J., Juliusdottir, T., Lowe, C.J., Freeman, R., Aronowicz, J., Kirschner, M., Lander, E.S., Thorndyke, M., Nakano, H., Kohn, A.B., Heyland, A., Moroz, L.L., Copley, R.R., and Telford, M.J. (2006). Deuterostome phylogeny reveals monophyletic chordates and the new phylum Xenoturbellida. *Nature* **444**, 85–88.
253. Bownds, M.D. (1999). "The Biology of Mind: Origins and Structures of Mind, Brain, and Consciousness." Fitzgerald Sci. Pr., Bethesda, Maryland.
254. Boy de la Tour, E. and Laemmli, U.K. (1988). The metaphase scaffold is helically folded: sister chromatids have predominantly opposite helical handedness. Cell **55**, 937–944.
255. Boyd, R. and Richerson, P.J. (1993). Culture and human evolution. *In* "The Origin and Evolution of Humans and Humanness" (D.T. Rasmussen, ed.). Jones & Bartlett: Boston, Mass., pp. 119–134.
256. Boyden, E.A. (1977). Development and growth of the airways. *In* "Development of the Lung" (W.A. Hodson, ed.). Marcel Dekker, New York, pp. 3–35.
257. Brace, C.L. (1963). Structural reduction in evolution. *Am. Nat.* **97**, 39–49.
258. Brachmann, C.B. and Cagan, R.L. (2003). Patterning the fly eye: the role of apoptosis. *Trends Genet.* **19**, 91–96.
259. Bradbury, J. (2005). Molecular insights into human brain evolution. *PLoS Biol.* **3** #3, 367–370 (e50).
260. Braddock, S.R., Jones, K.L., Bird, L.M., Villegas, I., and Jones, M.C. (1995). Anterior cervical hypertrichosis: a dominantly inherited isolated defect. *Am. J. Med. Genet.* **55**, 498–499.

261. Bradshaw, J.L. and Rogers, L.J. (1993). "The Evolution of Lateral Asymmetries, Language, Tool Use, and Intellect." Acad. Pr., New York.
262. Braendle, C. and Félix, M.-A. (2006). Sex determination: Ways to evolve a hermaphrodite. *Curr. Biol.* **16**, R468–R471.
263. Brainerd, E.L. (1994). The evolution of lung-gill bimodal breathing and the homology of vertebrate respiratory pumps. *Amer. Zool.* **34**, 289–299.
264. Brainerd, E.L. and Owerkowicz, T. (2006). Functional morphology and evolution of aspiration breathing in tetrapods. *Respir. Physiol. Neurobiol.* **154**, 73–88.
265. Brakefield, P.M. (2008). Prospects of evo-devo for linking pattern and process in the evolution of morphospace. *In* "Evolving Pathways: Key Themes in Evolutionary Developmental Biology", (A. Minelli and G. Fusco, eds.). Cambridge Univ. Pr., New York, pp. 62–79.
266. Brakefield, P.M. and French, V. (2004). How and why to spot fly wings. *Nature* **433**, 466–467.
267. Bramble, D.M. and Lieberman, D.E. (2004). Endurance running and the evolution of *Homo*. *Nature* **432**, 345–352.
268. Brandley, M.C., Huelsenbeck, J.P., and Wiens, J.J. (2008). Rates and patterns in the evolution of snake-like body form in squamate reptiles: evidence for repeated re-evolution of lost digits and long-term persistence of intermediate body forms. *Evolution* **62**, 2042–2064.
269. Brandman, O. and Meyer, T. (2008). Feedback loops shape cellular signals in space and time. *Science* **322**, 390–395. [See also (2008) *BioEssays* **30**, 542–555.]
270. Bratus, A. and Slota, E. (2006). DMRT1/Dmrt1, the sex determining or sex differentiating gene in Vertebrata. *Folia Biol. (Krakow)* **54**, 81–86.
271. Brawand, D., Wahli, W., and Kaessmann, H. (2008). Loss of egg yolk genes in mammals and the origin of lactation and placentation. *PLoS Biol.* **6** #3, 507–517 (e63).
272. Brennan, J. and Capel, B. (2004). One tissue, two fates: molecular genetic events that underlie testis versus ovary development. *Nature Rev. Genet.* **5**, 509–521.
273. Brennan, P.L.R., Prum, R.O., McCracken, K.G., Sorenson, M.D., Wilson, R.E., and Birkhead, T.R. (2007). Coevolution of male and female genital morphology in waterfowl. *PLoS ONE* **2** #5, e418.
274. Brenner, S. (1974). The genetics of *Caenorhabditis elegans*. *Genetics* **77**, 71–94.
275. Brenner, S. (1997). Centaur biology. *Curr. Biol.* **7**, R454.
276. Brent, A.E. (2005). Somite formation: where left meets right. *Curr. Biol.* **15**, R468–R470.
277. Breuker, C.J., Debat, V., and Klingenberg, C.P. (2006). Functional evo-devo. *Trends Ecol. Evol.* **21**, 488–492.
278. Brewer, S. and Williams, T. (2004). Finally, a sense of closure? Animal models of human ventral body wall defects. *BioEssays* **26**, 1307–1321.
279. Brezo, J., Klempan, T., and Turecki, G. (2008). The genetics of suicide: a critical review of molecular studies. *Psychiatr. Clin. North Am.* **31**, 179–203.
280. Brigham, P.A., Cappas, A., and Uno, H. (1988). The stumptailed macaque as a model for androgenetic alopecia: effects of topical minoxidil analyzed by use of the folliculogram. *Clinics Dermatol.* **6** #4, 177–187.
281. Brightwell, L.R. (1952). Some experiments with the common hermit crab (*Eupagurus bernhardus*) Linn., and transparent univalve shells. *Proc. Zool. Soc. Lond.* **121**, 279–283.
282. Brisson, J.A. and Stern, D.L. (2006). The pea aphid, *Acyrthosiphon pisum*: an emerging genomic model system for ecological, developmental and evolutionary studies. *BioEssays* **28**, 747–755.
283. Brito, J.M., Teillet, M.-A., and Le Douarin, N.M. (2008). Induction of mirror-image supernumerary jaws in chicken mandibular mesenchyme by Sonic hedgehog-producing cells. *Development* **135**, 2311–2319.
284. Brockes, J.P. and Kumar, A. (2005). Appendage regeneration in adult vertebrates and implications for regenerative medicine. *Science* **310**, 1919–1922.
285. Brockes, J.P., Kumar, A., and Velloso, C.P. (2001). Regeneration as an evolutionary variable. *J. Anat.* **199**, 3–11.
286. Bronner-Fraser, M. (2008). On the trail of the "new head" in Les Treilles. *Development* **135**, 2995–2999.
287. Brooijmans, N. and Kuntz, I.D. (2003). Molecular recognition and docking algorithms. *Annu. Rev. Biophys. Biomol. Struct.* **32**, 335–373.
288. Brown, C.D., Johnson, D.S., and Sidow, A. (2007). Functional architecture and evolution of transcriptional elements that drive gene coexpression. *Science* **317**, 1557–1560.
289. Brown, M.C., Southern, C.L., Anbarasu, A., Kaye, S.B., Fisher, A.C., Hagan, R.P., and Newman, W.D. (2006). Congenital absence of optic chiasm: demonstration of an uncrossed visual pathway using monocular flash visual evoked potentials. *Documenta Ophthalmol.* **113**, 1–4.
290. Brown, N.L., Patel, S., Brzezinski, J., and Glaser, T. (2001). *Math5* is required for retinal ganglion cell and optic nerve formation. *Development* **128**, 2497–2508.
291. Brown, R.H., Jr. (2008). Neuron research leaps ahead. *Science* **321**, 1169–1170.
292. Brown, S. (2008). Music of language or language of music? *Trends Cogn. Sci.* **12**, 246–247.
293. Brown, S. (2008). Top billing for platypus at end of evolution tree. *Nature* **453**, 138–139.
294. Brown, S. and Parsons, L.M. (2008). The neuroscience of dance. *Sci. Am.* **299** #1, 78–83.

295. Brown, S.L. (1994). Animals at play. *Nat. Geogr.* **186** #6, 2–35.
296. Bruce, N.W. and Wellstead, J.R. (1992). Spacing of fetuses and local competition in strains of mice with large, medium and small litters. *J. Reprod. Fertil.* **95**, 783–789.
297. Bruder, C.E.G., Piotrowski, A., Gijsbers, A.A.C.J., Andersson, R., Erickson, S., Diaz de Stahl, T., Menzel, U., Sandgren, J., von Tell, D., Poplawski, A., Crowley, M., Crasto, C., Partridge, E.C., Tiwari, H., Allison, D.B., Komorowski, J., van Ommen, G.-J.B., Boomsma, D.I., Pederson, N.L., den Dunnen, J.T., Wirdefeldt, K., and Dumanski, J.P. (2008). Phenotypically concordant and discordant monozygotic twins display different DNA copy-number-variation profiles. *Am. J. Human Genet.* **82**, 763–771.
298. Bruemmer, F. (1993). "The Narwhal: Unicorn of the Sea." Swan Hill Pr., Shrewsbury, U.K.
299. Brugmann, S.A., Tapadia, M.D., and Helms, J.A. (2006). The molecular origins of species-specific facial pattern. *Curr. Topics Dev. Biol.* **73**, 1–42.
300. Brunel, C.A., Madigan, S.J., Cassill, J.A., Edeen, P.T., and McKeown, M. (1998). *pcdr*, a novel gene with sexually dimorphic expression in the pigment cells of the *Drosophila* eye. *Dev. Genes Evol.* **208**, 327–335.
301. Brunet, J.-F. and Ghysen, A. (1999). Deconstructing cell determination: proneural genes and neuronal identity. *BioEssays* **21**, 313–318.
302. Bryant, P.J. (1978). Pattern formation in imaginal discs. *In* "The Genetics and Biology of Drosophila," Vol. 2c (M. Ashburner and T.R.F. Wright, eds.). Acad. Pr., New York, pp. 229–335.
303. Bryant, P.J. and Hsei, B.W. (1977). Pattern formation in asymmetrical and symmetrical imaginal discs of *Drosophila melanogaster*. *Amer. Zool.* **17**, 595–611.
304. Bryant, S.V., Bryant, P.J., and French, V. (1981). Distal regeneration and symmetry. *Science* **212**, 993–1002.
305. Bublitz, N. (2008). A face in the crowd. *Sci. Am. Mind* **19** #2, 58–65. [See also (2008) *Annu. Rev. Neurosci.* **31**, 411–437.]
306. Buchholtz, E.A. (2007). Modular evolution of the Cetacean vertebral column. *Evol. Dev.* **9**, 278–289.
307. Buck, C. and Bär, H. (1993). Investigations on the biomechanical significance of dermatoglyphic ridges. *In* "Hands of Primates" (H. Preuschoft and D.J. Chivers, eds.). Springer-Verlag: New York, pp. 285–306.
308. Buckingham, M., Meilhac, S., and Zaffran, S. (2005). Building the mammalian heart from two sources of myocardial cells. *Nature Rev. Genet.* **6**, 826–835.
309. Budd, G.E. (2006). On the origin and evolution of major morphological characters. *Biol. Rev.* **81**, 609–628.
310. Bull, J.J. (2000). Déjà vu. *Nature* **408**, 416–417.
311. Bull, J.J. and Charnov, E.L. (1985). On irreversible evolution. *Evolution* **39**, 1149–1155.
312. Bullock, T.H., Orkand, R., and Grinnell, A. (1977). "Introduction to Nervous Systems." W. H. Freeman, San Francisco, Calif.
313. Bulmer, M.G. (1970). "The Biology of Twinning in Man." Oxford Univ. Pr., New York.
314. Bunge, S.A., Dudukovic, N.M., Thomason, M.E., Vaidya, C.J., and Gabrieli, J.D.E. (2002). Immature frontal lobe contributions to cognitive control in children: evidence from fMRI. *Neuron* **33**, 301–311.
315. Bunge, S.A. and Wright, S.B. (2007). Neurodevelopmental changes in working memory and cognitive control. *Curr. Opin. Neurobiol.* **17**, 243–250.
316. Burd, M. (1986). Sexual selection and human evolution: all or none adaptation? *Amer. Anthrop.* **88**, 167–172.
317. Burgdorf, J. and Panksepp, J. (2001). Tickling induces reward in adolescent rats. *Physiol. Behav.* **72**, 167–173.
318. Burke, A.C., Nelson, C.E., Morgan, B.A., and Tabin, C. (1995). *Hox* genes and the evolution of vertebrate axial morphology. *Development* **121**, 333–346.
319. Burke, K.A., Franz, T.M., Miller, D.N., and Schoenbaum, G. (2008). The role of the orbitofrontal cortex in the pursuit of happiness and more specific rewards. *Nature* **454**, 340–344.
320. Burke, R.M., Rayan, S.S., Kasirajan, K., Chaikof, E.L., and Milner, R. (2006). Unusual case of right-sided May-Thurner syndrome and review of its management. *Vascular* **14**, 47–50.
321. Burn, J. (1991). Disturbance of morphological laterality in humans. *In* "Biological Asymmetry and Handedness," *Ciba Found. Symp. 162* (G.R. Bock and J. Marsh, eds.). Wiley, New York, pp. 282–299.
322. Burn, S.F., Boot, M.J., de Angelis, C., Doohan, R., Arques, C.G., Torres, M., and Hill, R.E. (2008). The dynamics of spleen morphogenesis. *Dev. Biol.* **318**, 303–311.
323. Burns, J.L., Soothill, P., and Hassan, A.B. (2007). Allometric growth ratios are independent of *Igf2* gene dosage during development. *Evol. Dev.* **9**, 155–164.
324. Burr, D. and Morrone, M.C. (2005). Eye movements: Building a stable world from glance to glance. *Curr. Biol.* **15**, R839–R840.
325. Burrows, A.M. (2008). The facial expression musculature in primates and its evolutionary significance. *BioEssays* **30**, 212–225.
326. Burt, A. and Trivers, R. (2006). "Genes in Conflict: The Biology of Selfish Genetic Elements." Harvard Univ. Pr., Cambridge, Mass.
327. Burtis, K.C. and Wolfner, M.F. (1992). The view from the bottom: sex-specific traits and their control in *Drosophila*. *Sems. Dev. Biol.* **3**, 331–340.
328. Burton, P.M. (2008). Insights from diploblasts: the evolution of mesoderm and muscle. *J. Exp. Zool. (Mol. Dev. Evol.)* **310B**, 5–14.
329. Bush, E.C. and Lahn, B.T. (2006). The evolution of word composition in metazoan promoter

sequence. *PLoS Comput. Biol.* **2** #11, 1343–1348 (e150).
330. Buss, D.M., Haselton, M.G., Shackelford, T.K., Bleske, A.L., and Wakefield, J.C. (1998). Adaptations, exaptations, and spandrels. *Am. Psychol.* **53**, 533–548.
331. Busser, B.W., Bulyk, M.L., and Michelson, A.M. (2008). Toward a systems-level understanding of developmental regulatory networks. *Curr. Opin. Gen. Dev.* **18**, 521–529.
332. Butler, A.B. and Hodos, W. (2005). "Comparative Vertebrate Neuroanatomy: Evolution and Adaptation," 2nd ed. Wiley, Hoboken, N.J.
333. Butler, P.M. (1978). The ontogeny of mammalian heterodonty. *J. Biol. Bucc.* **6**, 217–227.
334. Butts, T., Holland, P.W.H., and Ferrier, D.E.K. (2008). The Urbilaterian Super-Hox cluster. *Trends Genet.* **24**, 259–262.
335. Buxbaum, R.E. (1995). Biological levels. *Nature* **373**, 567–568.
336. Buzsáki, G. (2007). The structure of consciousness. *Nature* **446**, 267.
337. Byers, J.A. (1999). Play's the thing. *Nat. Hist.* **108** #6, 40–45.
338. Byrne, R.W. and Bates, L.A. (2007). Sociality, evolution and cognition. *Curr. Biol.* **17**, R714–R723.
339. Cahill, L. (2005). His brain, her brain. *Sci. Am.* **292** #5, 40–47.
340. Cai, J., Cho, S.-W., Kim, J.-Y., Lee, M.-J., Cha, Y.-G., and Jung, H.-S. (2007). Patterning the size and number of tooth and its cusps. *Dev. Biol.* **304**, 499–507. [See also (2009) *Science* **323**, 1232–1234.]
341. Calder, N. (1984). "Timescale: An Atlas of the Fourth Dimension." Chatto & Windus, London.
342. Calder, W.A., III (1978). The kiwi. *Sci. Am.* **239** #1, 132–142.
343. Calder, W.A., III (1979). The kiwi and egg design: evolution as a package deal. *BioScience* **29** #8, 461–467.
344. Call, M.K., Grogg, M.W., and Tsonis, P.A. (2005). Eye on regeneration. *Anat. Rec.* **287B**, 42–48.
345. Callaerts, P., Halder, G., and Gehring, W.J. (1997). *Pax-6* in development and evolution. *Annu. Rev. Neurosci.* **20**, 483–532.
346. Calvin, W.H. and Ojemann, G.A. (1994). "Conversations with Neil's Brain: The Neural Nature of Thought and Language." Addison-Wesley, New York.
347. Cameron, R.A., Rowen, L., Nesbitt, R., Bloom, S., Rast, J.P., Berney, K., Arenas-Mena, C., Martinez, P., Lucas, S., Richardson, P.M., Davidson, E.H., Peterson, K.J., and Hood, L. (2006). Unusual gene order and organization of the sea urchin Hox cluster. *J. Exp. Zool. (Mol. Dev. Evol.)* **306B**, 45–58.
348. Camon, J., Sabate, D., Degollada, E., and Lopezbejar, M.A. (1991). Vascular anatomy of a dicephalic cat. *Anat. Embryol.* **184**, 507–515.
349. Campbell, J. (1959). "The Masks of God. Vol. 1. Primitive Mythology." Viking Penguin, New York.
350. Campbell, J. (1962). "The Masks of God. Vol. 2. Oriental Mythology." Viking Penguin, New York.
351. Campbell, J. (1964). "The Masks of God. Vol. 3. Occidental Mythology." Viking Penguin, New York.
352. Campbell, J. (1968). "The Masks of God. Vol. 4. Creative Mythology." Viking Penguin, New York.
353. Campbell, N.A., Reece, J.B., Taylor, M.R., and Simon, E.J. (2006). "Biology: Concepts and Connections," 5th ed. Pearson/Benjamin Cummings, San Francisco, Calif.
354. Campeau, P.M., Foulkes, W.D., and Tischkowitz, M.D. (2008). Hereditary breast cancer: new genetic developments, new therapeutic avenues. *Hum. Genet.* **124**, 31–42.
355. Campione, M., Ros, M.A., Icardo, J.M., Piedra, E., Christoffels, V.M., Schweickert, A., Blum, M., Franco, D., and Moorman, A.F.M. (2001). *Pitx2* expression defines a left cardiac lineage of cells: evidence for atrial and ventricular molecular isomerism in the *iv/iv* mice. *Dev. Biol.* **231**, 252–264.
356. Candiani, S., Holland, N.D., Oliveri, D., Parodi, M., and Pestarino, M. (2008). Expression of the amphioxus Pit-1 gene (*AmphiPOU1F1/Pit-1*) exclusively in the developing preoral organ, a putative homolog of the vertebrate adenohypophysis. *Brain Res. Bull.* **75**, 324–330.
357. Cañestro, C. and Postlethwait, J.H. (2007). Development of a chordate anterior–posterior axis without classical retinoic acid signaling. *Dev. Biol.* **305**, 522–538.
358. Cañestro, C., Yokoi, H., and Postlethwait, J.H. (2007). Evolutionary developmental biology and genomics. *Nature Rev. Genet.* **8**, 932–942.
359. Cannon, B. and Nedergaard, J. (2008). Neither fat nor flesh. *Nature* **454**, 947–948.
360. Cant, J.G.H. (1981). Hypothesis for the evolution of human breasts and buttocks. *Am. Nat.* **117**, 199–204.
361. Cantalupo, C. and Hopkins, W.D. (2001). Asymmetric Broca's area in great apes. *Nature* **414**, 505.
362. Canto-Soler, M.V. and Adler, R. (2006). Optic cup and lens development requires Pax6 expression in the early optic vesicle during a narrow time window. *Dev. Biol.* **294**, 119–132.
363. Canup, R.M. and Asphaug, E. (2001). Origin of the Moon in a giant impact near the end of the Earth's formation. *Nature* **412**, 708–712.
364. Capellini, I., Barton, R.A., McNamara, P., Preston, B.T., and Nunn, C.L. (2008). Phylogenetic analysis of the ecology and evolution of mammalian sleep. *Evolution* **62**, 1764–1776.
365. Caporael, L.R. (1994). Of myth and science: origin stories and evolutionary scenarios. *Social Sci. Info.* **33** #1, 9–23.
366. Caporale, L. (2005). Darwin in the genome. *BioEssays* **27**, 984.

367. Caporale, L.H. (2003). "Darwin in the Genome: Molecular Strategies in Biological Evolution." McGraw-Hill, New York.
368. Capozzoli, N.J. (1995). Why are vertebrate nervous systems crossed? *Med. Hypotheses* **45**, 471–475.
369. Capozzoli, N.J. (1999). Why do we speak with the left hemisphere? *Med. Hypotheses* **52**, 497–503.
370. Capra, F. (2007). "The Science of Leonardo: Inside the Mind of the Great Genius of the Renaissance." Doubleday, New York.
371. Carapuço, M., Nóvoa, A., Bobola, N., and Mallo, M. (2005). *Hox* genes specify vertebral types in the presomitic mesoderm. *Genes Dev.* **19**, 2116–2121.
372. Cardoso, W.V. and Lü, J. (2006). Regulation of early lung morphogenesis: questions, facts and controversies. *Development* **133**, 1611–1624.
373. Carey, M. (1998). The enhanceosome and transcriptional synergy. *Cell* **92**, 5–8.
374. Cariani, P. (2001). Symbols and dynamics in the brain. *Biosystems* **60**, 59–83.
375. Carlborg, Ö., Jacobsson, L., Åhgren, P., Siegel, P., and Andersson, L. (2005). Epistasis and the release of genetic variation during long-term selection. *Nature Genet.* **38**, 418–420.
376. Carlson, B.M. (1994). "Human Embryology and Developmental Biology." Mosby, St. Louis, Mo.
377. Carmeliet, P. and Tessier-Lavigne, M. (2005). Common mechanisms of nerve and blood vessel wiring. *Nature* **436**, 193–200.
378. Carmona, F.D., Motokawa, M., Tokita, M., Tsuchiya, K., Jiménez, R., and Sánchez-Villagra, M.R. (2008). The evolution of female mole ovotestes evidences high plasticity of mammalian gonad development. *J. Exp. Zool. (Mol. Dev. Evol.)* **310B**, 259–266.
379. Caronia, G., Goodman, F.R., McKeown, C.M.E., Scambler, P.J., and Zappavigna, V. (2003). An I47L substitution in the HOXD13 homeodomain causes a novel human limb malformation by producing a selective loss of function. *Development* **130**, 1701–1712.
380. Carroll, L. and Gardner, M. (1960). "The Annotated Alice: Alice's Adventures in Wonderland & Through the Looking Glass." Meridian, New York.
381. Carroll, R.L. and Holmes, R.B. (2007). Evolution of the appendicular skeleton of amphibians. *In* "Fins into Limbs: Evolution, Development, and Transformation" (B.K. Hall, ed.). Univ. Chicago Pr.: Chicago, Ill., pp. 185–224.
382. Carroll, S.B. (2000). Endless forms: the evolution of gene regulation and morphological diversity. *Cell* **101**, 577–580.
383. Carroll, S.B. (2001). Chance and necessity: the evolution of morphological complexity and diversity. *Nature* **409**, 1102–1109.
384. Carroll, S.B. (2002). Stephen Jay Gould (1941–2002): a wonderful life. *Dev. Cell* **3**, 21–23.
385. Carroll, S.B. (2003). Genetics and the making of *Homo sapiens*. *Nature* **422**, 849–857.
386. Carroll, S.B. (2005). "Endless Forms Most Beautiful: The New Science of Evo Devo and the Making of the Animal Kingdom." Norton, New York.
387. Carroll, S.B. (2005). Evolution at two levels: On genes and form. *PLoS Biol.* **3** #7, 1159–1166 (e245).
388. Carroll, S.B. (2005). The origins of form. *Nat. Hist.* **114** #9, 58–63.
389. Carroll, S.B. (2006). "The Making of the Fittest: DNA and the Ultimate Forensic Record of Evolution." W.W. Norton, New York.
390. Carroll, S.B. (2008). Evo-devo and an expanding evolutionary synthesis: a genetic theory of morphological evolution. *Cell* **134**, 25–36.
391. Carroll, S.B., Grenier, J.K., and Weatherbee, S.D. (2001). "From DNA to Diversity: Molecular Genetics and the Evolution of Animal Design," Blackwell Science, Malden, Mass.
392. Carroll, S.B., Grenier, J.K., and Weatherbee, S.D. (2005). "From DNA to Diversity: Molecular Genetics and the Evolution of Animal Design," 2nd ed. Blackwell, Malden, Mass.
393. Carroll, S.B., Prud'homme, B., and Gompel, N. (2008). Regulating evolution. *Sci. Am.* **298** #5, 60–67.
394. Cartwright, J.H.E., Piro, O., and Tuval, I. (2004). Fluid-dynamical basis of the embryonic development of left-right asymmetry in vertebrates. *PNAS* **101** #19, 7234–7239.
395. Cartwright, P. and Collins, A. (2007). Fossils and phylogenies: integrating multiple lines of evidence to investigate the origin of early major metazoan lineages. *Integr. Comp. Biol.* **47**, 744–751.
396. Casellas, J. and Medrano, J.F. (2008). Within-generation mutation variance for litter size in inbred mice. *Genetics* **179**, 2147–2155.
397. Casey, B. (1998). Two rights make a wrong: human left-right malformations. *Hum. Molec. Genet.* **7**, 1565–1571.
398. Castelli-Gair, J. and Akam, M. (1995). How the Hox gene *Ultrabithorax* specifies two different segments: the significance of spatial and temporal regulation within metameres. *Development* **121**, 2973–2982.
399. Catchpole, C.K. and Slater, P.J.B. (1995). "Bird Song: Biological Themes and Variations." Cambridge Univ. Pr., New York.
400. Catz, D.S., Fischer, L.M., and Kelley, D.B. (1995). Androgen regulation of a laryngeal-specific myosin heavy chain mRNA isoform whose expression is sexually differentiated. *Dev. Biol.* **171**, 448–457.
401. Cavener, D.R. (1992). Transgenic animal studies on the evolution of genetic regulatory circuitries. *BioEssays* **14**, 237–244.
402. Cayouette, M. and Raff, M. (2003). The orientation of cell division influences cell-fate choice in the developing mammalian retina. *Development* **130**, 2329–2339.

403. Cazemajor, M., Joly, D., and Montchamp-Moreau, C. (2000). Sex-ratio meiotic drive in *Drosophila simulans* is related to equational nondisjunction of the Y chromosome. *Genetics* **154**, 229–236.

404. Cebrià, F. (2007). Regenerating the central nervous system: how easy for planarians! *Dev. Genes Evol.* **217**, 733–748.

405. Cebrià, F., Guo, T., Jopek, J., and Newmark, P.A. (2007). Regeneration and maintenance of the planarian midline is regulated by a *slit* orthologue. *Dev. Biol.* **307**, 394–406.

406. Cela-Conde, C.J. (1996). Bipedal/Savanna/Cladogeny Model. Can it still be held? *Hist. Phil. Life Sci.* **18**, 213–224.

407. Cela-Conde, J. and Ayala, F.J. (2007). "Human Evolution: Trails from the Past." Oxford Univ. Pr., New York.

408. Cellini, A. and Offidani, A. (1992). Familial supernumerary nipples and breasts. *Dermatology* **185**, 56–58.

409. Cerny, R., Lwigale, P., Ericsson, R., Meulemans, D., Epperlein, H.-H., and Bronner-Fraser, M. (2004). Developmental origins and evolution of jaws: new interpretation of "maxillary" and "mandibular." *Dev. Biol.* **276**, 225–236.

410. Chadwick, D. and Goode, J., eds. (2002). "The Genetics and Biology of Sex Determination." *Novartis Found. Symp. 244.* Wiley, New York.

411. Chae, T.H. and Walsh, C.A. (2007). Genes that control the size of the cerebral cortex. *In* "Cortical Development: Genes and Genetic Abnormalities," *Novartis Found. Symp. 288* (G. Bock and J. Goode, eds.). Wiley, Hoboken, N.J., pp. 79–95.

412. Chai, Y., Jiang, X., Ito, Y., Bringas, P., Jr., Han, J., Rowitch, D.H., Soriano, P., McMahon, A.P., and Sucov, H.M. (2000). Fate of the mammalian cranial neural crest during tooth and mandibular morphogenesis. *Development* **127**, 1671–1679.

413. Chai, Y. and Maxson, R.E., Jr. (2006). Recent advances in craniofacial morphogenesis. *Dev. Dynamics* **235**, 2353–2375.

414. Chakravarti, A. (2008). Victor Almon McKusick (1921–2008). *Nature* **455**, 46.

415. Chamberlain, S.R., Menzies, L., Hampshire, A., Suckling, J., Fineberg, N.A., del Campo, N., Aitken, M., Craig, K., Owen, A.M., Bullmore, E.T., Robbins, T.W., and Sahakian, B.J. (2008). Orbitofrontal dysfunction in patients with obsessive-compulsive disorder and their unaffected relatives. *Science* **321**, 421–422.

416. Chambeyron, S. and Bickmore, W.A. (2004). Chromatin decondensation and nuclear reorganization of the *HoxB* locus upon induction of transcription. *Genes Dev.* **18**, 1119–1130.

417. Chan, A.W.S., Dominko, T., Luetjens, C.M., Neuber, E., Martinovich, C., Hewitson, L., Simerly, C.R., and Schatten, G.P. (2000). Clonal propagation of primate offspring by embryo splitting. *Science* **287**, 317–319.

418. Chapais, B. (2008). "Primeval Kinship: How Pair-bonding Gave Birth to Human Society." Harvard Univ. Pr., Cambridge, Mass.

419. Chapman, A.J. (1975). Humorous laughter in children. *J. Pers. Soc. Psychol.* **31**, 42–49.

420. Chapman, S.C., Sawitzke, A.L., Campbell, D.S., and Schoenwolf, G.C. (2005). A three-dimensional atlas of pituitary gland development in the zebrafish. *J. Comp. Neurol.* **487**, 428–440.

421. Chapouton, P., Jagasia, R., and Bally-Cuif, L. (2007). Adult neurogenesis in non-mammalian vertebrates. *BioEssays* **29**, 745–757.

422. Charité, J., de Graaff, W., Consten, D., Reijnen, M.J., Korving, J., and Deschamps, J. (1998). Transducing positional information to the *Hox* genes: critical interaction of cdx gene products with position-sensitive regulatory elements. *Development* **125**, 4349–4358.

423. Charlesworth, B. (1996). The evolution of chromosomal sex determination and dosage compensation. *Curr. Biol.* **6**, 149–162.

424. Charman, W.N. (1991). The vertebrate dioptric apparatus. *In* "Evolution of the Eye and Visual System," (J.R. Cronly-Dillon and R.L. Gregory, eds.). CRC Pr., Boston, Mass., pp. 82–117.

425. Charoenkul, V. and Jimarkon, P. (1978). Gigantic bilateral aberrant axillary breasts: a case report. *Mt. Sinai J. Med.* **45**, 455–459.

426. Chatterjee, S. and Templin, R.J. (2004). Posture, locomotion, and paleoecology of pterosaurs. *Geol. Soc. Amer. Spec. Ppr.* **376**, 1–64.

427. Chea, H.K., Wright, C.V., and Swalla, B.J. (2005). Nodal signaling and the evolution of deuterostome gastrulation. *Dev. Dynamics* **234**, 269–278.

428. Chen, C.-H., Cretekos, C.J., Rasweiler, J.J., IV, and Behringer, R.R. (2005). *Hoxd13* expression in the developing limbs of the short-tailed fruit bat, *Carollia perspicillata. Evol. Dev.* **7**, 130–141.

429. Chen, E.H., Christiansen, A.E., and Baker, B.S. (2005). Allocation and specification of the genital disc precursor cells in *Drosophila. Dev. Biol.* **281**, 270–285.

430. Chen, F., Desai, T.J., Qian, J., Niederreither, K., Lü, J., and Cardoso, W.V. (2007). Inhibition of Tgfβ signaling by endogenous retinoic acid is essential for primary lung bud induction. *Development* **134**, 2969–2979.

431. Chen, J.-Y., Bottjer, D.J., Oliveri, P., Dornbos, S.Q., Gao, F., Ruffins, S., Chi, H., Li, C.-W., and Davidson, E.H. (2004). Small bilaterian fossils from 40 to 55 million years before the Cambrian. *Science* **305**, 218–222.

432. Chen, Y., Knezevic, V., Ervin, V., Hutson, R., Ward, Y., and Mackem, S. (2004). Direct interaction with Hoxd proteins reverses Gli3-repressor function

to promote digit formation downstream of Shh. *Development* **131**, 2339–2347.

433. Chen, Z., Gore, B.B., Long, H., Ma, L., and Tessier-Lavigne, M. (2008). Alternative splicing of the Robo3 axon guidance receptor governs the midline switch from attraction to repulsion. *Neuron* **58**, 325–332.
434. Chen, Z.-L., Yu, W.-M., and Strickland, S. (2007). Peripheral regeneration. *Annu. Rev. Neurosci.* **30**, 209–233.
435. Cheng, C.W., Yan, C.H.M., Hui, C.-c., Strähle, U., and Chang, S.H. (2006). The homeobox gene *irx1a* is required for the propagation of the neurogenic waves in the zebrafish retina. *Mechs. Dev.* **123**, 252–263.
436. Chernoff, E.A.G., Sato, K., Corn, A., and Karcavich, R.E. (2002). Spinal cord regeneration: intrinsic properties and emerging mechanisms. *Sems. Cell Dev. Biol.* **13**, 361–368.
437. Chevone, B.I. and Richards, A.G. (1976). Ultrastructure of the atypical muscles associated with terminalial inversion in male *Aedes aegypti* (L). *Biol. Bull.* **151**, 283–296.
438. Chi, N.C., Shaw, R.M., De Val, S., Kang, G., Jan, L.Y., Black, B.L., and Stainier, D.Y.R. (2008). Foxn4 directly regulates *tbx2b* expression and atrioventricular canal formation. *Genes Dev.* **22**, 734–739.
439. Chiang, C., Litingtung, Y., Harris, M.P., Simandl, B.K., Li, Y., Beachy, P.A., and Fallon, J.F. (2001). Manifestation of the limb prepattern: limb development in the absence of Sonic hedgehog function. *Dev. Biol.* **236**, 421–435.
440. Child, C.M. (1941). "Patterns and Problems of Development." Univ. of Chicago Pr., Chicago.
441. Chin, A.J., Tsang, M., and Weinberg, E.S. (2000). Heart and gut chiralities are controlled independently from initial heart position in the developing zebrafish. *Dev. Biol.* **227**, 403–421.
442. Chipman, A.D. (2001). Developmental exaptation and evolutionary change. *Evol. Dev.* **3**, 299–301.
443. Chipman, A.D. (2008). Thoughts and speculations on the ancestral arthropod segmentation pathway. *In* "Evolving Pathways: Key Themes in Evolutionary Developmental Biology" (A. Minelli and G. Fusco, eds.). Cambridge Univ. Pr., New York, pp. 343–358.
444. Chippindale, P.T., Bonett, R.M., Baldwin, A.S., and Wiens, J.J. (2004). Phylogenetic evidence for a major reversal of life-history evolution in plethodontid salamanders. *Evolution* **58**, 2809–2822.
445. Chitty, D. (1996). "Do Lemmings Commit Suicide?" Oxford Univ. Pr., New York.
446. Chivers, D.J. (1992). Diet and guts. *In* "The Cambridge Encyclopedia of Human Evolution" (S. Bunney, S. Jones, R. Martin, and D. Pilbeam, eds.). Cambridge Univ. Pr., New York, pp. 60–64.
447. Cho, K.-W., Kim, J.-Y., Song, S.-J., Farrell, E., Eblaghie, M.C., Kim, H.-J., Tickle, C., and Jung, H.-S. (2006). Molecular interactions between *Tbx3* and *Bmp4* and a model for dorsoventral positioning of mammary gland development. *PNAS* **103** #45, 16788–16793.
448. Cho, S., Huang, Z.Y., and Zhang, J. (2007). Sex-specific splicing of the honeybee *doublesex* gene reveals 300 million years of evolution at the bottom of the insect sex-determination pathway. *Genetics* **177**, 1733–1741.
449. Cholfin, J.A. and Rubenstein, J.L.R. (2007). Genetic regulation of prefrontal cortex development and function. *In* "Cortical Development: Genes and Genetic Abnormalities," *Novartis Found. Symp. 288* (G. Bock and J. Goode, eds.). Wiley, Hoboken, N.J., pp. 165–177.
450. Chothia, C. (1991). Asymmetry in protein structures. *In* "Biological Asymmetry and Handedness," *Ciba Found. Symp. 162* (G.R. Bock and J. Marsh, eds.). Wiley, New York, pp. 36–57.
451. Chow, R.L., Altmann, C.R., Lang, R.A., and Hemmati-Brivanlou, A. (1999). Pax6 induces ectopic eyes in a vertebrate. *Development* **126**, 4213–4222.
452. Christ, B., Jacob, H.J., Brand-Saberi, B., and Grim, M. (1993). On the development of the human hand. *In* "Hands of Primates" (H. Preuschoft and D.J. Chivers, eds.). Springer-Verlag, New York, pp. 405–421.
453. Christel, M. (1993). Grasping techniques and hand preferences in Hominoidea. *In* "Hands of Primates" (H. Preuschoft and D.J. Chivers, eds.). Springer-Verlag, New York, pp. 91–108.
454. Christen, H.J., Hanefeld, F., Kruse, E., Imhäuser, S., Ernst, J.P., and Finkenstaedt, M. (2000). Foix-Chavany-Marie (anterior-operculum) syndrome in childhood: a reappraisal of Worster-Drought syndrome. *Dev. Med. Child Neurol.* **42**, 122–132.
455. Christensen, K., Johnson, T.E., and Vaupel, J.W. (2006). The quest for genetic determinants of human longevity: challenges and insights. *Nature Rev. Genet.* **7**, 436–448.
456. Christiaen, L., Jaszczyszyn, Y., Kerfant, M., Kano, S., Thermes, V., and Joly, J.-S. (2007). Evolutionary modification of mouth position in deuterostomes. *Sems. Cell Dev. Biol.* **18**, 502–511.
457. Christiansen, A.E., Keisman, E.L., Ahmad, S.M., and Baker, B.S. (2002). Sex comes in from the cold: the integration of sex and pattern. *Trends Genet.* **18**, 510–516.
458. Christiansen, P. (2008). Evolution of skull and mandible shape in cats (Carnivora: Felidae). *PLoS ONE* **3** #7, e2807.
459. Chung, K.W. (2005). "Gross Anatomy (Board Review Series)," 5th ed. Lippincott, Williams, & Wilkins, Baltimore, MD.
460. Cirelli, C. and Tononi, G. (2008). Is sleep essential? *PLoS Biol.* **6** #8, 1605–1611 (e216).
461. Clark, D.A., Mitra, P.P., and Wang, S.S.-H. (2001). Scalable architecture in mammalian brains. *Nature* **411**, 189–193.

462. Clark, R.B. (1964). "Dynamics in Metazoan Evolution: The Origin of the Coelom and Segments." Oxford Univ. Pr., New York.
463. Clarke, P.G.H. (1981). Chance, repetition, and error in the development of normal nervous systems. *Persp. Biol. Med.* **25**, 2–19. [See also (2009) *PLoS Biol.* 7 #2, 265–277.]
464. Claxton, G. (1975). Why can't we tickle ourselves? *Percept. Mot. Skills* **41**, 335–338.
465. Clements, R., Liew, T.-S., Vermeulen, J.J., and Schilthuizen, M. (2008). Further twists in gastropod shell evolution. *Biol. Lett.* **4**, 179–182.
466. Cline, T.W. (1989). The affairs of *daughterless* and the promiscuity of developmental regulators. *Cell* **59**, 231–234.
467. Cline, T.W. (2005). Reflections on a path to sexual commitment. *Genetics* **169**, 1179–1185.
468. Cloud, P. (1983). The biosphere. *Sci. Am.* **249** #3, 176–189.
469. Clutton-Brock, T. (2007). Sexual selection in males and females. *Science* **318**, 1882–1885.
470. Clyne, J.D. and Miesenböck, G. (2008). Sex-specific control and tuning of the pattern generator for courtship song in *Drosophila*. *Cell* **133**, 354–363.
471. Coates, M. and Ruta, M. (2000). Nice snake, shame about the legs. *Trends Ecol. Evol.* **15**, 503–507.
472. Coates, M.I. (1994). The origin of vertebrate limbs. *Development* **1994 Suppl.**, 169–180.
473. Coates, M.I. and Cohn, M.J. (1998). Fins, limbs, and tails: outgrowths and axial patterning in vertebrate evolution. *BioEssays* **20**, 371–381.
474. Coates, M.I. and Ruta, M. (2007). Skeletal changes in the transition from fins to limbs. *In* "Fins into Limbs: Evolution, Development, and Transformation" (B.K. Hall, ed.). Univ. Chicago Pr.: Chicago, Ill., pp. 15–38.
475. Cobb, J. and Duboule, D. (2005). Comparative analysis of genes downstream of the Hoxd cluster in developing digits and external genitalia. *Development* **132**, 3055–3067.
476. Cobourne, M.T. and Mitsiadis, T. (2006). Neural crest cells and patterning of the mammalian dentition. *J. Exp. Zool. (Mol. Dev. Evol.)* **306B**, 251–260.
477. Cobourne, M.T. and Sharpe, P.T. (2003). Tooth and jaw: molecular mechanisms of patterning in the first branchial arch. *Arch. Oral Biol.* **48**, 1–14.
478. Cochard, L.R. (2002). "Netter's Atlas of Human Embryology." Icon Learning Systems, Teterboro, N.J.
479. Cock, A.G. (1966). Genetical aspects of metrical growth and form in animals. *Q. Rev. Biol.* **41**, 131–190.
480. Coen, E. (1999). "The Art of Genes: How Organisms Make Themselves." Oxford Univ. Pr., New York.
481. Cohen, E.D. and Morrisey, E.E. (2008). A house with many rooms: how the heart got its chambers with *foxn4*. *Genes Dev.* **22**, 706–710.
482. Cohen, J. (1978). "Food Webs and Niche Space." Princeton Univ. Pr., Princeton, N.J.
483. Cohen, J. (1995). Getting all turned around over the origins of life on earth. *Science* **267**, 1265–1266.
484. Cohen, J. (2007). The endangered lab chimp. *Science* **315**, 450–452.
485. Cohen, M.M. (2006). "Perspectives on the Face." Oxford Univ. Pr., New York.
486. Cohen, M.M., Jr. (2001). Asymmetry: molecular, biologic, embryopathic, and clinical perspectives. *Am. J. Med. Genet.* **101**, 292–314.
487. Cohen, M.M., Jr. (2002). Malformations of the craniofacial region: evolutionary, embryonic, genetic, and clinical perspectives. *Am. J. Med. Genet. (Semin. Med. Genet.)* **115**, 245–268.
488. Cohen, M.M., Jr. (2006). Holoprosencephaly: clinical, anatomic, and molecular dimensions. *Birth Defects Res. (Pt. A): Clin. Mol. Teratol.* **76**, 658–673.
489. Cohn, M.J., Izpisúa-Belmonte, J.-C., Abud, H., Heath, J.K., and Tickle, C. (1995). Fibroblast growth factors induce additional limb development from the flank of chick embryos. *Cell* **80**, 739–746.
490. Cohn, M.J. and Tickle, C. (1999). Developmental basis of limblessness and axial patterning in snakes. *Nature* **399**, 474–479.
491. Cole, N.J. and Currie, P.D. (2007). Insights from sharks: evolutionary and developmental models of fin development. *Dev. Dynamics* **236**, 2421–2431.
492. Collin, R. and Cipriani, R. (2003). Dollo's law and the re-evolution of shell coiling. *Proc. Roy. Soc. Lond. B* **270**, 2551–2555.
493. Collins, S., de Meaux, J., and Acquisti, C. (2007). Adaptive walks toward a moving optimum. *Genetics* **176**, 1089–1099.
494. Colucci-D'Amato, L. and di Porzio, U. (2008). Neurogenesis in adult CNS: from denial to opportunities and challenges for therapy. *BioEssays* **30**, 135–145.
495. Compton, R.N. and Pagni, R.M. (2002). The chirality of biomolecules. *Adv. Atomic Molec. Optical Phys.* **48**, 219–261.
496. Concha, M.L. and Wilson, S.W. (2001). Asymmetry in the epithalamus of vertebrates. *J. Anat.* **199**, 63–84.
497. Condous, G. (2006). Ectopic pregnancy—risk factors and diagnosis. *Aust. Fam. Physician* **35**, 854–857.
498. Conlon, I. and Raff, M. (1999). Size control in animal development. *Cell* **96**, 235–244.
499. Conlon, R.A. (1995). Retinoic acid and pattern formation in vertebrates. *Trends Genet.* **11**, 314–319.
500. Connerney, J., Andreeva, V., Leshem, Y., Mercado, M.A., Dowell, K., Yang, X., Lindner, V., Friesel, R.E., and Spicer, D.B. (2008). Twist1 homodimers enhance FGF responsiveness of the cranial sutures and promote suture closure. *Dev. Biol.* **318**, 323–334.

501. Conniff, R. (2008). That great beast of a town. *Nat. Hist.* **117** #2, 44–49.
502. Connolly, G., Naidoo, C., Conroy, R.M., Byrne, P., and McKenna, P. (2003). A new predictor of cephalopelvic disproportion? *J. Obstet. Gynecol.* **23**, 27–29.
503. Connor, C.E., Brincat, S.L., and Pasupathy, A. (2007). Transformation of shape information in the ventral pathway. *Curr. Opin. Neurobiol.* **17**, 140–147.
504. Constantine-Paton, M. and Law, M.I. (1978). Eye-specific termination bands in tecta of three-eyed frogs. *Science* **202**, 639–641.
505. Conway, A., Jarrold, C., Kane, M., Miyake, A., and Towse, J., eds. (2007). "Variation in Working Memory." Oxford Univ. Pr., New York.
506. Conway-Morris, S. (2003). The Cambrian "explosion" of metazoans and molecular biology: would Darwin be satisfied? *Int. J. Dev. Biol.* **47**, 505–515.
507. Cook, N.D. and Hayashi, T. (2008). The psychoacoustics of harmony perception. *Am. Sci.* **96**, 311–319.
508. Cook, T. (2003). Cell diversity in the retina: more than meets the eye. *BioEssays* **25**, 921–925.
509. Cook, T. and Desplan, C. (2001). Photoreceptor subtype specification: from flies to humans. *Sems. Cell Dev. Biol.* **12**, 509–518.
510. Cook, T.A. (1914). "The Curves of Life." Constable, London.
511. Cooke, J. (2004). Developmental mechanism and evolutionary origin of vertebrate left/right asymmetries. *Biol. Rev.* **79**, 377–407.
512. Cooke, J. (2004). The evolutionary origins and significance of vertebrate left–right organisation. *BioEssays* **26**, 413–421.
513. Cooper, K.L. and Tabin, C.J. (2008). Understanding of bat wing evolution takes flight. *Genes Dev.* **22**, 121–124.
514. Copp, A.J., Greene, N.D.E., and Murdoch, J.N. (2003). The genetic basis of mammalian neurulation. *Nature Rev. Genet.* **4**, 784–793.
515. Coppens, Y. (1991). The origin and evolution of man. *Diogenes* **39** #155, 111–134.
516. Coppens, Y. (1995). Brain, locomotion, diet, and culture: how a primate, by chance, became a man. *In* "Origins of the Human Brain" (J.-P. Changeux and J. Chavaillon, eds.). Clarendon Pr., Oxford, pp. 104–115.
517. Corballis, M.C. (1989). Laterality and human evolution. *Psychol. Rev.* **96**, 492–505.
518. Corballis, M.C. (2007). The uniqueness of human recursive thinking. *Am. Sci.* **95**, 240–248.
519. Corballis, M.C. and Morgan, M.J. (1978). On the biological basis of human laterality. I. Evidence for a maturational left-right gradient. *Behav. Brain Sci.* **2**, 261–269.
520. Corballis, M.C. and Morgan, M.J. (1978). On the biological basis of human laterality. II. The mechanisms of inheritance. *Behav. Brain Sci.* **2**, 270–336.
521. Cordes, R., Schuster-Gossler, K., Serth, K., and Gossier, A. (2004). Specification of vertebral identity is coupled to Notch signalling and the segmentation clock. *Development* **131**, 1221–1233.
522. Coren, S. and Porac, C. (1977). Fifty centuries of right-handedness: the historical record. *Science* **198**, 631–632.
523. Cornwell, R.E., Law Smith, M.J., Boothroyd, L.G., Moore, F.R., Davis, H.P., Stirrat, M., Tiddeman, B., and Perrett, D.I. (2006). Reproductive strategy, sexual development, and attraction to facial characteristics. *Phil. Trans. Roy. Soc. B* **361**, 2143–2154.
524. Corpa, J.M. (2006). Ectopic pregnancy in animals and humans. *Reproduction* **131**, 631–640.
525. Correia, H., Balseiro, S., and De Areia, M. (2005). Sexual dimorphism in the human pelvis: testing a new hypothesis. *HOMO (J. Comp. Hum. Biol.)* **56**, 153–160.
526. Coschigano, K.T. and Wensink, P.C. (1993). Sex-specific transcriptional regulation by the male and female doublesex proteins of *Drosophila. Genes Dev.* **7**, 42–54.
527. Costa, M.M.R., Fox, S., Hanna, A.I., Baxter, C., and Coen, E. (2005). Evolution of regulatory interactions controlling floral asymmetry. *Development* **132**, 5093–5101.
528. Coutelis, J.B., Petzoldt, A.G., Spéder, P., Suzanne, M., and Noselli, S. (2008). Left–right asymmetry in *Drosophila. Sems. Cell Dev. Biol.* **19**, 252–262.
529. Cowan, M.J., Gladwin, M.T., and Shelhamer, J.H. (2001). Disorders of ciliary motility. *Am. J. Med. Sci.* **321**, 3–10.
530. Coyne, J.A., Kay, E.H., and Pruett-Jones, S. (2007). The genetic basis of sexual dimorphism in birds. *Evolution* **62**, 214–219.
531. Coyne, J.A. and Orr, H.A. (2004). "Speciation." Sinauer, Sunderland, Mass.
532. Cranford, T.W., Amundin, M., and Norris, K.S. (1996). Functional morphology and homology in the odontocete nasal complex: implications for sound generation. *J. Morph.* **228**, 223–285.
533. Crelin, E.S. (1987). "The Human Vocal Tract." Vantage Pr., New York.
534. Crick, F. and Koch, C. (2003). A framework for consciousness. *Nature Neurosci.* **6**, 119–126.
535. Crick, F.H.C. (1968). The origin of the genetic code. *J. Mol. Biol.* **38**, 367–379.
536. Crickmore, M.A. and Mann, R.S. (2006). Hox control of organ size by regulation of morphogen production and mobility. *Science* **313**, 63–68.
537. Crickmore, M.A. and Mann, R.S. (2008). The control of size in animals: insights from selector genes. *BioEssays* **30**, 843–853.
538. Crofts, D.R. (1955). Muscle morphogenesis in primitive gastropods and its relation to torsion. *Proc. Zool. Soc. Lond.* **125**, 711–750.
539. Cronly-Dillon, J.R. (1991). Origin of invertebrate and vertebrate eyes. *In* "Evolution of the Eye and

Visual System" (J.R. Cronly-Dillon and R.L. Gregory, eds.). CRC Pr., Boston, Mass., pp. 15–51.

540. Cross, A., Collard, M., and Nelson, A. (2008). Body segment differences in surface area, skin temperature and 3D displacement and the estimation of heat balance during locomotion in hominins. *PLoS ONE* 3 #6, e2464.
541. Crotwell, P.L. and Mabee, P.M. (2007). Gene expression patterns underlying proximal-distal skeletal segmentation in late-stage zebrafish, *Danio rerio*. *Dev. Dynamics* **236**, 3111–3128.
542. Crow, J.F. (1991). Why is Mendelian segregation so exact? *BioEssays* **13**, 305–312.
543. Cummins, H. and Midlo, C. (1943). "Finger Prints, Palms and Soles. An Introduction to Dermatoglyphics." Dover, New York.
544. Cunningham, C.W., Blackstone, N.W., and Buss, L.W. (1992). Evolution of king crabs from hermit crab ancestors. *Nature* **355**, 539–542.
545. Curcio, C.A., Sloan, K.R., Jr., Packer, O., Hendrickson, A.E., and Kalina, R.E. (1987). Distribution of cones in human and monkey retina: individual variability and radial asymmetry. *Science* **236**, 579–582.
546. Curtiss, J. and Mlodzik, M. (2000). Morphogenetic furrow initiation and progression during eye development in *Drosophila*: the roles of *decapentaplegic, hedgehog*, and *eyes absent*. *Development* **127**, 1325–1336.
547. Cutter, A.D., Wasmuth, J.D., and Washington, N.L. (2008). Patterns of molecular evolution in Caenorhabditis preclude ancient origins of selfing. *Genetics* **178**, 2093–2104.
548. Cvekl, A. and Duncan, M.K. (2007). Genetic and epigenetic mechanisms of gene regulation during lens development. *Prog. Retinal Eye Res.* **26**, 555–597.
549. Cvekl, A. and Piatigorsky, J. (1996). Lens development and crystallin gene expression: many roles for Pax-6. *BioEssays* **18**, 621–630.
550. Cvekl, A. and Tamm, E.R. (2004). Anterior eye development and ocular mesenchyme: new insights from mouse models and human disease. *BioEssays* **26**, 374–386.
551. da Vinci, L. (1983). "Leonardo on the Human Body." Dover, New York.
552. Dacey, D.M., Liao, H.-W., Peterson, B.B., Robinson, F.R., Smith, V.C., Pokorny, J., Yau, K.-W., and Gamlin, P.D. (2005). Melanopsin-expressing ganglion cells in primate retina signal colour and irradiance and project to the LGN. *Nature* **433**, 749–754.
553. Daeschler, E.B., Shubin, N.H., and Jenkins, F.A., Jr. (2006). A Devonian tetrapod-like fish and the evolution of the tetrapod body plan. *Nature* **440**, 757–763.
554. Dagle, J.M., Sabel, J.L., Littig, J.L., Sutherland, L.B., Kolker, S.J., and Weeks, D.L. (2003). Pitx2c attenuation results in cardiac defects and abnormalities of intestinal orientation in developing *Xenopus laevis*. *Dev. Biol.* **262**, 268–281.
555. Dahn, R.D. and Fallon, J.F. (2000). Interdigital regulation of digit identity and homeotic transformation by modulated BMP signaling. *Science* **289**, 438–441.
556. Dalzell, B. (1974). Exotic bestiary for vicarious space voyagers. *Smithsonian* **5** #7, 84–91.
557. Damen, W.G.M. (2007). Evolutionary conservation and divergence of the segmentation process in arthropods. *Dev. Dynamics* **236**, 1379–1391.
558. Damen, W.G.M., Saridaki, T., and Averof, M. (2002). Diverse adaptations of an ancestral gill: a common evolutionary origin for wings, breathing organs, and spinnerets. *Curr. Biol.* **12**, 1711–1716.
559. Darwin, C. (1859). "On the Origin of Species by Means of Natural Selection, or the Preservation of Favoured Races in the Struggle for Life." John Murray, London.
560. Darwin, C. (1868). "The Variation of Animals and Plants Under Domestication." John Murray, London.
561. Darwin, C. (1871). "The Descent of Man, and Selection in Relation to Sex." John Murray, London.
562. Darwin, C. (1872). "The Expression of the Emotions in Man and Animals." John Murray, London.
563. Darwin, C. (1877). "The Various Contrivances by which Orchids are Fertilised by Insects," 2nd (revised) ed. John Murray, London.
564. Darwin, C. (1882). "The Descent of Man, and Selection in Relation to Sex," 2nd ed. John Murray, London.
565. Darwin, F., ed. (1925). "The Life and Letters of Charles Darwin, Including an Autobiographical Chapter." Vol. **2**. Appleton, New York.
566. Dassule, H.R., Lewis, P., Bei, M., Maas, R., and McMahon, A.P. (2000). Sonic hedgehog regulates growth and morphogenesis of the tooth. *Development* **127**, 4775–4785.
567. Dassule, H.R. and McMahon, A.P. (1998). Analysis of epithelial-mesenchymal interactions in the initial morphogenesis of the mammalian tooth. *Dev. Biol.* **202**, 215–227.
568. Davenport, J.R. and Yoder, B.K. (2005). An incredible decade for the primary cilium: a look at a once-forgotten organelle. *Am. J. Physiol. Renal Physiol.* **289**, F1159–F1169.
569. Davidson, B.P., Kinder, S.J., Steiner, K., Schoenwolf, G.C., and Tam, P.P.L. (1999). Impact of node ablation on the morphogenesis of the body axis and the lateral asymmetry of the mouse embryo during early organogenesis. *Dev. Biol.* **211**, 11–26.
570. Davidson, E.H. (1990). How embryos work: a comparative view of diverse modes of cell fate specification. *Development* **108**, 365–389.
571. Davidson, E.H. (2006). "The Regulatory Genome: Gene Regulatory Networks in Development and Evolution." Acad. Pr., New York.

572. Davidson, E.H. and Erwin, D.H. (2006). Gene regulatory networks and the evolution of animal body plans. *Science* **311**, 796–800.
573. Davidson, K. (1999). "Carl Sagan: A Life." Wiley, New York.
574. Davidson, R.J. and Hugdahl, K., eds. (1995). "Brain Asymmetry." M.I.T. Pr., Cambridge, Mass.
575. Davies, J.A., ed. (2005). "Branching Morphogenesis." Springer, Berlin.
576. Davies, J.A. (2005). "Mechanisms of Morphogenesis." Elsevier Acad. Pr., Burlington, Mass.
577. Davies, J.A. (2005). Watching tubules glow and branch. *Curr. Opin. Gen. Dev.* **15**, 364–370.
578. Davis, D.D. (1964). "The Giant Panda: A Morphological Study of Evolutionary Mechanisms." *Fieldiana: Zoology Memoirs, Vol. 3,* Chicago Nat. Hist. Mus., Chicago, Ill.
579. Davis, D.D. (1966). Non-functional anatomy. *Folia Biotheor.* **6**, 5–8.
580. Davis, E.E., Brueckner, M., and Katsanis, N. (2006). The emerging complexity of the vertebrate cilium: new functional roles for an ancient organelle. *Dev. Cell* **11**, 9–19.
581. Davis, M.C., Dahn, R.D., and Shubin, N.H. (2007). An autopodial-like pattern of Hox expression in the fins of a basal actinopterygian fish. *Nature* **447**, 473–476.
582. Davis, N.M., Kurpios, N.A., Sun, X., Gros, J., Martin, J.F., and Tabin, C.J. (2008). The chirality of gut rotation derives from left-right asymmetric changes in the architecture of the dorsal mesentery. *Dev. Cell* **15**, 134–145.
583. Davis, R.J., Harding, M., Moayedi, Y., and Mardon, G. (2008). Mouse *Dach1* and *Dach2* are redundantly required for Müllerian duct development. *Genesis* **46**, 205–213.
584. Davis-Silberman, N. and Ashery-Padan, R. (2008). Iris development in vertebrates: genetic and molecular considerations. *Brain Res.* **1192**, 17–28.
585. Davison, A. (2006). The ovotestis: an underdeveloped organ of evolution. *BioEssays* **28**, 642–650.
586. Davison, A., Chiba, S., Barton, N.H., and Clarke, B. (2005). Speciation and gene flow between snails of opposite chirality. *PLoS Biol.* **3** #9, 1559–1571 (e282).
587. Dawkins, R. (1982). "The Extended Phenotype." Oxford Univ. Pr., New York.
588. Dawkins, R. (1986). "The Blind Watchmaker: Why the Evidence of Evolution Reveals a Universe Without Design." Norton, New York.
589. Dawkins, R. (1989). The evolution of evolvability. *In* "Artificial Life," *Santa Fe Institute Studies in the Sciences of Complexity,* vol. 6, (C.G. Langton, ed.). Addison-Wesley, New York, pp. 201–220.
590. Dawkins, R. (1989). "The Selfish Gene," 2nd ed. Oxford Univ. Pr., New York.
591. Dawkins, R. (1994). The eye in a twinkling. *Nature* **368**, 690–691.
592. Dawkins, R. (1995). The evolved imagination: Animals as models of their world. *Nat. Hist.* **104** #9, 8–24.
593. Dawkins, R. (1996). "Climbing Mount Improbable." Norton, New York.
594. Dawkins, R. (1998). "Unweaving the Rainbow: Science, Delusion and the Appetite for Wonder." Houghton Mifflin, New York.
595. Dawkins, R. (2004). "The Ancestor's Tale: A Pilgrimage to the Dawn of Evolution." Houghton Mifflin, New York.
596. Dawkins, R. (2005). The illusion of design. *Nat. Hist.* **114** #9, 35–37.
597. de Arsuaga, J.L. (1986). Longueur du col du fémur et largeur du canal obstétrical dans l'évolution des hominidés. *L'Anthrop. (Paris)* **90**, 567–577.
598. de Beer, G. (1958). "Embryos and Ancestors," 3rd ed. Clarendon Pr., Oxford.
599. de Beer, G.R. (1924). The evolution of the pituitary. *J. Exp. Biol.* **1**, 271–291.
600. De Jong, M.C.M. and Sabelis, M.W. (1991). Limits to runaway sexual selection: the wallflower paradox. *J. Evol. Biol.* **4**, 637–655.
601. de Joussineau, C., Soulé, J., Martin, M., Anguille, C., Montcourrier, P., and Alexandre, D. (2003). Delta-promoted filopodia mediate long-range lateral inhibition in *Drosophila. Nature* **426**, 555–559.
602. De Leyn, P., Bedert, L., Delcroix, M., Depuydt, P., Lauwers, G., Sokolov, Y., Van Meerhaeghe, A., and Van Schil, P. (2007). Tracheotomy: clinical review and guidelines. *Eur. J. Cardiothorac. Surg.* **32**, 412–421.
603. de Perera, T.B. and Braithwaite, V.A. (2005). Laterality in a non-visual sensory modality—the lateral line of fish. *Curr. Biol.* **15**, R241.
604. De Robertis, E.M. (2008). Evo-devo: variations on ancestral themes. *Cell* **132**, 185–195.
605. De Robertis, E.M., Fainsod, A., Gont, L.K., and Steinbeisser, H. (1994). The evolution of vertebrate gastrulation. *Development* **1994 Suppl.**, 117–124.
606. De Robertis, E.M. and Sasai, Y. (1996). A common plan for dorsoventral patterning in Bilateria. *Nature* **380**, 37–40.
607. de Waal, F. (1997). Bonobo dialogs. *Nat. Hist.* **106** #4, 22–25.
608. de Waal, F. (2005). "Our Inner Ape." Riverhead Bks., N.Y.
609. de Waal, F.B.M. and Parish, A.R. (2000). The other "closest living relative": How bonobos (*Pan paniscus*) challenge traditional assumptions about females, dominance, intra- and intersexual interactions, and hominid evolution. *Ann. N. Y. Acad. Sci.* **907**, 97–113.
610. Deacon, T.W. (1990). Rethinking mammalian brain evolution. *Amer. Zool.* **30**, 629–705.
611. Deacon, T.W. (1992). The human brain. *In* "The Cambridge Encyclopedia of Human Evolution" (S. Bunney, S. Jones, R. Martin, and D. Pilbeam, eds.). Cambridge Univ. Pr., New York, pp. 115–123.

612. Deacon, T.W. (1997). "The Symbolic Species: The Co-Evolution of Language and the Brain." Norton, New York.
613. Dean, A.M. and Thornton, J.W. (2007). Mechanistic approaches to the study of evolution: the functional synthesis. *Nature Rev. Genet.* **8**, 675–688.
614. Dean, C. (1992). Jaws and teeth. *In* "The Cambridge Encyclopedia of Human Evolution" (S. Bunney, S. Jones, R. Martin, and D. Pilbeam, eds.). Cambridge Univ. Pr., New York, pp. 56–59.
615. Deban, S.M. and Olson, W.M. (2002). Suction feeding by a tiny predatory tadpole. *Nature* **420**, 41–42.
616. Decembrini, S., Andreazzoli, M., Vignali, R., Barsacchi, G., and Cremisi, F. (2006). Timing the generation of distinct retinal cells by homeobox proteins. *PLoS Biol.* **4** #9, 1562–1571 (e272).
617. Deeming, D.C. and Ferguson, M.W.J. (1989). The mechanism of temperature dependent sex determination in crocodilians: a hypothesis. *Amer. Zool.* **29**, 973–985.
618. DeFalco, T.J., Verney, G., Jenkins, A.B., McCaffery, J.M., Russell, S., and Van Doren, M. (2003). Sex-specific apoptosis regulates sexual dimorphism in the *Drosophila* embryonic gonad. *Dev. Cell* **5**, 205–216.
619. Degabriele, R. (1980). The physiology of the koala. *Sci. Am.* **243** #1, 110–117.
620. Tyson, N.D. and Goldsmith, D. (2004). "Origins: Fourteen Billion Years of Cosmic Evolution." Norton, New York.
621. Deininger, W., Fuhrmann, M., and Hegemann, P. (2000). Opsin evolution: out of wild green yonder? *Trends Genet.* **16**, 158–159.
622. del Corral, R.D., Olivera-Martinez, I., Goriely, A., Gale, E., Maden, M., and Storey, K. (2003). Opposing FGF and retinoid pathways control ventral neural pattern, neuronal differentiation, and segmentation during body axis extension. *Neuron* **40**, 65–79.
623. DeLaurier, A., Schweitzer, R., and Logan, M. (2006). *Pitx1* determines the morphology of muscle, tendon, and bones of the hindlimb. *Dev. Biol.* **299**, 22–34.
624. Delbrück, M. (1986). "Mind From Matter? An Essay on Evolutionary Epistemology." Palo Alto, Calif., Blackwell.
625. Delezoide, A.-L., Narcy, F., and Larroche, J.-C. (1990). Cerebral midline developmental anomalies: spectrum and associated features. *Genet. Couns.* **1**, 197–210.
626. Delph, L.F. (2001). Mutated into oblivion. *Science* **292**, 2437.
627. Delsuc, F., Brinkmann, H., Chourrout, D., and Philippe, H. (2006). Tunicates and not cephalochordates are the closest relatives of vertebrates. *Nature* **439**, 965–968.
628. deMenocal, P.B. (2004). African climate change and faunal evolution during the Pliocene-Pleistocene. *Earth Planetary Sci. Ltrs.* **220**, 3–24.
629. Denes, A.S., Jékely, G., Steinmetz, P.R.H., Raible, F., Snyman, H., Prud'homme, B., Ferrier, D.E.K., Balavoine, G., and Arendt, D. (2007). Molecular architecture of annelid nerve cord supports common origin of nervous system centralization in Bilateria. *Cell* **129**, 277–288.
630. Denis, H. (1994). A parallel between development and evolution: germ cell recruitment by the gonads. *BioEssays* **16**, 933–938.
631. Dennis, C. (2004). A mole in hand.... *Nature* **432**, 142–143.
632. Denver, R.J. (2008). Chordate metamorphosis: ancient control by iodothyronines. *Curr. Biol.* **18**, R567–569.
633. Depew, M.J., Lufkin, T., and Rubenstein, J.L.R. (2002). Specification of jaw subdivisions by Dlx genes. *Science* **298**, 381–385.
634. Depew, M.J. and Olsson, L. (2008). Symposium on the evolution and development of the vertebrate head. *J. Exp. Zool. (Mol. Dev. Evol.)* **310B**, 287–293.
635. Depew, M.J. and Simpson, C.A. (2006). 21st century neontology and the comparative development of the vertebrate skull. *Dev. Dynamics* **235**, 1256–1291.
636. Depew, M.J., Simpson, C.A., Morasso, M., and Rubenstein, J.L.R. (2005). Reassessing the *Dlx* code: the genetic regulation of branchial arch skeletal pattern and development. *J. Anat.* **207**, 501–561.
637. Deppmann, C.D., Mihalas, S., Sharma, N., Lonze, B.E., Niebur, E., and Ginty, D.D. (2008). A model for neuronal competition during development. *Science* **320**, 369–373.
638. DeSalle, R. and Tattersall, I. (2008). "Human Origins: What Bones and Genomes Tell Us About Ourselves." Texas A&M Univ. Pr., College Station, Texas.
639. Deschamps, J. (2007). Ancestral and recently recruited global control of the *Hox* genes in development. *Curr. Opin. Gen. Dev.* **17**, 422–427.
640. Deschamps, J. (2008). Ancestral and recently recruited global control of the *Hox* genes in development. *Curr. Opin. Gen. Dev.* **17**, 422–427.
641. Deschamps, J. (2008). Tailored *Hox* gene transcription and the making of the thumb. *Genes Dev.* **22**, 293–296.
642. Deschamps, J. and van Nes, J. (2005). Developmental regulation of the Hox genes during axial morphogenesis in the mouse. *Development* **132**, 2931–2942.
643. DeSilva, J. and Lesnik, J. (2006). Chimpanzee neonatal brain size: implications for brain growth in *Homo erectus*. *J. Human Evol.* **51**, 207–212.
644. Dessaud, E., McMahon, A.P., and Briscoe, J. (2008). Pattern formation in the vertebrate neural tube: a sonic hedgehog morphogen-regulated transcriptional network. *Development* **135**, 2489–2503.
645. Dessaud, E., Yang, L.L., Hill, K., Cox, B., Ulloa, F., Ribeiro, A., Mynett, A., Novitch, B.G., and Briscoe, J. (2007). Interpretation of the sonic hedgehog

morphogen gradient by a temporal adaptation mechanism. *Nature* **450**, 717–720.

646. Deutsch, J.S. and Lopez, P. (2008). Are transposition events at the origin of the bilaterian Hox complexes? *In* "Evolving Pathways: Key Themes in Evolutionary Developmental Biology" (A. Minelli and G. Fusco, eds.). Cambridge Univ. Pr., New York, pp. 239–260.
647. Deutsch, S. and Deutsch, A. (1993). "Understanding the Nervous System: An Engineering Perspective." IEEE Pr., New York.
648. Devlin, J.T. (2006). Are we dancing apes? *Science* **314**, 926–927.
649. Dewing, P., Chiang, C.W.K., Sinchak, K., Sim, H., Fernagut, P.-O., Kelly, S., Chesselet, M.F., Micevych, P.E., Albrecht, K.H., Harley, V.R., and Vilain, E. (2006). Distinct regulation of adult brain function by the male-specific factor SRY. *Curr. Biol.* **16**, 415–420.
650. Diamond, J. (1992). "The Third Chimpanzee: The Evolution and Future of the Human Animal." HarperCollins, New York.
651. Diamond, J. (1997). "Why Is Sex Fun? The Evolution of Human Sexuality." Basic Books, New York.
652. Diamond, J.M. (1987). Aristotle's theory of mammalian teat number is confirmed. *Nature* **325**, 200.
653. Dichtel-Danjoy, M.-L. and Félix, M.-A. (2004). Phenotypic neighborhood and micro-evolvability. *Trends Genet.* **20**, 268–276.
654. Dicke, U. and Roth, G. (2008). Intelligence evolved. *Sci. Am. Mind* **19** #4, 70–77.
655. Dickinson, M.H., Farley, C.T., Full, R.J., Koehl, M.A.R., Kram, R., and Lehman, S. (2000). How animals move: an integrative view. *Science* **288**, 100–106.
656. Dickinson, W.J. (1988). On the architecture of regulatory systems: evolutionary insights and implications. *BioEssays* **8**, 204–208.
657. Dickman, S. (1997). HOX gene links limb, genital defects. *Science* **275**, 1568.
658. Dickson, B.J. and Gilestro, G.F. (2006). Regulation of commissural axon pathfinding by Slit and its Robo receptors. *Annu. Rev. Cell Dev. Biol.* **22**, 651–675.
659. Diekamp, B., Regolin, L., Güntürkün, O., and Vallortigara, G. (2005). A left-sided visuospatial bias in birds. *Curr. Biol.* **15**, R372–R373.
660. Dierick, H.A. (2008). Fly fighting: Octopamine modulates aggression. *Curr. Biol.* **18**, R161–R163.
661. Dietrich, M.R. (2000). From hopeful monsters to homeotic effects: Richard Goldschmidt's integration of development, evolution, and genetics. *Amer. Zool.* **40**, 738–747.
662. Dietrich, M.R. (2003). Richard Goldschmidt: hopeful monsters and other "heresies." *Nature Rev. Genet.* **4**, 68–74.
663. Dinstein, I., Thomas, C., Behrmann, M., and Heeger, D.J. (2008). A mirror up to nature. *Curr. Biol.* **18**, R13–R18.
664. Dobzhansky, T. (1974). Chance and creativity in evolution. *In* "Studies in the Philosophy of Biology: Reduction and Related Problems" (F.J. Ayala and T. Dobzhansky, eds.). Univ. Calif. Pr., Berkeley, Calif., pp. 307–338.
665. Dodt, E. and Meissl, H. (1982). The pineal and parietal organs of lower vertebrates. *Experientia* **38**, 996–1000.
666. Doi, O., Hutson, J.M., Myers, N.A., and McKelvie, P.A. (1988). Branchial remnants: a review of 58 cases. *J. Pediatr. Surg.* **23**, 789–792.
667. Dolan, R.J. (2002). Emotion, cognition, and behavior. *Science* **298**, 1191–1194.
668. Dong, J., Feldmann, G., Huang, J., Wu, S., Zhang, N., Comerford, S.A., Gayyed, M.F., Anders, R.A., Maitra, A., and Pan, D. (2007). Elucidation of a universal size-control mechanism in *Drosophila* and mammals. *Cell* **130**, 1120–1133.
669. Donjacour, A.A. and Cunha, G.R. (1988). The effect of androgen deprivation on branching morphogenesis in the mouse prostate. *Dev. Biol.* **128**, 1–14.
670. Donner, A.L., Lachke, S.A., and Maas, R.L. (2006). Lens induction in vertebrates: variations on a conserved theme of signaling events. *Sems. Cell Dev. Biol.* **17**, 676–685.
671. Donoghue, P.C.J., Graham, A., and Kelsh, R.N. (2008). The origin and evolution of the neural crest. *BioEssays* **30**, 530–541.
672. Dorit, R. (2007). The undiscovered country. *Am. Sci.* **95**, 398–401.
673. Dornan, A.J. and Goodwin, S.F. (2008). Fly courtship song: Triggering the light fantastic. *Cell* **133**, 210–212.
674. Doss, R.P. (1978). Handedness in duckweed: double flowering fronds produce right- and left-handed lineages. *Science* **199**, 1465–1466.
675. Douglas, R.H., Partridge, J.C., Dulai, K., Hunt, D., Mullineaux, C.W., Tauber, A.Y., and Hynninen, P.H. (1998). Dragon fish see using chlorophyll. *Nature* **393**, 423–424.
676. Dover, G. (2000). How genomic and developmental dynamics affect evolutionary processes. *BioEssays* **22**, 1153–1159.
677. Dreger, A.D. (2004). "One of Us: Conjoined Twins and the Future of Normal." Harvard Univ. Pr., Cambridge, Mass.
678. Drimmer, F. (1973). "Very Special People." Amjon, New York.
679. Drögemüller, C., Karlsson, E.K., Hytönen, M.K., Perloski, M., Dolf, G., Sainio, K., Lohi, H., Lindblad-Toh, K., and Leeb, T. (2008). A mutation in hairless dogs implicates *FOXI3* in ectodermal development. *Science* **321**, 1462.
680. Duboc, V. and Lepage, T. (2008). A conserved role for the nodal signaling pathway in the establishment of dorso-ventral and left–right axes in deuterostomes. *J. Exp. Zool. (Mol. Dev. Evol.)* **310B**, 41–53.

681. Duboc, V., Röttinger, E., Lapraz, F., Besnardeau, L., and Lepage, T. (2005). Left–right asymmetry in the sea urchin embryo is regulated by Nodal signaling on the right side. *Dev. Cell* **9**, 147–158.
682. Duboule, D. (2007). The rise and fall of Hox gene clusters. *Development* **134**, 2549–2560.
683. Duboule, D. and Wilkins, A.S. (1998). The evolution of "bricolage." *Trends Genet.* **14**, 54–59.
684. Dubrulle, J. and Pourquié, O. (2004). Coupling segmentation to axis formation. *Development* **131**, 5783–5793.
685. Dudek, R.W. and Fix, J.D. (2005). "Embryology (Board Review Series)," 3rd ed. Lippincott, Williams, & Wilkins, Baltimore, Md.
686. Dulac, C. and Kimchi, T. (2007). Neural mechanisms underlying sex-specific behaviors in vertebrates. *Curr. Opin. Neurobiol.* **17**, 675–683.
687. Dunbar, M.E., Dann, P.R., Robinson, G.W., Hennighausen, L., Zhang, J.-P., and Wysolmerski, J.J. (1999). Parathyroid hormone-related protein signaling is necessary for sexual dimorphism during embryonic mammary development. *Development* **126**, 3485–3493.
688. Dunbar, R. (1996). "Grooming, Gossip, and the Evolution of Language." Harvard Univ. Pr., Cambridge, Mass.
689. Dunbar, R. (2004). From spears to speech: Could throwing spears have laid the foundations for language acquisition? *Nature* **427**, 783.
690. Dupé, V. and Lumsden, A. (2001). Hindbrain patterning involves graded responses to retinoic acid signalling. *Development* **128**, 2199–2208.
691. Durand, J.-B., Nelissen, K., Joly, O., Wardak, C., Todd, J.T., Norman, J.F., Janssen, P., Vanduffel, W., and Orban, G.A. (2007). Anterior regions of monkey parietal cortex process visual 3D shape. *Neuron* **55**, 493–505.
692. Dürr, V., Schmitz, J., and Cruse, H. (2004). Behaviour-based modelling of hexapod locomotion: linking biology and technical application. *Arthropod Struct. Dev.* **33**, 237–250.
693. Dworkin, I. (2005). Canalization, cryptic variation, and developmental buffering: a critical examination and analytical perspective. *In* "Variation: A Central Concept in Biology" (B. Hallgrímsson and B.K. Hall, eds.). Elsevier Acad. Pr., New York, pp. 131–158.
694. Eakin, R.M. (1973). "The Third Eye." Univ. Calif. Pr., Berkeley, Calif.
695. Eakin, R.M. and Brandenberger, J.L. (1980). Unique eye of probable evolutionary significance. *Science* **211**, 1189–1190.
696. Eberhard, W.G. (1985). "Sexual Selection and Animal Genitalia." Harvard Univ. Pr., Cambridge, Mass.
697. Eberhard, W.G. (1987). Runaway sexual selection. *Nat. Hist.* **96** #12, 4–8.
698. Eberhard, W.G. (1990). Animal genitalia and female choice. *Am. Sci.* **78**, 134–141.
699. Eberhard, W.G. (2001). Species-specific genitalic copulatory courtship in sepsid flies (Diptera, Sepsidae, *Microsepsis*) and theories of genitalic evolution. *Evolution* **55**, 93–102.
700. Eberhart, J.K., Swartz, M.E., Crump, J.G., and Kimmel, C.B. (2006). Early Hedgehog signaling from neural to oral epithelium organizes anterior craniofacial development. *Development* **133**, 1069–1077.
701. Eckstein, G. (1948). "Everyday Miracle." Harper, New York.
702. Economides, K.D., Zeltser, L., and Capecchi, M.R. (2003). *Hoxb13* mutations cause overgrowth of caudal spinal cord and tail vertebrae. *Dev. Biol.* **256**, 317–330.
703. Edelsteinkeshet, L. and Ermentrout, G.B. (1990). Contact response of cells can mediate morphogenetic pattern-formation. *Differentiation* **45**, 147–159.
704. Edwards, A.W.F. (2008). Commemorating Galton's intellectual legacy. *BioEssays* **30**, 919.
705. Eglen, S.J. and Willshaw, D.J. (2002). Influence of cell fate mechanisms upon retinal mosaic formation: a modelling study. *Development* **129**, 5399–5408.
706. Ehrman, L., Thompson, J.N., Jr., Perelle, I., and Hisey, B.N. (1978). Some approaches to the question of *Drosophila* laterality. *Genet. Res. Camb.* **32**, 231–238.
707. Eibl-Eibesfeldt, I. (1989). "Human Ethology." Aldine de Gruyter, New York.
708. Eibner, C., Pittlik, S., Meyer, A., and Begemann, G. (2008). An organizer controls the development of the "sword," a sexually selected trait in swordtail fish. *Evol. Dev.* **10**, 403–412.
709. Eichmann, A., Makinen, T., and Alitalo, K. (2005). Neural guidance molecules regulate vascular remodeling and vessel navigation. *Genes Dev.* **19**, 1013–1021.
710. Eisenberg, J.F. (1981). "The Mammalian Radiations: An Analysis of Trends in Evolution, Adaptation, and Behavior." Univ. Chicago Pr., Chicago, Ill.
711. Ekström, P. and Meissl, H. (2003). Evolution of photosensory pineal organs in new light: the fate of neuroendocrine photoreceptors. *Phil. Trans. Roy. Soc. Lond. B* **358**, 1679–1700.
712. El-Hawrani, A., Sohn, M., Noga, M., and El-Hakim, H. (2006). The face does predict the brain–midline facial and forebrain defects uncovered during the investigation of nasal obstruction and rhinorrhea. Case report and a review of holoprosencephaly and its classifications. *Int. J. Pediatr. Otorhinolaryngology* **70**, 935–940.
713. Ellegren, H. and Parsch, J. (2007). The evolution of sex-biased genes and sex-biased gene expression. *Nature Rev. Genet.* **8**, 689–698.
714. Ellegren, H. and Sheldon, B.C. (2008). Genetic basis of fitness differences in natural populations. *Nature* **452**, 169–175.

715. Emerson, S.B. (1985). Jumping and leaping. *In* "Functional Vertebrate Morphology" (M. Hildebrand, D.M. Bramble, K.F. Liem, and D.B. Wake, eds.). Harvard Univ. Pr., Cambridge, Mass., pp. 58–72.
716. Emlen, D.J. and Nijhout, H.F. (2000). The development and evolution of exaggerated morphologies in insects. *Annu. Rev. Entomol.* **45**, 661–708.
717. Enard, W., Przeworski, M., Fisher, S.E., Lai, C.S.L., Wiebe, V., Kitano, T., Monaco, A.P., and Pääbo, S. (2002). Molecular evolution of *FOXP2*, a gene involved in speech and language. *Nature* **418**, 869–872.
718. Endo, H., Yamagiwa, D., Hayashi, Y., Koie, H., Yamaya, Y., and Kimura, J. (1999). Role of the giant panda's "pseudo-thumb." *Nature* **397**, 309–310.
719. England, M.A. (1996). "Life before Birth," 2nd ed. Mosby-Wolfe, New York.
720. England, S.J., Blanchard, G.B., Mahadevan, L., and Adams, R.J. (2006). A dynamic fate map of the forebrain shows how vertebrate eyes form and explains two causes of cyclopia. *Development* **133**, 4613–4617.
721. Epper, F. (1981). Morphological analysis and fate map of the intersexual genital disc of the mutant doublesex-dominant in *Drosophila melanogaster*. *Dev. Biol.* **88**, 104–114.
722. Epper, F. and Nöthiger, R. (1982). Genetic and developmental evidence for a repressed genital primordium in Drosophila melanogaster. *Dev. Biol.* **94**, 163–175.
723. Epstein, C.J., Erickson, R.P., and Wynshaw-Boris, A., eds. (2004). "Inborn Errors of Development: The Molecular Basis of Clinical Disorders of Morphogenesis." *Oxford Monographs on Medical Genetics, No. 49*, Oxford Univ. Pr., New York.
724. Erclik, T., Hartenstein, V., Lipshitz, H.D., and McInnes, R.R. (2008). Conserved role of the Vsx genes supports a monophyletic origin for bilaterian visual systems. *Curr. Biol.* **18**, 1278–1287. [See also (2008) *Development* **135**, 805–811.]
725. Erdman, S.E. and Burtis, K.C. (1993). The *Drosophila* doublesex proteins share a novel zinc finger DNA binding domain. *EMBO J.* **12**, 527–535.
726. Erickson, J.W. (2001). Sex and the neighboring cell. *Dev. Cell* **1**, 156–158.
727. Erickson, J.W. and Quintero, J.J. (2007). Indirect effects of ploidy suggest X chromosome dose, not the X:A ratio, signals sex in *Drosophila*. *PLoS Biol.* **5** #2, 2821–2830 (e332).
728. Erskine, L. and Herrera, E. (2007). The retinal ganglion cell axon's journey: insights into molecular mechanims of axon guidance. *Dev. Biol.* **308**, 1–14.
729. Erwin, D.H. (2006). Evolutionary contingency. *Curr. Biol.* **16**, R825–R826.
730. Erwin, D.H. (2006). "Extinction: How Life on Earth Nearly Ended 250 Million Years Ago." Princeton Univ. Pr., Princeton, N.J.
731. Erwin, D.H. and Davidson, E.H. (2002). The last common bilaterian ancestor. *Development* **129**, 3021–3032.
732. Essner, J.J., Branford, W.W., Zhang, J., and Yost, H.J. (2000). Mesendoderm and left-right brain, heart and gut development are differentially regulated by *pitx2* isoforms. *Development* **127**, 1081–1093.
733. Estrada, B., Casares, F., and Sánchez-Herrero, E. (2003). Development of the genitalia in *Drosophila melanogaster*. *Differentiation* **71**, 299–310.
734. Estrada, B. and Sánchez-Herrero, E. (2001). The Hox gene *Abdominal-B* antagonizes appendage development in the genital disc of *Drosophila*. *Development* **128**, 331–339.
735. Euston, D.R., Tatsuno, M., and NcNaughton, B.L. (2007). Fast-forward playback of recent memory sequences in prefrontal cortex during sleep. *Science* **318**, 1147–1150.
736. Extavour, C.G.M. (2007). Evolution of the bilaterian germ line: lineage origin and modulation of specification mechanisms. *Integr. Comp. Biol.* **47**, 770–785.
737. Extavour, C.G.M. (2008). Urbisexuality: the evolution of bilaterian germ cell specification and reproductive systems. *In* "Evolving Pathways: Key Themes in Evolutionary Developmental Biology" (A. Minelli and G. Fusco, eds.). Cambridge Univ. Pr., New York, pp. 321–342.
738. Eyre-Walker, A. and Keightley, P.D. (2007). The distribution of fitness effects of new mutations. *Nature Rev. Genet.* **8**, 610–618.
739. Ezaz, T., Stiglec, R., Veyrunes, F., and Graves, J.A.M. (2006). Relationships between vertebrate ZW and XY sex chromosome systems. *Curr. Biol.* **16**, R736–R743.
740. Ezzell, C. (2003). Why? The neuroscience of suicide. *Sci. Am.* **288** #2, 44–51.
741. Fagen, R. (1981). "Animal Play Behavior." Oxford Univ. Pr., New York.
742. Fairbairn, D.J., Blanckenhorn, W.U., and Székely, T., eds. (2007). "Sex, Size, and Gender Roles: Evolutionary Studies of Sexual Size Dimorphism." Oxford Univ. Pr., New York.
743. Falk, D. (1992). "Braindance." Henry Holt, New York.
744. Falk, D. (2001). The evolution of sex differences in primate brains. *In* "Evolutionary Anatomy of the Primate Cerebral Cortex" (D. Falk and K.R. Gibson, eds.). Cambridge Univ. Pr.: New York, pp. 98–112.
745. Falk, D. and Gibson, K.R., eds. (2001). "Evolutionary Anatomy of the Primate Cerebral Cortex." Cambridge Univ. Pr., New York.
746. Farlow, J.O. and Brett-Surman, M.K., eds. (1997). "The Complete Dinosaur." Indiana Univ. Pr., Bloomington, Ind.
747. Farnum, C.E. (2007). Postnatal growth of fins and limbs through endochondral ossification. *In* "Fins

into Limbs: Evolution, Development, and Transformation," (B.K. Hall, ed.). Univ. Chicago Pr., Chicago, Ill., pp. 118–151.

748. Farrer, L.A., Arnos, K.S., Asher, J.H., Jr., Baldwin, C.T., Diehl, S.R., Friedman, T.B., Greenberg, J., Grundfast, K.M., Hoth, C., Lalwani, A.K., Landa, B., Leverton, K., Milunsky, A., Morell, R., Nance, W.E., Newton, V., Ramesar, R., Rao, V.S., Reynolds, J.E., San Agustin, T.B., Wilcox, E.R., Winship, I., and Read, A.P. (1994). Locus heterogeneity for Waardenburg syndrome is predictive of clinical subtypes. *Am. J. Hum. Genet.* **55**, 728–737.
749. Farzan, S.F., Ascano, M., Jr., Ogden, S.K., Sanial, M., Brigui, A., Plessis, A., and Robbins, D.J. (2008). Costal2 functions as a kinesin-like protein in the Hedgehog signal transduction pathway. *Curr. Biol.* **18**, 1215–1220.
750. Faurie, C. and Raymond, M. (2004). Handedness frequency over more than ten thousand years. *Proc. Roy. Soc. Lond. B (Suppl.)* **271**, S43–S45.
751. Favor, J., Gloeckner, C.J., Neuhäuser-Klaus, A., Pretsch, W., Sandulache, R., Saule, S., and Zaus, I. (2008). Relationship of *Pax6* activity levels to the extent of eye development in the mouse, *Mus musculus*. *Genetics* **179**, 1345–1355.
752. Fawcett, J.W. and Willshaw, D.J. (1982). Compound eyes project stripes on the optic tectum in *Xenopus*. *Nature* **296**, 350–352.
753. Featherstone, D.E. and Broadie, K. (2002). Wrestling with pleiotropy: genomic and topological analysis of the yeast gene expression network. *BioEssays* **24**, 267–274.
754. Fedak, T., Franz-Odendaal, T., Hall, B., and Vickaryous, M. (2004). Epigenetics: the context of development. *Paleont. Assoc. Newsletter* **56**, 44–49.
755. Fehilly, C.B., Willadsen, S.M., and Tucker, E.M. (1984). Interspecific chimaerism between sheep and goat. *Nature* **307**, 634–636.
756. Feistel, K. and Blum, M. (2006). Three types of cilia including a novel 9+4 axoneme on the notochordal plate of the rabbit embryo. *Dev. Dynamics* **235**, 3348–3358.
757. Fekete, D.M. (2004). Development of the ear. *In* "Inborn Errors of Development: The Molecular Basis of Clinical Disorders of Morphogenesis," *Oxford Monographs on Medical Genetics, No. 49* (C.J. Epstein, R.P. Erickson, and A. Wynshaw-Boris, eds.). Oxford Univ. Pr.: New York, pp. 89–106.
758. Feldhamer, G.A., Drickamer, L.C., Vessey, S.H., Merritt, J.F., and Krajewski, C. (2007). "Mammalogy: Adaptation, Diversity, Ecology," 3rd ed. Johns Hopkins Univ. Pr., Baltimore, Maryland.
759. Félix, M.-A. and Barrière, A. (2005). Evolvability of cell specification mechanisms. *J. Exp. Zool. (Mol. Dev. Evol.)* **304B**, 536–547.
760. Fenno, L.E., Ptaszek, L.M., and Cowan, C.A. (2008). Human embryonic stem cells: emerging technologies and practical applications. *Curr. Opin. Gen. Dev.* **18**, 324–329.
761. Ferguson, C.A., Tucker, A.S., and Sharpe, P.T. (2000). Temporospatial cell interactions regulating mandibular and maxillary arch patterning. *Development* **127**, 403–412.
762. Ferguson-Smith, M. (2007). The evolution of sex chromosomes and sex determination in vertebrates and the key role of *DMRT1*. *Sex. Dev.* **1**, 2–11.
763. Fernald, R.D. (2006). Casting a genetic light on the evolution of eyes. *Science* **313**, 1914–1918.
764. Ferrier, D.E.K. (2008). When is a Hox gene not a Hox gene? The importance of gene nomenclature. *In* "Evolving Pathways: Key Themes in Evolutionary Developmental Biology" (A. Minelli and G. Fusco, eds.). Cambridge Univ. Pr., New York, pp. 175–193.
765. Figuera, L.E., Pandolfo, M., Dunne, P.W., Cantú, J.M., and Patel, P.I. (1995). Mapping of the congenital generalized hypertrichosis locus to chromosome Xq24–q27.1. *Nature Genet.* **10**, 202–207.
766. Filler, A.G. (2007). Homeotic evolution in the Mammalia: Diversification of Therian axial seriation and the morphogenetic basis of human origins. *PLoS ONE* **2** #10, e1019.
767. Findlay, J.M., Walker, R., and Kentridge, R.W., eds. (1995). "Eye Movement Research: Mechanisms, Processes and Application." Elsevier, Amsterdam.
768. Fini, M.E., Strissel, K.J., and West-Mays, J.A. (1997). Perspectives on eye development. *Dev. Genet.* **20**, 175–185.
769. Finlay, B.L. and Darlington, R.B. (1995). Linked regularities in the development and evolution of mammalian brains. *Science* **268**, 1578–1584.
770. Finlay, B.L., Darlington, R.B., and Nicastro, N. (2001). Developmental structure in brain evolution. *Behav. Brain Sci.* **24**, 263–308.
771. Finn, C.A. (1998). Menstruation: a nonadaptive consequence of uterine evolution. *Q. Rev. Biol.* **73**, 163–173.
772. Fischer, A., Viebahn, C., and Blum, M. (2002). FGF8 acts as a right determinant during establishment of the left-right axis in the rabbit. *Curr. Biol.* **12**, 1807–1816.
773. Fischer, J.A. (2000). Molecular motors and developmental asymmetry. *Curr. Opin. Gen. Dev.* **10**, 489–496.
774. Fischer, W.W. (2008). Life before the rise of oxygen. *Nature* **455**, 1051–1052.
775. Fisher, R.A. (1930). "The Genetical Theory of Natural Selection." Clarendon Pr., Oxford.
776. Fisher, S.E. and Marcus, G.F. (2006). The eloquent ape: genes, brains and the evolution of language. *Nature Rev. Genet.* **7**, 9–20.
777. Fishman, M.C. and Olson, E.N. (1997). Parsing the heart: genetic modules for organ assembly. *Cell* **91**, 153–156.
778. Fitch, D.H.A. and Sudhaus, W. (2002). One small step for worms, one giant leap for "Bauplan"? *Evol. Dev.* **4**, 243–246.
779. Fitch, W.T. (2000). The evolution of speech: a comparative review. *Trends Cognitive Sci.* **4**, 258–267.

780. Fitch, W.T. (2005). Dancing to Darwin's tune. *Nature* **438**, 288.
781. Fitch, W.T. (2005). The evolution of language: a comparative review. *Biol. Philos.* **20**, 193–230.
782. Fitch, W.T., Hauser, M.D., and Chomsky, N. (2005). The evolution of the language faculty: clarifications and implications. *Cognition* **97**, 179–210.
783. Fitch, W.T. and Reby, D. (2001). The descended larynx is not uniquely human. *Proc. Roy. Soc. Lond. B* **268**, 1669–1675.
784. Flatt, T. (2005). The evolutionary genetics of canalization. *Q. Rev. Biol.* **80**, 287–316.
785. Flatt, T. and Promislow, D.E.L. (2007). Still pondering an age-old question. *Science* **318**, 1255–1256.
786. Fleagle, J.G. (1992). Primate locomotion and posture. *In* "The Cambridge Encyclopedia of Human Evolution" (S. Bunney, S. Jones, R. Martin, and D. Pilbeam, eds.). Cambridge Univ. Pr., New York, pp. 75–79.
787. Fliesler, S.J. and Kisselev, O., eds. (2008). "Signal Transduction in the Retina." CRC Pr., Boca Raton, Fla.
788. Flinn, M.V., Geary, D.C., and Ward, C.V. (2005). Ecological dominance, social competition, and coalitionary arms races: Why humans evolved extraordinary intelligence. *Evol. Human Behav.* **26**, 10–46.
789. Foley, R. (2001). The evolutionary consequences of increased carnivory in hominids. *In* "Meat-eating and Human Evolution" (C.B. Stanford and H.T. Bunn, eds.). Oxford Univ. Pr., New York, pp. 305–331.
790. Foltys, H., Krings, T., Meister, I.G., Sparing, R., Boroojerdi, B., Thron, A., and Topper, R. (2003). Motor representation in patients rapidly recovering after stroke: a functional magnetic resonance imaging and transcranial magnetic stimulation study. *Clin. Neurophysiol.* **114**, 2404–2415.
791. Fomenou, M.D., Scaal, M., Stockdale, F.E., Christ, B., and Huang, R. (2005). Cells of all somitic compartments are determined with respect to segmental identity. *Dev. Dynamics* **233**, 1386–1393.
792. Fondon, J.W., III and Garner, H.R. (2004). Molecular origins of rapid and continuous morphological evolution. *PNAS* **101** #52, 18058–18063.
793. Foote, M. (1997). The evolution of morphological diversity. *Annu. Rev. Ecol. Syst.* **28**, 129–152.
794. Force, A., Lynch, M., Pickett, F.B., Amores, A., Yan, Y.-l., and Postlethwait, J. (1999). Preservation of duplicate genes by complementary, degenerative mutations. *Genetics* **151**, 1531–1545.
795. Ford, J. and Ford, D. (1986). Narwhal: unicorn of the Arctic seas. *Nat. Geogr.* **169** #3, 354–363.
796. Forey, P. and Janvier, P. (1993). Agnathans and the origin of jawed vertebrates. *Nature* **361**, 129–134.
797. Forlani, S., Lawson, K.A., and Deschamps, J. (2003). Acquisition of Hox codes during gastrulation and axial elongation in the mouse embryo. *Development* **130**, 3807–3819.
798. Foster, R.G. and Hankins, M.W. (2007). Circadian vision. *Curr. Biol.* **17**, R746–R751.
799. Fox, C.W., Scheibly, K.L., and Reed, D.H. (2008). Experimental evolution of the genetic load and its implications for the genetic basis of inbreeding depression. *Evolution* **62**, 2236–2249.
800. Fraga, M.F., Ballestar, E., Paz, M.F., Ropero, S., Setien, F., Ballestar, M.L., Heine-Suñer, D., Cigudosa, J.C., Urioste, M., Benitez, J., Boix-Chornet, M., Sanchez-Aguilera, A., Ling, C., Carlsson, E., Poulsen, P., Vaag, A., Stephan, Z., Spector, T.D., Wu, Y.-Z., Plass, C., and Esteller, M. (2005). Epigenetic differences arise during the lifetime of monozygotic twins. *PNAS* **102** #30, 10604–10609.
801. Francis, C.M., Anthony, E.L.P., Brunton, J.A., and Kunz, T.H. (1994). Lactation in male fruit bats. *Nature* **367**, 691–692.
802. Frank, L.G. (1997). Evolution of genital masculinization: why do female hyaenas have such a large "penis"? *Trends Ecol. Evol.* **12**, 58–62.
803. Frank, L.G., Weldele, M.L., and Glickman, S.E. (1995). Masculinization costs in hyaenas. *Nature* **377**, 584–585.
804. Frankel, J. (1989). "Pattern Formation: Ciliate Studies and Models." Oxford Univ. Pr., New York.
805. Frankel, J. (1990). Positional order and cellular handedness. *J. Cell Sci.* **97**, 205–211.
806. Frankel, J. (2008). What do genic mutations tell us about the structural patterning of a complex single–celled organism?. *Eukaryotic Cell* **7**, 1617–1639.
807. Franz-Odendaal, T.A. and Hall, B.K. (2006). Modularity and sense organs in the blind cavefish, Astyanax mexicanus. *Evol. Dev.* **8**, 94–100.
808. Fraser, J.A., Diezmann, S., Subaran, R.L., Allen, A., Lengeler, K.B., Dietrich, F.S., and Heitman, J. (2004). Convergent evolution of chromosomal sex-determining regions in the animal and fungal kingdoms. *PLoS Biol.* **2** #2, 2243–2255 (e384).
809. Frayer, D.W. and Wolpoff, M.H. (1985). Sexual dimorphism. *Annu. Rev. Anthropol.* **14**, 429–473.
810. Freake, M.J. (1999). Evidence for orientation using the e-vector direction of polarised light in the sleepy lizard *Tiliqua rugosa*. *J. Exp. Biol.* **202**, 1159–1166.
811. Freckleton, R.P. and Harvey, P.H. (2006). Detecting non-Brownian trait evolution in adaptive radiations. *PLoS Biol.* **4** #11, 2104–2111 (e373).
812. Freedman, M.S., Lucas, R.J., Soni, B., von Schantz, M., Muñoz, M., David-Gray, Z., and Foster, R. (1999). Regulation of mammalian circadian behavior by non-rod, non-cone, ocular photoreceptors. *Science* **284**, 502–504.
813. Freeman, G. and Lundelius, J.W. (1982). The developmental genetics of dextrality and sinistrality in the gastropod *Lymnaea peregra*. *W. Roux's Arch.* **191**, 69–83.
814. Freeman, M. (2000). Feedback control of intercellular signalling in development. *Nature* **408**, 313–319.

815. Freeman, S. and Herron, J.C. (2007). "Evolutionary Analysis," 4th ed. Pearson Prentice Hall, Upper Saddle River, N.J.
816. Freitas, R., Zhang, G., and Cohn, M.J. (2007). Biphasic *Hoxd* gene expression in shark paired fins reveals an ancient origin of the distal limb domain. *PLoS ONE* **2** #8, e754.
817. Freitas, R., Zhang, G.J., and Cohn, M.J. (2006). Evidence that mechanisms of fin development evolved in the midline of early vertebrates. *Nature* **442**, 1033–1037.
818. French, V. (1983). Development and evolution of the insect segment. *In* "Development and Evolution," *Symp. Brit. Soc. Dev. Biol.*, Vol. **6** (B.C. Goodwin, N. Holder, and C.C. Wylie, eds.). Cambridge Univ. Pr.: Cambridge, pp. 161–193.
819. French, V., Bryant, P.J., and Bryant, S.V. (1976). Pattern regulation in epimorphic fields. *Science* **193**, 969–981.
820. Fridlund, A.J. and Loftis, J.M. (1990). Relations between tickling and humorous laughter: preliminary support for the Darwin-Hecker hypothesis. *Biol. Psychol.* **30**, 141–150.
821. Fried, I., Wilson, C.L., MacDonald, K.A., and Behnke, E.J. (1998). Electric current stimulates laughter. *Nature* **391**, 650.
822. Friedman, M. (2008). The evolutionary origin of flatfish asymmetry. *Nature* **454**, 209–212.
823. Friedman, M., Coates, M.I., and Anderson, P. (2007). First discovery of a primitive coelacanth fin fills a major gap in the evolution of lobed fins and limbs. *Evol. Dev.* **9**, 329–337.
824. Frith, C.D. (2008). Social cognition: Hi there! Here's something interesting. *Curr. Biol.* **18**, R524–R525.
825. Fröbisch, N.B., Carroll, R.L., and Schoch, R.R. (2007). Limb ossification in the Paleozoic branchiosaurid *Apateon* (Temnospondyli) and the early evolution of preaxial dominance in tetrapod limb development. *Evol. Dev.* **9**, 69–75.
826. Fry, C.J. and Peterson, C.L. (2001). Chromatin remodeling enzymes: who's on first? *Curr. Biol.* **11**, R185–R197.
827. Fujii, N. (2002). D-amino acids in living higher organisms. *Orig. Life Evol. Biosph.* **32**, 103–127.
828. Fujisawa, K., Wrana, J.L., and Culotti, J.G. (2007). The Slit receptor EVA-1 coactivates a SAX-3/Robo-mediated guidance signal in C. elegans. *Science* **317**, 1934–1938.
829. Fujiwara, S. and Kawamura, K. (2003). Acquisition of retinoic acid signaling pathway and innovation of the chordate body plan. *Zool. Sci.* **20**, 809–818.
830. Fukushige, T., Brodigan, T.M., Schriefer, L.A., Waterston, R.H., and Krause, M. (2006). Defining the transcriptional redundancy of early bodywall muscle development in C. elegans: evidence for a unified theory of animal muscle development. *Genes Dev.* **20**, 3395–3406.
831. Furlow, B. (2000). The uses of crying and begging. *Nat. Hist.* **109** #8, 62–67.
832. Fusco, G. (2001). How many processes are responsible for phenotypic evolution? *Evol. Dev.* **3**, 279–286.
833. Futuyma, D.J., Edwards, S.V., and True, J.R. (2005). "Evolution." Sinauer, Sunderland, Mass.
834. Gad, J.M. and Tam, P.P.L. (1999). Axis development: the mouse becomes a dachshund. *Curr. Biol.* **9**, R783–R786.
835. Gage, P.J., Suh, H., and Camper, S.A. (1999). Dosage requirement of *Pitx2* for development of multiple organs. *Development* **126**, 4643–4651.
836. Gailey, D.A., Billeter, J.-C., Liu, J.H., Bauzon, F., Allendorfer, J.B., and Goodwin, S.F. (2006). Functional conservation of the *fruitless* male sex-determination gene across 250 Myr of insect evolution. *Mol. Biol. Evol.* **23**, 633–643.
837. Gailey, D.A., Ohshima, S., Santiago, S.J.-M., Montez, J.M., Arellano, A.R., Robillo, J., Villarimo, C.A., Roberts, L., Fine, E., Villella, A., and Hall, J.C. (1997). The muscle of Lawrence in *Drosophila*: a case of repeated evolutionary loss. *Proc. Natl. Acad. Sci. USA* **94**, 4543–4547.
838. Galaburda, A.M., LoTurco, J., Ramus, F., Fitch, R.H., and Rosen, G.D. (2006). From genes to behavior in developmental dyslexia. *Nature Neurosci.* **9**, 1213–1217.
839. Galbreath, G.J. (1985). The evolution of monozygotic polyembryony in *Dasypus*. *In* "Ecology of Armadillos, Sloths, and Vermilinguas" (G.G. Montgomery, ed.). Smithsonian Inst. Pr., Washington, D.C., pp. 243–246.
840. Galis, F. (1999). Why do almost all mammals have seven cervical vertebrae? Developmental constraints, *Hox* genes, and cancer. *J. Exp. Zool. (Mol. Dev. Evol.)* **285**, 19–26. [See also (2009) *Evol. Dev.* **11**, 69–79.]
841. Galis, F., Kundrát, M., and Metz, J.A.J. (2005). Hox genes, digit identities and the theropod/bird transition. *J. Exp. Zool. (Mol. Dev. Evol.)* **304B**, 198–205.
842. Galis, F. and Metz, J.A.J. (2003). Anti-cancer selection as a source of developmental and evolutionary constraints. *BioEssays* **25**, 1035–1039.
843. Galis, F. and Metz, J.A.J. (2007). Evolutionary novelties: the making and breaking of pleiotropic constraints. *Integr. Comp. Biol.* **47**, 409–419.
844. Galis, F., van Alphen, J.J.M., and Metz, J.A.J. (2001). Why five fingers? Evolutionary constraints on digit numbers. *Trends Ecol. Evol.* **16**, 637–646.
845. Galis, F., Van Dooren, T.J.M., Feuth, J.D., Metz, J.A.J., Witkam, A., Ruinard, S., Steigenga, M.J., and Wijnaendts, L.C.D. (2006). Extreme selection in humans against homeotic transformations of cervical vertebrae. *Evolution* **60**, 2643–2654.
846. Galis, F., van Dooren, T.J.M., and Metz, J.A.J. (2002). Conservation of the segmented germband stage: robustness or pleiotropy? *Trends Genet.* **18**, 504–509.
847. Galis, F., Wagner, G.P., and Jockusch, E.L. (2003). Why is limb regeneration possible in amphibians but not in reptiles, birds, and mammals? *Evol. Dev.* **5**, 208–220.

848. Galli, D., Domínguez, J.N., Zaffran, S., Munk, A., Brown, N.A., and Buckingham, M.E. (2008). Atrial myocardium derives from the posterior region of the second heart field, which acquires left-right identity as Pitx2c is expressed. *Development* **135**, 1157–1167.
849. Galli-Resta, L. (1998). Patterning the vertebrate retina: the early appearance of retinal mosaics. *Sems. Cell Dev. Biol.* **9**, 279–284.
850. Galloway, J. (1987). Evolution of helicity: a cause for reflection? *Nature* **330**, 204–205.
851. Galloway, J. (1989). Ciliate through the looking glass. *Nature* **340**, 16–17.
852. Galloway, J.W. (1989). Reflections on the ambivalent helix. *Experientia* **45**, 859–872.
853. Galloway, J.W. (1991). Macromolecular asymmetry. *In* "Biological Asymmetry and Handedness," *Ciba Found. Symp. 162* (G.R. Bock and J. Marsh, eds.). Wiley, New York, pp. 16–35.
854. Galton, F. (1869). "Hereditary Genius: An Inquiry into Its Laws and Consequences." MacMillan, London.
855. Galton, F. (1889). "Natural Inheritance." MacMillan, London.
856. Gannon, P.J., Kheck, N.M., and Hof, P.R. (2001). Language areas of the hominoid brain: a dynamic communicative shift on the upper east side planum. *In* "Evolutionary Anatomy of the Primate Cerebral Cortex" (D. Falk and K.R. Gibson, eds.). Cambridge Univ. Pr., New York, pp. 216–240.
857. Garavelli, L., Zanacca, C., Caselli, G., Banchini, G., Dubourg, C., David, V., Odent, S., Gurrieri, F., and Neri, G. (2004). Solitary median maxillary central incisor syndrome: clinical case with a novel mutation of Sonic Hedgehog. *Am. J. Med. Genet.* **127A**, 93–95.
858. Garcia, L.R., LeBoeuf, B., and Koo, P. (2007). Diversity in mating behavior of hermaphroditic and male-female Caenorhabditis nematodes. *Genetics* **175**, 1761–1771.
859. García-Bellido, A. (1975). Genetic control of wing disc development in *Drosophila. In* "Cell Patterning," *Ciba Found. Symp.*, Vol. 29 (R. Porter and J. Rivers, eds.). Elsevie, Amsterdam, pp. 161–182.
860. García-Bellido, A. (1977). Homoeotic and atavic mutations in insects. *Amer. Zool.* **17**, 613–629.
861. García-Bellido, A., Lawrence, P.A., and Morata, G. (1979). Compartments in animal development. *Sci. Am.* **241** #1, 102–110.
862. Garcia-Cruz, D., Figuera, L.E., and Cantu, J.M. (2002). Inherited hypertrichoses. *Clin. Genet.* **61**, 321–329.
863. Garcia-Fernàndez, J. (2005). The genesis and evolution of homeobox gene clusters. *Nature Rev. Genet.* **6**, 881–892.
864. Gardiner, D.M. and Bryant, S.V. (2007). Tetrapod limb regeneration. *In* "Fins into Limbs: Evolution, Development, and Transformation" (B.K. Hall, ed.). Univ. Chicago Pr., Chicago, Ill., pp. 163–182.
865. Gardner, M. (1952). Is nature ambidextrous? *Philos. Phenom. Res.* **13**, 200–211.
866. Garvie, C.W. and Wolberger, C. (2001). Recognition of specific DNA sequences. *Molec. Cell* **8**, 937–946.
867. Garwin, L. (2007). Harmony of the hemispheres. *Nature* **449**, 977–978.
868. Gatesy, S.M. and Middleton, K.M. (2007). Skeletal adaptations for flight. *In* "Fins into Limbs: Evolution, Development, and Transformation" (B.K. Hall, ed.). Univ. Chicago Pr., Chicago, Ill., pp. 269–283.
869. Gavrilets, S. (2004). "Fitness Landscapes and the Origin of Species." Princeton Univ. Pr., Princeton, N.J.
870. Gavrilets, S. and Vose, A. (2005). Dynamic patterns of adaptive radiation. *PNAS* **102** #50, 18040–18045. [See also (2009) *Science* **323**, 732–737.]
871. Gayon, J. (2000). History of the concept of allometry. *Amer. Zool.* **40**, 748–758.
872. Gazzaniga, M.S. (1998). The split brain revisited. *Sci. Am.* **279** #1, 51–55.
873. Gebo, D.L. (1987). Functional anatomy of the tarsier foot. *Am. J. Phys. Anthrop.* **73**, 9–31.
874. Gee, H. (2008). The amphioxus unleashed. *Nature* **453**, 999–1000.
875. Gehring, W.J. (1998). "Master Control Genes in Development and Evolution: The Homeobox Story." Yale Univ. Pr., New Haven.
876. Gehring, W.J. (2002). The genetic control of eye development and its implications for the evolution of the various eye-types. *Int. J. Dev. Biol.* **46**, 65–73.
877. Gehring, W.J., Affolter, M., and Bürglin, T. (1994). Homeodomain proteins. *Annu. Rev. Biochem.* **63**, 487–526.
878. Gehring, W.J. and Ikeo, K. (1999). *Pax6*: mastering eye morphogenesis and eye evolution. *Trends Genet.* **15**, 371–377.
879. Gehring, W.J., Qian, Y.Q., Billeter, M., Furukubo-Tokunaga, K., Schier, A.F., Resendez-Perez, D., Affolter, M., Otting, G., and Wüthrich, K. (1994). Homeodomain-DNA recognition. *Cell* **78**, 211–223.
880. Gendron-Maguire, M., Mallo, M., Zhang, M., and Gridley, T. (1993). *Hoxa-2* mutant mice exhibit homeotic transformation of skeletal elements derived from cranial neural crest. *Cell* **75**, 1317–1331.
881. Georgiev, P., Tikhomirova, T., Yelagin, V., Belenkaya, T., Gracheva, E., Parshikov, A., Evgen'ev, M.B., Samarina, O.P., and Corces, V.G. (1997). Insertions of hybrid *P* elements in the *yellow* gene of *Drosophila* cause a large variety of mutant phenotypes. *Genetics* **146**, 583–594.
882. Gerber, S., Eble, G.J., and Neige, P. (2008). Allometric space and allometric disparity: a developmental perspective in the macroevolutionary analysis of morphological disparity. *Evolution* **62**, 1450–1457.
883. Gerhart, J. (1999). Signaling pathways in development (1998 Warkany lecture). *Teratology* **60**, 226–239.

884. Gerhart, J. and Kirschner, M. (1997). "Cells, Embryos, and Evolution." Blackwell Science, Malden, Mass.
885. Gerhart, J. and Kirschner, M. (2007). The theory of facilitated variation. *PNAS* **104 Suppl.1**, 8582–8589.
886. Gerhart, J., Lowe, C., and Kirschner, M. (2005). Hemichordates and the origin of chordates. *Curr. Opin. Gen. Dev.* **15**, 461–467.
887. Gervais, M. and Wilson, D.S. (2005). The evolution and functions of laughter and humor: a synthetic approach. *Q. Rev. Biol.* **80**, 395–430.
888. Gesta, S., Blüher, M., Yamamoto, Y., Norris, A.W., Berndt, J., Kralisch, S., Boucher, J., Lewis, C., and Kahn, C.R. (2006). Evidence for a role of developmental genes in the origin of obesity and body fat distribution. *PNAS* **103** #17, 6676–6681.
889. Ghazanfar, A.A. and Rendall, D. (2008). Evolution of human vocal production. *Curr. Biol.* **18**, R457–R460.
890. Ghazanfar, A.A., Turesson, H.K., Maier, J.X., van Dinther, R., Patterson, R.D., and Logothetis, N.K. (2007). Vocal-tract resonances as indexical cues in rhesus monkeys. *Curr. Biol.* **17**, 425–430.
891. Gherman, A., Chen, P.E., Teslovich, T.M., Stankiewicz, P., Withers, M., Kashuk, C.S., Chakravarti, A., Lupski, J.R., Cutler, D.J., and Katsanis, N. (2007). Population bottlenecks as a potential major shaping force of human genome architecture. *PLoS Genet.* **3** #7, 1223–1231 (e119).
892. Ghiselin, M.T. (2006). Sexual selection in hermaphrodites: where did our ideas come from? *Integr. Comp. Biol.* **46**, 368–372.
893. Ghysen, A. (2003). The origin and evolution of the nervous system. *Int. J. Dev. Biol.* **47**, 555–562.
894. Gibbons, A. (2008). Brainy babies and risky births for Neandertals. *Science* **312**, 1429.
895. Gibbons, A. (2008). Millennium ancestor gets its walking papers. *Science* **319**, 1599–1601.
896. Gibson, G. (1999). Developmental evolution: Going beyond the "just so." *Curr. Biol.* **9**, R942–R945.
897. Gibson, G. and Honeycutt, E. (2002). The evolution of developmental regulatory pathways. *Curr. Opin. Gen. Dev.* **12**, 695–700.
898. Gibson, G. and Russell, I. (2006). Flying in tune: Sexual recognition in mosquitoes. *Curr. Biol.* **16**, 1311–1316.
899. Gibson, G. and Wagner, G. (2000). Canalization in evolutionary genetics: a stablilizing theory? *BioEssays* **22**, 372–380.
900. Gibson, K.R., Rumbaugh, D., and Beran, M. (2001). Bigger is better: primate brain size in relationship to cognition. *In* "Evolutionary Anatomy of the Primate Cerebral Cortex" (D. Falk and K.R. Gibson, eds.). Cambridge Univ. Pr., New York, pp. 79–97.
901. Gieler, U. and Walter, B. (2008). Scratch this! *Sci. Am. Mind* **19** #3, 52–57.
902. Gierer, A. (1974). Molecular models and combinatorial principles in cell differentiation and morphogenesis. *Cold Spring Harb. Symp. Quant. Biol.* **38**, 951–961.
903. Gierer, A. (2008). Brain, mind and limitations of a scientific theory of human consciousness. *BioEssays* **30**, 499–505.
904. Gilbert, A.N. (1986). Mammary number and litter size in Rodentia: the "one-half rule." *Proc. Natl. Acad. Sci. USA* **83**, 4828–4830.
905. Gilbert, S.F. (2000). Diachronic biology meets evo-devo: C. H. Waddington's approach to evolutionary developmental biology. *Amer. Zool.* **40**, 729–737.
906. Gilbert, S.F. (2003). Opening Darwin's black box: teaching evolution through developmental genetics. *Nature Rev. Genet.* **4**, 735–741.
907. Gilbert, S.F. (2004). General principles of differentiation and morphogenesis. *In* "Inborn Errors of Development: The Molecular Basis of Clinical Disorders of Morphogenesis," Oxford *Monographs on Medical Genetics, No. 49* (C.J. Epstein, R.P. Erickson, and A. Wynshaw-Boris, eds.). Oxford Univ. Pr., New York, pp. 10–24.
908. Gilbert, S.F. and Singer, S.R. (2006). "Developmental Biology," 8th ed. Sinauer, Sunderland, Mass.
909. Gilbert, S.J. and Burgess, P.W. (2008). Executive function. *Curr. Biol.* **18**, R110–R114.
910. Gilbert, S.L., Dobyns, W.B., and Lahn, B.T. (2005). Genetic links between brain development and brain evolution. *Nature Rev. Genet.* **6**, 581–590.
911. Gilbert-Barness, E., Debich-Spicer, D., and Opitz, J.M. (2003). Conjoined twins: morphogenesis of the heart and a review. *Am. J. Med. Genet.* **120A**, 568–582.
912. Gilburn, A.S. and Day, T.H. (1996). The evolution of female choice when the preference and the preferred trait are linked to the same inversion system. *Heredity* **76**, 19–27.
913. Giles, J. (2004). Change of mind. *Nature* **430**, 14.
914. Gilissen, E. (2001). Structural symmetries and asymmetries in human and chimpanzee brains. *In* "Evolutionary Anatomy of the Primate Cerebral Cortex" (D. Falk and K.R. Gibson, eds.). Cambridge Univ. Pr.: New York, pp. 187–215.
915. Gill, F.B. (2007). "Ornithology," 3rd ed. W. H. Freeman, New York.
916. Gillham, N.W. (2001). Evolution by jumps: Francis Galton and William Bateson and the mechanism of evolutionary change. *Genetics* **159**, 1383–1392.
917. Girton, J.R. and Bryant, P.J. (1980). The use of cell lethal mutations in the study of *Drosophila* development. *Dev. Biol.* **77**, 233–243.
918. Gislén, A., Dacke, M., Kröger, R.H.H., Abrahamsson, M., Nilsson, D.-E., and Warrant, E.J. (2003). Superior underwater vision in a human population of sea gypsies. *Curr. Biol.* **13**, 833–836.
919. Gleichauf, R. (1936). Anatomie und Variabilität des Geschlechtsapparates von *Drosophila melanogaster* (Meigen). *Z. wiss. Zool.* **148**, 1–66.
920. Glick, S.D., ed. (1985). "Cerebral Lateralization in Nonhuman Species." Acad. Pr., New York.
921. Glickman, S.E., Cunha, G.R., Drea, C.M., Conley, A.J., and Place, N.J. (2006). Mammalian sexual

differentiation: lessons from the spotted hyena. *Trends Endocr. Metab.* **17**, 349–356.

922. Glucksmann, A. (1974). Sexual dimorphism in mammals. *Biol. Rev.* **49**, 423–475.
923. Godt, D. and Tepass, U. (2003). Organogenesis: keeping in touch with the germ cells. *Curr. Biol.* **13**, R683–R685. [See also (2007) *Development* **134**, 3401–3411.]
924. Goldberger, A.L., Rigney, D.R., and West, B.J. (1990). Chaos and fractals in human physiology. *Sci. Am.* **262** #2, 42–49.
925. Goldbeter, A., Gonze, D., and Pourquié, O. (2007). Sharp developmental thresholds defined through bistability by antagonistic gradients of retinoic acid and FGF signaling. *Dev. Dynamics* **236**, 1495–1508.
926. Goldensohn, E. (2002). Waiting to inhale. *Nat. Hist.* **111** #7, 8.
927. Goldman, A.S. (2002). Evolution of the mammary gland defense system and the ontogeny of the immune system. *J. Mammary Gland Biol. Neoplasia* **7**, 277–289.
928. Goldman, T.D. and Arbeitman, M.N. (2007). Genomic and functional studies of *Drosophila* sex hierarchy regulated gene expression in adult head and nervous system tissues. *PLoS Genet.* **3** #11, 2278–2295 (e216).
929. Goldschmidt, R. (1938). "Physiological Genetics." McGraw-Hill, New York.
930. Goldschmidt, R. (1940). "The Material Basis of Evolution." Yale Univ. Pr., New Haven.
931. Goldschmidt, R.B. (1952). Homoeotic mutants and evolution. *Acta Biotheor.* **10**, 87–104.
932. Goldsmith, T.H. (1990). Optimization, constraint, and history in the evolution of eyes. *Q. Rev. Biol.* **65**, 281–322.
933. Goller, F. and Suthers, R.A. (1995). Implications for lateralization of bird song from unilateral gating of bilateral motor patterns. *Nature* **373**, 63–66.
934. Golovnin, A., Birukova, I., Romanova, O., Silicheva, M., Parshikov, A., Savitskaya, E., Pirrotta, V., and Georgiev, P. (2003). An endogenous Su(Hw) insulator separates the *yellow* gene from the *Achaete-scute* gene complex in *Drosophila*. *Development* **130**, 3249–3258.
935. Gomez, C., Özbudak, E.M., Wunderlich, J., Baumann, D., Lewis, J., and Pourquié, O. (2008). Control of segment number in vertebrate embryos. *Nature* **454**, 335–339.
936. Gómez-Skarmeta, J.L., Rodríguez, I., Martínez, C., Culí, J., Ferrés-Marcó, D., Beamonte, D., and Modolell, J. (1995). *Cis*-regulation of *achaete* and *scute*: shared enhancer-like elements drive their coexpression in proneural clusters of the imaginal discs. *Genes Dev.* **9**, 1869–1882.
937. Gomis-Rüth, S., Wierenga, C.J., and Bradke, F. (2008). Plasticity of polarization: changing dendrites into axons in neurons integrated in neuronal circuits. *Curr. Biol.* **18**, 992–1000.
938. Gommery, D. (2005). Vertebral column, bipedalism and freedom of the hands. *In* "From Tools to Symbols: From Early Hominids to Modern Humans" (F. d'Errico and L. Backwell, eds.). Wits Univ. Pr., Johannesburg, South Africa, pp. 183–197.
939. Gompel, N., Prud'homme, B., Wittkopp, P.J., Kassner, V.A., and Carroll, S.B. (2005). Chance caught on the wing: *cis*-regulatory evolution and the origin of pigment patterns in *Drosophila*. *Nature* **433**, 481–487.
940. Gonzalez, F., Duboule, D., and Spitz, F. (2007). Transgenic analysis of *Hoxd* gene regulation during digit development. *Dev. Biol.* **306**, 847–859.
941. Goodman, F.R. and Scambler, P.J. (2004). *HOXD13* and synpolydactyly. *In* "Inborn Errors of Development: The Molecular Basis of Clinical Disorders of Morphogenesis," *Oxford Monographs on Medical Genetics, No. 49* (C.J. Epstein, R.P. Erickson, and A. Wynshaw-Boris, eds.). Oxford Univ. Pr., New York, pp. 521–528.
942. Goodwin, B. (1994). "How the Leopard Changed Its Spots." Charles Scribner's Sons, New York. [See also (2009) *Sems. Cell Dev. Biol.* **20**, 82–89.]
943. Goodwin, B.C. (1984). Changing from an evolutionary to a generative paradigm in biology. *In* "Evolutionary Theory: Paths into the Future" (J.W. Pollard, ed.). Wiley, New York, pp. 99–120.
944. Gordo, I. and Campos, P.R.A. (2008). Sex and deleterious mutations. *Genetics* **179**, 621–626.
945. Gordo, I., Navarro, A., and Charlesworth, B. (2002). Muller's ratchet and the pattern of variation at a neutral locus. *Genetics* **161**, 835–848.
946. Gordon, M.S. (1968). "Animal Function: Principles and Adaptations." MacMillan, New York.
947. Gorfinkiel, N., Sánchez, L., and Guerrero, I. (1999). *Drosophila* terminalia as an appendage-like structure. *Mechs. Dev.* **86**, 113–123.
948. Gorfinkiel, N., Sánchez, L., and Guerrero, I. (2003). Development of the *Drosophila* genital disc requires interactions between its segmental primordia. *Development* **130**, 295–305.
949. Goss, R.J. (1969). "Principles of Regeneration." Acad. Pr., New York.
950. Gosse, N.J., Nevin, L.M., and Baier, H. (2008). Retinotopic order in the absence of axon competition. *Nature* **452**, 892–895. [See also (2008) *Development* **135**, 1833–1841.]
951. Gostling, N.J. and Shimeld, S.M. (2003). Protochordate Zic genes define primitive somite compartments and highlight molecular changes underlying neural crest evolution. *Evol. Dev.* **5**, 136–144.
952. Gould, S.J. (1966). Allometry and size in ontogeny and phylogeny. *Biol. Rev.* **41**, 587–640.
953. Gould, S.J. (1971). D'Arcy Thompson and the science of form. *New Lit. Hist.* **2**, 229–258.
954. Gould, S.J. (1975). On the scaling of tooth size in mammals. *Amer. Zool.* **15**, 351–362.
955. Gould, S.J. (1977). "Ever since Darwin: Reflections in Natural History." W.W. Norton, New York.

956. Gould, S.J. (1977). "Ontogeny and Phylogeny." Harvard Univ. Pr., Cambridge, Mass.
957. Gould, S.J. (1977). The return of hopeful monsters. *Nat. Hist.* **86** #6, 22–30.
958. Gould, S.J. (1978). The panda's thumb and the orchids' trap. *New Sci.* **80**, 700–701.
959. Gould, S.J. (1980). The evolutionary biology of constraint. *Proc. Amer. Acad. Arts Sci.* **109** #2, 39–52.
960. Gould, S.J. (1980). Is a new and general theory of evolution emerging? *Paleobiol.* **6**, 119–130.
961. Gould, S.J. (1980). "The Panda's Thumb: More Reflections in Natural History." Norton, New York.
962. Gould, S.J. (1982). The oddball human male. *Nat. Hist.* **91** #7, 14–22.
963. Gould, S.J. (1983). "Hen's Teeth and Horse's Toes." W.W. Norton, New York.
964. Gould, S.J. (1984). Human equality is a contingent fact of history. *Nat. Hist.* **93** #11, 26–33.
965. Gould, S.J. (1984). Only his wings remained. *Nat. Hist.* **93** #9, 10–18.
966. Gould, S.J. (1985). To be a platypus. *Nat. Hist.* **94** #8, 10–15.
967. Gould, S.J. (1986). Archetype and adaptation. *Nat. Hist.* **95** #10, 16–27.
968. Gould, S.J. (1986). The egg-a-day barrier. *Nat. Hist.* **95** #7, 16–24.
969. Gould, S.J. (1986). Of kiwi eggs and the Liberty Bell. *Nat. Hist.* **95** #11, 20–29.
970. Gould, S.J. (1986). Play it again, life. *Nat. Hist.* **95** #2, 18–26.
971. Gould, S.J. (1987). Bushes all the way down. *Nat. Hist.* **96** #6, 12–19.
972. Gould, S.J. (1987). Freudian slip. *Nat. Hist.* **96** #2, 14–21.
973. Gould, S.J. (1991). Exaptation: a crucial tool for an evolutionary psychology. *J. Social Issues* **47** #3, 43–65.
974. Gould, S.J. (1991). What the immaculate pigeon teaches the burdened mind. *Nat. Hist.* **100** #4, 12–21.
975. Gould, S.J. (1994). Common pathways of illumination. *Nat. Hist.* **103** #12, 10–20.
976. Gould, S.J. (1995). Left snails and right minds: Were the earliest conchologists choosing sides when they held the mirror up to nature? *Nat. Hist.* **104** #4, 10–18.
977. Gould, S.J. (1997). As the worm turns. *Nat. Hist.* **106** #1, 24–27 and 68–73.
978. Gould, S.J. (1997). Unanswerable questions. *In* "A Glorious Accident" (W. Kayzer, ed.). W. H. Freeman, New York, pp. 75–104.
979. Gould, S.J. (2000). The narthex of San Marco and the pangenetic paradigm. *Nat. Hist.* **109** #6, 24–37.
980. Gould, S.J. (2000). Of coiled oysters and big brains: how to rescue the terminology of heterochrony, now gone astray. *Evol. Dev.* **2**, 241–248.
981. Gould, S.J. (2001). Size matters and function counts. *In* "Evolutionary Anatomy of the Primate Cerebral Cortex" (D. Falk and K.R. Gibson, eds.). Cambridge Univ. Pr., New York, pp. xiii–xvii.
982. Gould, S.J. and Lewontin, R.C. (1979). The spandrels of San Marco and the Panglossian paradigm: a critique of the adaptationist programme. *Proc. Roy. Soc. Lond. B* **205**, 581–598.
983. Gould, S.J. and Vrba, E.S. (1982). Exaptation—a missing term in the science of form. *Paleobiology* **8**, 4–15.
984. Gould, S.J. and Young, N.D. (1985). The consequences of being different: sinistral coiling in *Cerion*. *Evolution* **39**, 1364–1379.
985. Govind, C.K. (1989). Asymmetry in lobster claws. *Am. Sci.* **77**, 468–474.
986. Govind, C.K. and Pearce, J. (1986). Differential reflex activity determines claw and closer muscle asymmetry in developing lobsters. *Science* **233**, 354–356.
987. Govind, C.K. and Pearce, J. (1989). Delayed determination of claw laterality in lobsters following loss of target. *Development* **107**, 547–551.
988. Graf, W. and Baker, R. (1983). Adaptive changes of the vestibulo-ocular reflex in flatfish are achieved by reorganization of central nervous pathways. *Science* **221**, 777–779.
989. Graham, A. (2001). The development and evolution of the pharyngeal arches. *J. Anat.* **199**, 133–141.
990. Graham, A., Okabe, M., and Quinlan, R. (2005). The role of the endoderm in the development and evolution of the pharyngeal arches. *J. Anat.* **207**, 479–487.
991. Graham, J.B. (1994). An evolutionary perspective for bimodal respiration: a biological synthesis of fish air breathing. *Amer. Zool.* **34**, 229–237.
992. Graham, J.H., Freeman, D.C., and Emlen, J.M. (1994). Antisymmetry, directional asymmetry, and dynamic morphogenesis. *In* "Developmental Instability: Its Origins and Evolutionary Implications" (T.A. Markow, ed.). Kluwer, London, pp. 123–139.
993. Graham, P., Penn, J.K.M., and Schedl, P. (2003). Masters change, slaves remain. *BioEssays* **25**, 1–4.
994. Grainger, R.M. (1996). New perspectives on embryonic lens induction. *Sems. Cell Dev. Biol.* **7**, 149–155.
995. Grammatopoulos, G.A., Bell, E., Toole, L., Lumsden, A., and Tucker, A.S. (2000). Homeotic transformation of branchial arch identity after *Hoxa2* overexpression. *Development* **127**, 5355–5365.
996. Grammer, K., Fink, B., Moller, A.P., and Thornhill, R. (2003). Darwinian aesthetics: sexual selection and the biology of beauty. *Biol. Rev.* **78**, 385–407.
997. Grant, P.R. (1981). Speciation and the adaptive radiation of Darwin's finches. *Am. Sci.* **69**, 653–663.
998. Grant, P.R. and Grant, B.R. (2008). "How and Why Species Multiply: The Radiation of Darwin's Finches." Princeton Univ. Pr., Princeton, N.J.
999. Grant, S., Waller, W., Bhalla, A., and Kennard, C. (2003). Normal chiasmatic routing of uncrossed projections from the ventrotemporal retina in

albino *Xenopus* frogs. *J. Comp. Neurol.* **458**, 425–439.

1000. Graves, J.A.M. (2001). Of course sex matters. *Cell* **107**, 285–287. [See also (2008) *Nature Rev. Genet.* **9**, 911–922]
1001. Graves, J.A.M. and Schmidt, M.M. (1992). Mammalian sex chromosomes: design or accident? *Curr. Opin. Gen. Dev.* **2**, 890–901.
1002. Graveson, A.C., Smith, M.M., and Hall, B.K. (1997). Neural crest potential for tooth development in a urodele amphibian: developmental and evolutionary significance. *Dev. Biol.* **188**, 34–42.
1003. Gray, G.W. (1948). The great raveled knot. *Sci. Am.* **179** #10, 26–39.
1004. Gray, H. (1977). "Anatomy, Descriptive and Surgical." 15th ed. Bounty Bks., New York.
1005. Gray, M.A. and Critchley, H.D. (2007). Interoceptive basis to craving. *Neuron* **54**, 183–186.
1006. Graybiel, A.M. and Saka, E. (2002). A genetic basis for obsessive grooming. *Neuron* **33**, 1–2.
1007. Graze, R.M., Barmina, O., Tufts, D., Naderi, E., Harmon, K.L., Persianinova, M., and Nuzhdin, S.V. (2007). New candidate genes for sex-comb divergence between *Drosophila mauritiana* and *Drosophila simulans*. *Genetics* **176**, 2561–2576.
1008. Grbic, M. (2000). "Alien" wasps and evolution of development. *BioEssays* **22**, 920–932.
1009. Green, H. and Thomas, J. (1978). Pattern formation by cultured human epidermal cells: development of curved ridges resembling dermatoglyphics. *Science* **200**, 1385–1388.
1010. Greer, J.M. and Capecchi, M.R. (2002). *Hoxb8* is required for normal grooming behavior in mice. *Neuron* **33**, 23–34.
1011. Greer, J.M., Puetz, J., Thomas, K.R., and Capecchi, M.R. (2000). Maintenance of functional equivalence during paralogous Hox gene evolution. *Nature* **403**, 661–665.
1012. Gridley, T. (2006). The long and short of it: somite formation in mice. *Dev. Dynamics* **235**, 2330–2336.
1013. Griffith, A. (2007). Healing broken nerves. *Sci. Am.* **297** #3, 28–30.
1014. Griffith, D.R. (1981). "The Question of Animal Awareness: Evolutionary Continuity of Mental Experience." Rockefeller Univ. Pr., New York.
1015. Grigoryan, T., Wend, P., Klaus, A., and Birchmeier, W. (2008). Deciphering the function of canonical Wnt signals in development and disease: conditional loss- and gain-of-function mutations of β-catenin in mice. *Genes Dev.* **22**, 2308–2341.
1016. Grimaldi, D. and Engel, M.S. (2005). "Evolution of the Insects." Cambridge Univ. Pr., New York.
1017. Gritli-Linde, A. (2007). Molecular control of secondary palate development. *Dev. Biol.* **301**, 309–326.
1018. Gritli-Linde, A., Hallberg, K., Harfe, B.D., Reyahi, A., Kannius-Janson, M., Nilsson, J., Cobourne, M.T., Sharpe, P.T., McMahon, A.P., and Linde, A. (2007). Abnormal hair development and apparent follicular transformation to mammary gland in the absence of Hedgehog signaling. *Dev. Cell* **12**, 99–112.
1019. Gröbner, G., Burnett, I.J., Glaubitz, C., Choi, G., Mason, A.J., and Watts, A. (2000). Observations of light-induced structural changes of retinal within rhodopsin. *Nature* **405**, 810–813.
1020. Grogg, M.W., Call, M.K., and Tsonis, P.A. (2006). Signaling during lens regeneration. *Sems. Cell Dev. Biol.* **17**, 753–758.
1021. Gröning, J. and Hochkirch, A. (2008). Reproductive interference between animal species. *Q. Rev. Biol.* **83**, 257–282. [See also (2007) *Curr. Biol.* **17**, 1943–1947.]
1022. Gross, L. (2008). A molecular link between albinism and visual deficits. *PLoS Biol.* **6** #9, 1813–1814 (e248).
1023. Grossl, N.A. (2000). Supernumerary breast tissue: historical perspectives and clinical features. *South Med. J.* **93**, 29–32.
1024. Grzeschik, K.-H. (2002). Human limb malformations: an approach to the molecular basis of development. *Int. J. Dev. Biol.* **46**, 983–991.
1025. Guerin, M.B., McKernan, D.P., O'Brien, C.J., and Cotter, T.G. (2006). Retinal ganglion cells: dying to survive. *Int. J. Dev. Biol.* **50**, 665–674.
1026. Guillery, R.W. (1974). Visual pathways in albinos. *Sci. Am.* **230** #5, 44–54.
1027. Guillery, R.W. (1996). Why do albinos and other hypopigmented mutants lack normal binocular vision, and what else is abnormal in their central visual pathways? *Eye* **10**, 217–221.
1028. Guioli, S. and Lovell-Badge, R. (2007). PITX2 controls asymmetric gonadal development in both sexes of the chick and can rescue the degeneration of the right ovary. *Development* **134**, 4199–4208.
1029. Güler, A.D., Ecker, J.L., Lall, G.S., Haq, S., Altimus, C.M., Liao, H.-W., Barnard, A.R., Cahill, H., Badea, T.C., Zhao, H., Hankins, M.W., Berson, D.M., Lucas, R.J., Yau, K.-W., and Hattar, S. (2008). Melanopsin cells are the principal conduits for rod-cone input to non-image-forming vision. *Nature* **453**, 102–105.
1030. Gura, T. (2008). Addicted to starvation. *Sci. Am. Mind* **19** #3, 60–65.
1031. Gurdon, J.B. (2005). Sinistral snails and gentlemen scientists. *Cell* **123**, 751–753.
1032. Gurevich, I., Tamir, H., Arango, V., Dwork, A.J., Mann, J.J., and Schmauss, C. (2002). Altered editing of serotonin 2C receptor pre-mRNA in the prefrontal cortex of depressed suicide victims. *Neuron* **34**, 349–356.
1033. Gurnett, C.A., Bowcock, A.M., Dietz, F.R., Morcuende, J.A., Murray, J.C., and Dobbs, M.B. (2007). Two novel point mutations in the long-range SHH enhancer in three families with triphalangeal thumb and preaxial polydactyly. *Am. J. Med. Genet. A* **143A**, 27–32.

1034. Gutmann, M.F. (1993). The constructional preconditions of the basic organization of the tetrapod limb. *In* "Hands of Primates" (H. Preuschoft and D.J. Chivers, eds.). Springer-Verlag, New York, pp. 309–321.

1035. Guttmacher, A.F. (1967). Biographical notes on some famous conjoined twins. *In* "Conjoined Twins," *Birth Defects Orig. Article. Ser., Vol. 3, No. 1* (D. Bergsma, R.J. Blattner, B.L. Nichols, and A.J. Rudolph, eds.). Natl. Found. March of Dimes:, New York, pp. 10–17.

1036. Guttmacher, A.F. and Nichols, B.L. (1967). Teratology of conjoined twins. *In* "Conjoined Twins," *Birth Defects Orig. Article. Ser., Vol. 3, No. 1* (D. Bergsma, R.J. Blattner, B.L. Nichols, and A.J. Rudolph, eds.). Natl. Found. March of Dimes, New York, pp. 3–9.

1037. Guyton, A.C. (1969). "Function of the Human Body." W. B. Saunders, Philadelphia.

1038. Haag, E.S. (2005). Echinoderm rudiments, rudimentary bilaterians, and the origin of the chordate CNS. *Evol. Dev.* **7**, 280–281.

1039. Haag, E.S. and Doty, A.V. (2005). Sex determination across evolution: connecting the dots. *PLoS Biol.* **3** #1, 21–24 (e21).

1040. Haag, E.S. and True, J.R. (2001). From mutants to mechanisms? Assessing the candidate gene paradigm in evolutionary biology. *Evolution* **55**, 1077–1084.

1041. Hadley, M.E. (1972). Functional significance of vertebrate integumental pigmentation. *Amer. Zool.* **12**, 63–76.

1042. Haesler, S. (2007). Programmed for speech. *Sci. Am. Mind* **18** #3, 66–71.

1043. Haesler, S., Rochefort, C., Georgi, B., Licznerski, P., Osten, P., and Scharff, C. (2007). Incomplete and innacurate vocal imitation after knockdown of *FoxP2* in songbird basal ganglia nucleus area X. *PLoS Biol.* **5** #12, 2885–2897.

1044. Hafen, E. and Stocker, H. (2003). How are the sizes of cells, organs, and bodies controlled? *PLoS Biol.* **1** #3, 319–322 (e86).

1045. Hagmann, M. (2000). Why chicks aren't all thumbs. *Science* **289**, 372–373.

1046. Hagmann, P., Cammoun, L., Gigandet, X., Meuli, R., Honey, C.J., Wedeen, V.J., and Sporns, O. (2008). Mapping the structural core of the human cerebral cortex. *PLoS Biol.* **6** #7, 1479–1493 (e159).

1047. Hahn, J.S. and Plawner, L.L. (2004). Evaluation and management of children with holoprosencephaly. *Pediatr. Neurol.* **31**, 79–88.

1048. Halanych, K.M. (2004). The new view of animal phylogeny. *Annu. Rev. Ecol. Evol. Syst.* **35**, 229–256.

1049. Haldane, J.B.S. (1928). On being the right size. *In* "Possible Worlds and Other Papers," Harper & Bros., New York, pp. 20–28.

1050. Haldane, J.B.S. (1928). "Possible Worlds and Other Papers." Harper & Bros., New York.

1051. Halder, G., Callaerts, P., and Gehring, W.J. (1995). Induction of ectopic eyes by targeted expression of the *eyeless* gene in *Drosophila*. *Science* **267**, 1788–1792.

1052. Hall, A.J. (1984). Man and manatee: Can we live together? *Nat. Geogr.* **166** #3, 400–413.

1053. Hall, B.K. (1984). Developmental mechanisms underlying the formation of atavisms. *Biol. Rev.* **59**, 89–124.

1054. Hall, B.K. (1992). Waddington's legacy in development and evolution. *Amer. Zool.* **32**, 113–122.

1055. Hall, B.K. (2000). The neural crest as a fourth germ layer and vertebrates as quadroblastic not triploblastic. *Evol. Dev.* **2**, 3–5.

1056. Hall, B.K. (2003). Descent with modification: the unity underlying homology and homoplasy as seen through an analysis of development and evolution. *Biol. Rev.* **78**, 409–433.

1057. Hall, B.K. (2003). *Evo-devo*: evolutionary developmental mechanisms. *Int. J. Dev. Biol.* **47**, 491–495.

1058. Hall, B.K. (2005). Betrayed by *Balanoglossus*: William Bateson's rejection of evolutionary embryology as the basis for understanding evolution. *J. Exp. Zool. (Mol. Dev. Evol.)* **304B**, 1–17.

1059. Hall, B.K., ed. (2007). "Fins into Limbs: Evolution, Development, and Transformation." Univ. Chicago Pr., Chicago, Ill.

1060. Hall, B.K. (2008). Vertebrate origins: riding the crest of a new wave, or the wave of a new crest? *Evol. Dev.* **10**, 261–262.

1061. Hall, B.K. and Olson, W.M., eds. (2003). "Keywords and Concepts in Evolutionary Developmental Biology." Harvard Univ. Pr., Cambridge, Mass.

1062. Hall, J.C. (1995). Tripping along the trail to the molecular mechanisms of biological clocks. *Trends Neurosci.* **18**, 230–240.

1063. Hall, S.S. (2008). Last of the Neanderthals. *Nat. Geogr.* **214** #4, 34–59.

1064. Hallgrímsson, B., Brown, J.J.Y., and Hall, B.K. (2005). The study of phenotypic variability: an emerging research agenda for understanding the developmental-genetic architecture underlying phenotypic variation. *In* "Variation: A Central Concept in Biology" (B. Hallgrímsson and B.K. Hall, eds.). Elsevier Acad. Pr., New York, pp. 525–551.

1065. Hallgrímsson, B. and Hall, B.K. (2005). "Variation: A Central Concept in Biology." Elsevier Acad. Pr., New York.

1066. Hallsson, J.H., Haflidadóttir, B.S., Stivers, C., Odenwald, W., Arnheiter, H., Pignoni, F., and Steingrímsson, E. (2004). The basic helix-loop-helix leucine zipper transcription factor *Mitf* is conserved in *Drosophila* and functions in eye development. *Genetics* **167**, 233–241.

1067. Hamada, H., Meno, C., Watanabe, D., and Saijoh, Y. (2002). Establishment of vertebrate left-right asymmetry. *Nature Rev. Gen.* **3**, 103–113.

1068. Hamilton, C.R. and Vermeire, B.A. (1988). Complementary hemispheric specialization in monkeys. *Science* **242**, 1691–1694.

1069. Hamilton, W.D. (1967). Extraordinary sex ratios. *Science* **156**, 477–488.

1070. Hampton, A.N., Adolphs, R., Tyszka, M.J., and O'Doherty, J.P. (2007). Contributions of the amygdala to reward expectancy and choice signals in human prefrontal cortex. *Neuron* **55**, 545–555.

1071. Hamrick, M.W. (2001). Primate origins: evolutionary change in digital ray patterning and segmentation. *J. Human Evol.* **40**, 339–351.

1072. Han, M., Yang, X., Lee, J., Allan, C.H., and Muneoka, K. (2008). Development and regeneration of the neonatal digit tip in mice. *Dev. Biol.* **315**, 125–135.

1073. Hanafy, A. and Peterson, C.M. (1997). Twin-reversed arterial perfusion (TRAP) sequence: case reports and review of literature. *Aust. N. Z. J. Obstet. Gynaecol.* **37**, 187–191.

1074. Hancock, J.M. (2005). Gene factories, microfunctionalization and the evolution of gene families. *Trends Genet.* **21**, 591–595.

1075. Hand, E. (2008). The hole at the bottom of the Moon. *Nature* **453**, 1160–1163.

1076. Handrigan, G.R. (2003). *Concordia discors*: duality in the origin of the vertebrate tail. *J. Anat.* **202**, 255–267.

1077. Handrigan, G.R., Haas, A., and Wassersug, R.J. (2007). Bony-tailed tadpoles: the development of supernumerary caudal vertebrae in larval megophryids (Anura). *Evol. Dev.* **9**, 190–202.

1078. Handrigan, G.R. and Wassersug, R.J. (2007). The anuran *Bauplan*: a review of the adaptive, developmental, and genetic underpinnings of frog and tadpole morphology. *Biol. Rev.* **82**, 1–25.

1079. Hanken, J. and Gross, J.B. (2005). Evolution of cranial development and the role of neural crest: insights from amphibians. *J. Anat.* **207**, 437–446.

1080. Hannula-Jouppi, K., Kaminen-Ahola, N., Taipale, M., Eklund, R., Nopola-Hemmi, J., Käärläinen, H., and Kere, J. (2006). The axon guidance receptor gene *ROBO1* is a candidate gene for developmental dyslexia. *PLoS Genet.* **1** #4, 467–474 (e50).

1081. Hansen, T.F., Pienaar, J., and Orzack, S.H. (2008). A comparative method for studying adaptation to a randomly evolving environment. *Evolution* **62**, 1965–1977.

1082. Hanson, I.M. (2001). Mammalian homologues of the *Drosophila* eye specification genes. *Sems. Cell Dev. Biol.* **12**, 475–484.

1083. Happle, R. (1993). Mosaicism in human skin: understanding the patterns and mechanisms. *Arch. Dermatol.* **129**, 1460–1470.

1084. Harada, T., Harada, C., and Parada, L.F. (2007). Molecular regulation of visual system development: more than meets the eye. *Genes Dev.* **21**, 367–378.

1085. Harada, Y., Hosoiri, Y., and Kuroda, R. (2004). Isolation and evaluation of dextral-specific and dextral-enriched cDNA clones as candidates for the handedness-determining gene in a freshwater gastropod, *Lymnaea stagnalis*. *Dev. Genes Evol.* **214**, 159–169.

1086. Harada, Y., Takagaki, Y., Sunagawa, M., Saito, T., Yamada, L., Taniguchi, H., Shoguchi, E., and Sawada, H. (2008). Mechanism of self-sterility in a hermaphroditic chordate. *Science* **320**, 548–550.

1087. Harbison, S.T. and Sehgal, A. (2008). Quantitative genetic analysis of sleep in *Drosophila melanogaster*. *Genetics* **178**, 2341–2360.

1088. Harcourt-Smith, W.E.H. and Aiello, L.C. (2004). Fossils, feet and the evolution of human bipedal locomotion. *J. Anat.* **204**, 403–416.

1089. Harden, N. (2005). Of grainy heads and broken skins. *Science* **308**, 364–365.

1090. Hardie, R.C. (1985). Functional organization of the fly retina. *In* "Progress in Sensory Physiology," Vol. **5**(D. Ottoson, ed.). Springer-Verlag, Berlin, pp. 1–79.

1091. Hardie, R.C. (1986). The photoreceptor array of the dipteran retina. *Trends Neurosci.* **9**, 419–423.

1092. Hardy, A., Chan, C.L.H., and Cohen, C.R.G. (2005). The surgical management of haemorrhoids—a review. *Dig. Surg.* **22**, 26–33.

1093. Hare, B., Plyusnina, I., Ignacio, N., Schepina, O., Stepika, A., Wrangham, R., and Trut, L. (2005). Social cognitive evolution in captive foxes is a correlated by-product of experimental domestication. *Curr. Biol.* **15**, 226–230.

1094. Harfe, B.D., Scherz, P.J., Nissim, S., Tian, H., McMahon, A.P., and Tabin, C.J. (2004). Evidence for an expansion-based temporal Shh gradient in specifying vertebrate digit identities. *Cell* **118**, 517–528.

1095. Harma, M., Harma, M., Mil, Z., and Oksuzler, C. (2005). Vaginal delivery of dicephalic parapagus conjoined twins: case report and literature review. *Tohoku J. Exp. Med.* **205**, 179–185.

1096. Harnad, S.R., Steklis, H.D., and Lancaster, J., eds. (1976). "Origins and Evolution of Language and Speech." *Ann. N. Y. Acad. Sci., Vol. 280*, N. Y. Acad. Sci., New York.

1097. Harris, C.R. (1999). The mystery of ticklish laughter. *Am. Sci.* **87**, 344–351.

1098. Harris, J.M. (2007). Bones from the tar pits. *Nat. Hist.* **116** #5, 18–22.

1099. Harris, M.L. and Erickson, C.A. (2007). Lineage specification in neural crest cell pathfinding. *Dev. Dynamics* **236**, 1–19.

1100. Harris, M.P., Hasso, S.M., Ferguson, M.W.J., and Fallon, J.F. (2006). The development of archosaurian first-generation teeth in a chicken mutant. *Curr. Biol.* **16**, 371–377.

1101. Harrison, D.A. (2007). Sex determination: controlling the master. *Curr. Biol.* **17**, R328–R330.

1102. Harrison, L.G. (1993). "Kinetic Theory of Living Pattern." *Developmental and Cell Biology Series*, Vol. 28. Cambridge Univ. Pr., Cambridge.

1103. Harrison, R.J. and Montagna, W. (1969). "Man." Appleton-Century-Crofts, New York.

1104. Hartenstein, V., Younossi-Hartenstein, A., and Lekven, A. (1994). Delamination and division in the *Drosophila* neurectoderm: spatiotemporal pattern, cytoskeletal dynamics, and common control by neurogenic and segment polarity genes. *Dev. Biol.* **165**, 480–499.

1105. Hartline, D.K. and Colman, D.R. (2007). Rapid conduction and the evolution of giant axons and myelinated fibers. *Curr. Biol.* **17**, R29–R35.

1106. Hartman, C.G. (1927). A case of supernumerary nipple in *Macacus rhesus*, with remarks upon the biology of polymastia and polythelia. *J. Mammalogy* **8**, 96–106.

1107. Harvey, A.W. (1998). Genes for asymmetry easily overruled. *Nature* **392**, 345–346.

1108. Harvey, P.H. and Bennett, P.M. (1983). Brain size, energetics, ecology and life history patterns. *Nature* **306**, 314–315.

1109. Harvey, P.H. and Krebs, J.R. (1990). Comparing brains. *Science* **249**, 140–146.

1110. Harvey, R.J., Topf, M., Harvey, K., and Rees, M.I. (2008). The genetics of hyperekplexia: more than startle! *Trends Genet.* **24**, 439–447.

1111. Hashimoto, H., Mizuta, A., Okada, N., Suzuki, T., Tagawa, M., Tabata, K., Yokoyama, Y., Sakaguchi, M., Tanaka, M., and Toyohara, H. (2002). Isolation and characterization of a Japanese flounder clonal line, *reversed*, which exhibits reversal of metamorphic left-right asymmetry. *Mechs. Dev.* **111**, 17–24.

1112. Hashimoto, T. (2002). Molecular genetic analysis of left-right handedness in plants. *Phil. Trans. Roy. Soc. Lond. B* **357**, 799–808.

1113. Hasson, P., Del Buono, J., and Logan, M.P.O. (2007). *Tbx5* is dispensable for forelimb outgrowth. *Development* **134**, 85–92.

1114. Hastings, M.H., Vance, G., and Maywood, E. (1989). Some reflections on the phylogeny and function of the pineal. *Experientia* **45**, 903–909.

1115. Hattar, S., Liao, H.-W., Takao, M., Berson, D.M., and Yau, K.-W. (2002). Melanopsin-containing retinal ganglion cells: architecture, projections, and intrinsic photosensitivity. *Science* **295**, 1065–1070.

1116. Hattar, S., Lucas, R.J., Mrosovsky, N., Thompson, S., Douglas, R.H., Hankins, M.W., Lem, J., Biel, M., Hofmann, F., Foster, R.G., and Yau, K.W. (2003). Melanopsin and rod-cone photoreceptive systems account for all major accessory visual functions in mice. *Nature* **424**, 76–81.

1117. Hauser, M.D. (1993). Right hemisphere dominance for the production of facial expression in monkeys. *Science* **261**, 475–477.

1118. Hauser, M.D., Chomsky, N., and Fitch, W.T. (2002). The faculty of language: What is it, who has it, and how did it evolve? *Science* **298**, 1569–1579.

1119. Hawkes, K. (1993). Why hunter-gatherers work: An ancient version of the problem of public goods. *Curr. Anthrop.* **34**, 341–361.

1120. Hawkes, K. (2006). Life history theory and human evolution: a chronicle of ideas and findings. *In* "The Evolution of Human Life History" (K. Hawkes and R.R. Paine, eds.). School of Amer. Res. Pr., Santa Fe, New Mexico, pp. 45–93.

1121. Hawkes, K. and O'Connell, J.F. (2005). How old is human longevity? *J. Human Evol.* **49**, 650–653.

1122. Hawkins, J.R. and Sinclair, A.H. (1991). Identification and isolation of *SRY*, a gene from the sex determining region of the Y chromosome. *Sems. Dev. Biol.* **2**, 251–258.

1123. Haworth, K.E., Wilson, J.M., Grevellec, A., Cobourne, M.T., Healy, C., Helms, J.A., Sharpe, P.T., and Tucker, A.S. (2007). Sonic hedgehog in the pharyngeal endoderm controls arch pattern via regulation of *Fgf8* in head ectoderm. *Dev. Biol.* **303**, 244–258.

1124. Hay, J.M., Subramanian, S., Millar, C.D., Mohandesan, E., and Lambert, D.M. (2008). Rapid molecular evolution in a living fossil. *Trends Genet.* **24**, 106–109.

1125. Hayashi, T., Mizuno, N., Takada, R., Takada, S., and Kondoh, H. (2006). Determinative role of Wnt signals in dorsal iris-derived lens regeneration in newt eye. *Mechs. Dev.* **123**, 793–800.

1126. Hayashi, T., Mizuno, N., Ueda, Y., Okamoto, M., and Kondoh, H. (2004). FGF2 triggers iris-derived lens regeneration in newt eye. *Mechs. Dev.* **121**, 519–526.

1127. Hebb, D.O. (1949). "The Organization of Behavior: A Neuropsychological Theory." Wiley, New York.

1128. Heberlein, U., Singh, C.M., Luk, A.Y., and Donohoe, T.J. (1995). Growth and differentiation in the *Drosophila* eye coordinated by *hedgehog*. *Nature* **373**, 709–711.

1129. Heckscher-Sørensen, J., Watson, R.P., Lettice, L.A., Serup, P., Eley, L., De Angelis, C., Ahlgren, U., and Hill, R.E. (2004). The splanchnic mesodermal plate directs spleen and pancreatic laterality, and is regulated by *Bapx1/Nkx3.2*. *Development* **131**, 4665–4675.

1130. Hegstrom, R.A. and Kondepudi, D.K. (1990). The handedness of the universe. *Sci. Am.* **262** #1, 108–115.

1131. Heimlich, H.J. (1975). A life-saving maneuver to prevent food-choking. *JAMA* **234**, 398–401.

1132. Heimlich, H.J. and Patrick, E.A. (1990). The Heimlich maneuver: Best technique for saving any choking victim's life. *Postgrad. Med.* **87**, 38–48.

1133. Heiss, H. (1957). Beiderseitige kongenitale daumenlose Fünffingerhand bei Mutter und Kind. *Z. Anat. Entw.-Gesch.* **120**, 226–231.

1134. Held, L.I., Jr. (1991). Bristle patterning in *Drosophila*. *BioEssays* **13**, 633–640.

1135. Held, L.I., Jr. (1992). "Models for Embryonic Periodicity." *Monographs in Developmental Biology*, Vol. 24. Karger, Basel.

1136. Held, L.I., Jr. (1995). Axes, boundaries and coordinates: the ABCs of fly leg development. *BioEssays* **17**, 721–732.

1137. Held, L.I., Jr. (2002). "Imaginal Discs: The Genetic and Cellular Logic of Pattern Formation." *Developmental and Cell Biology Series*, Vol. **39.** Cambridge Univ. Pr., New York.
1138. Held, L.I., Jr., Grimson, M.J., and Du, Z. (2004). The sex comb rotates at 16 to 24 hours after pupariation. *Drosophila Info. Serv.* **87**, 76–78. [See also (2009) *Evol. Dev.* **11**, 191–204.]
1139. Held, L.I., Jr. and Heup, M. (1996). Genetic mosaic analysis of *decapentaplegic* and *wingless* gene function in the *Drosophila* leg. *Dev. Genes Evol.* **206**, 180–194.
1140. Held, L.I., Jr., Heup, M.A., Sappington, J.M., and Peters, S.D. (1994). Interactions of *decapentaplegic*, *wingless*, and *Distal-less* in the *Drosophila* leg. *Roux's Arch. Dev. Biol.* **203**, 310–319.
1141. Helms, J.A. and Brugmann, S.A. (2007). The origins of species-specific facial morphology: the proof is in the pigeon. *Integr. Comp. Biol.* **47**, 338–342.
1142. Helms, J.A., Cordero, D., and Tapadia, M.D. (2005). New insights into craniofacial morphogenesis. *Development* **132**, 851–861.
1143. Hendrikse, J.L., Parsons, T.E., and Hallgrímsson, B. (2007). Evolvability as the proper focus of evolutionary developmental biology. *Evol. Dev.* **9**, 393–401.
1144. Hendry, A. (2007). The Elvis paradox. *Nature* **446**, 147–150.
1145. Hennighausen, L. and Robinson, G.W. (1998). Think globally, act locally: the making of a mouse mammary gland. *Genes Dev.* **12**, 449–455.
1146. Henriksen, T. (2008). The macrosomic fetus: a challenge in current obstetrics. *Acta Obstet. Gynecol. Scand.* **87**, 134–145.
1147. Henry, G.H. and Vidyasagar, T.R. (1991). Evolution of mammalian visual pathways. *In* "Evolution of the Eye and Visual System" (J.R. Cronly-Dillon and R.L. Gregory, eds.). CRC Pr., Boston, Mass., pp. 442–465.
1148. Hens, J.R., Dann, P., Zhang, J.-P., Harris, S., Robinson, G.W., and Wysolmerski, J. (2007). BMP4 and PTHrP interact to stimulate ductal outgrowth during embryonic mammary development and to inhibit hair follicle induction. *Development* **134**, 1221–1230.
1149. Herman, L.H. (2006). Intelligence and rational behaviour in the bottlenosed dolphin. *In* "Rational Animals?" (S. Hurley and M. Nudds, eds.). Oxford Univ. Pr., New York, pp. 439–468.
1150. Herrera, E. and Garcia-Frigola, C. (2008). Genetics and development of the optic chiasm. *Frontiers Biosci.* **13**, 1646–1653.
1151. Herring, S.W. (1993). Formation of the vertebrate face: epigenetic and functional influences. *Amer. Zool.* **33**, 472–483.
1152. Herring, S.W. and Rowlatt, U.F. (1981). Anatomy and embryology in cephalothoracopagus twins. *Teratology* **23**, 159–173.
1153. Hersh, B.M. and Carroll, S.B. (2005). Direct regulation of *knot* gene expression by Ultrabithorax and the evolution of cis-regulatory elements in *Drosophila*. *Development* **132**, 1567–1577.
1154. Hey, J., Fitch, W.M., and Ayala, F.J. (2005). Systematics and the origin of species: an introduction. *PNAS* **102** #Suppl. 1, 6515–6519.
1155. Heyning, J.E. and Lento, G.M. (2002). The evolution of marine mammals. *In* "Marine Mammal Biology: An Evolutionary Approach" (A.R. Hoelzel, ed.). Blackwell Science: Malden, Mass., pp. 38–72.
1156. Hierck, B.P., Witte, B., Poelmann, R.E., Gittenberger-de-Groot, A.C., and Gittenberger, E. (2005). Chirality in snails is determined by highly conserved asymmetry genes. *J. Molluscan Studies* **71**, 192–195. [See also (2009) *Nature* **457**, 1007–1011.]
1157. Higgie, M. and Blows, M.W. (2008). The evolution of reproductive character displacement conflicts with how sexual selection operates within a species. *Evolution* **62**, 1192–1203.
1158. Hikosaka, O., Bromberg-Martin, E., Hong, S., and Matsumoto, M. (2008). New insights on the subcortical representation of reward. *Curr. Opin. Neurobiol.* **18**, 203–208.
1159. Hildebrand, M. (1985). Walking and running. *In* "Functional Vertebrate Morphology" (M. Hildebrand, D.M. Bramble, K.F. Liem, and D.B. Wake, eds.). Harvard Univ. Pr., Cambridge, Mass., pp. 38–57.
1160. Hillis, D.M. (2007). Making evolution relevant and exciting to biology students. *Evolution* **61**, 1261–1264.
1161. Hilloowala, R.A. (1975). Comparative anatomical study of the hyoid apparatus in selected primates. *Am. J. Anat.* **142**, 367–384.
1162. Hilloowala, R.A. (1976). The primate hyolaryngeal apparatus and herbivorous modifications. *Acta Anat. (Basel)* **95**, 260–278.
1163. Hintz, M., Bartholmes, C., Nutt, P., Ziermann, J., Hameister, S., Neuffer, B., and Theissen, G. (2007). Catching a "hopeful monster": shepherd's purse (*Capsella bursa-pastoris*) as a model system to study the evolution of flower development. *J. Exp. Botany* **57**, 3531–3542.
1164. Hirokawa, N., Tanaka, Y., Okada, Y., and Takeda, S. (2006). Nodal flow and the generation of left-right asymmetry. *Cell* **125**, 33–45.
1165. Hirth, F., Kammermeier, L., Frei, E., Walldorf, U., Noll, M., and Reichert, H. (2003). An urbilaterian origin of the tripartite brain: developmental genetic insights from *Drosophila*. *Development* **130**, 2365–2373.
1166. Ho, M.-W., Bolton, E., and Saunders, P.T. (1983). Bithorax phenocopy and pattern formation. I. Spatiotemporal characteristics of the phenocopy response. *Exp. Cell Biol.* **51**, 282–290.
1167. Hoagland, M. and Dodson, B. (1995). "The Way Life Works." Random House, New York.
1168. Hobson, A. (2004). A model for madness? *Nature* **430**, 21.

1169. Hobson, J.A. (1999). "Dreaming as Delirium: How the Brain Goes Out of Its Mind." MIT Pr., Cambridge, Mass.
1170. Hobson, J.A. (2002). "Dreaming: An Introduction to the Science of Sleep." Oxford Univ. Pr., New York.
1171. Hochachka, P.W. and Guppy, M. (1987). "Metabolic Arrest and the Control of Biological Time." Harvard Univ. Pr., Cambridge, Mass.
1172. Hodgkin, J. (1983). Two types of sex determination in a nematode. *Nature* **304**, 267–268.
1173. Hodgkin, J. (1985). Males, hermaphrodites and females: sex determination in *Caenorhabditis elegans*. *Trends Genet.* **1**, 85–88.
1174. Hodgkin, J. (1992). Genetic sex determination mechanisms and evolution. *BioEssays* **14**, 253–261.
1175. Hodgkin, J. (1998). Seven types of pleiotropy. *Int. J. Dev. Biol.* **42**, 501–505.
1176. Hodin, J. (2000). Plasticity and constraints in development and evolution. *J. Exp. Zool. (Mol. Dev. Evol.)* **288**, 1–20.
1177. Hoekstra, H.E. (2006). Genetics, development and evolution of adaptive pigmentation in vertebrates. *Heredity* **97**, 222–234.
1178. Hoekstra, H.E. and Coyne, J.A. (2007). The locus of evolution: Evo devo and the genetics of adaptation. *Evolution* **61**, 995–1016.
1179. Hoelzel, A.R., ed. (2002). "Marine Mammal Biology: An Evolutionary Approach." Blackwell Science, Malden, Mass.
1180. Hoffmann, A.A. and McKenzie, J.A. (2005). Mutation and phenotypic variation: Where is the connection? Capacitors, stressors, phenotypic variability, and evolutionary change. *In* "Variation: A Central Concept in Biology" (B. Hallgrímsson and B.K. Hall, eds.). Elsevier Acad. Pr., New York, pp. 159–189.
1181. Hogan, B.L.M. (1999). Morphogenesis. *Cell* **96**, 225–233.
1182. Hogan, B.L.M. and Kolodziej, P.A. (2002). Molecular mechanisms of tubulogenesis. *Nature Rev. Genet.* **3**, 513–523.
1183. Holden, C. (1998). No last word on language origins. *Science* **282**, 1455–1458.
1184. Holden, C. (2003). Opposites attract. *Science* **302**, 49.
1185. Holden, C. (2008). Biologists change one cell type directly into another. *Science* **321**, 1143.
1186. Holden, C. (2008). Fire out of Africa. *Science* **321**, 1613.
1187. Holden, C. (2008). Rebuilding the injured warrior. *Science* **320**, 437.
1188. Holder, N. (1983). The vertebrate limb: patterns and constraints in development and evolution. *In* "Development and Evolution" *Symp. Brit. Soc. Dev. Biol.*, Vol. 6 (B.C. Goodwin, N. Holder, and C.C. Wylie, eds.). Cambridge Univ. Pr., Cambridge, pp. 399–425.
1189. Holland, L.Z. (2002). Heads or tails? Amphioxus and the evolution of anterior-posterior patterning in deuterostomes. *Dev. Biol.* **241**, 209–228.
1190. Holland, L.Z. (2005). Non-neural ectoderm is really neural: evolution of developmental patterning mechanisms in the non-neural ectoderm of chordates and the problem of sensory cell homologies. *J. Exp. Zool. (Mol. Dev. Evol.)* **304B**, 304–323.
1191. Holland, L.Z. and Holland, N.D. (2001). Evolution of the neural crest and placodes: amphioxus as a model for the ancestral vertebrate? *J. Anat.* **199**, 85–98.
1192. Holland, L.Z. and Holland, N.D. (2007). A revised fate map for amphioxus and the evolution of axial patterning in chordates. *Integr. Comp. Biol.* **47**, 360–372.
1193. Holland, P.W.H. (1990). Homeobox genes and segmentation: co-option, co-evolution, and convergence. *Sems. Dev. Biol.* **1**, 135–145.
1194. Holland, P.W.H. (1999). The future of evolutionary developmental biology. *Nature* **402 Suppl.**, C41–C44.
1195. Holloway, M. (2002). A promenade with prosimians: visiting lemurs and their next of kin at the Duke Primate Center. *Sci. Am.* **287** #3, 94–95.
1196. Holloway, R.L. (1995). Toward a synthetic theory of human brain evolution. *In* "Origins of the Human Brain" (J.-P. Changeux and J. Chavaillon, eds.). Clarendon Pr., Oxford, pp. 42–60.
1197. Holloway, R.L. (2001). Does allometry mask important brain structure residuals relevant to species-specific behavioral evolution? *Behav. Brain Sci.* **24** #2, 286–287.
1198. Holloway, R.L., Broadfield, D.C., and Yuan, M.S. (2001). Revisiting australopithecine visual striate cortex: newer data from chimpanzee and human brains suggest it could have been reduced during australopithecine times. *In* "Evolutionary Anatomy of the Primate Cerebral Cortex" (D. Falk and K.R. Gibson, eds.). Cambridge Univ. Pr., New York, pp. 177–186.
1199. Hondo, E., Phichitrasilp, T., Kokubu, K., Kusakabe, K., Nakamuta, N., Oniki, H., and Kiso, Y. (2007). Distribution patterns of uterine glands and embryo spacing in the mouse. *Anat. Histol. Embryol.* **36**, 157–159.
1200. Hong, C.-S., Park, B.-Y., and Saint-Jeannet, J.-P. (2007). The function of *Dmrt* genes in vertebrate development: it is not just about sex. *Dev. Biol.* **310**, 1–9. [See also (2008) *Development* **135**, 311–321.]
1201. Hoogaars, W.M.H., Engel, A., Brons, J.F., Verkerk, A.O., de Lange, F.J., Wong, L.Y.E., Bakker, M.L., Clout, D.E., Wakker, V., Barnett, P., Ravesloot, J.H., Moorman, A.F.M., Verheijck, E.E., and Christoffels, V.M. (2007). Tbx3 controls the sinoatrial node gene program and imposes pacemaker function on the atria. *Genes Dev.* **21**, 1098–1112.

1202. Horabin, J.I. (2005). Splitting the Hedgehog signal: sex and patterning in *Drosophila. Development* **132**, 4801–4810.

1203. Hori, M. (1993). Frequency-dependent natural selection in the handedness of scale-eating cichlid fish. *Science* **260**, 216–219.

1204. Horne-Badovinac, S., Rebagliati, M., and Stainier, D.Y.R. (2003). A cellular framework for gut-looping morphogenesis in zebrafish. *Science* **302**, 662–665.

1205. Hornett, E.A., Charlat, S., Duplouy, A.M.R., Davies, N., Roderick, G.K., Wedell, N., and Hurst, G.D.D. (2006). Evolution of male-killer suppression in a natural population. *PLoS Biol.* **4** #9, 1643–1648 (e283).

1206. Hornett, E.A., Duplouy, A.M.R., Davies, N., Roderick, G.K., Wedell, N., Hurst, G.D.D., and Charlat, S. (2008). You can't keep a good parasite down: evolution of a male-killer suppressor uncovers cytoplasmic incompatibility. *Evolution* **62**, 1258–1263.

1207. Horridge, G.A. (1991). Evolution of visual processing. *In* "Evolution of the Eye and Visual System" (J.R. Cronly-Dillon and R.L. Gregory, eds.). CRC Pr., Boston, Mass., pp. 229–270.

1208. Horton, A.C., Mahadevan, N.R., Minguillon, C., Osoegawa, K., Rokhsar, D.S., Ruvinsky, I., de Jong, P.J., Logan, M.P., and Gibson-Brown, J.J. (2008). Conservation of linkage and evolution of developmental function within the *Tbx2/3/4/5* subfamily of T-box genes: implications for the origin of vertebrate limbs. *Dev. Genes Evol.* **218**, 613–628.

1209. Hosken, D.J. (2008). Clitoral variation says nothing about female orgasm. *Evol. Dev.* **10**, 393–395.

1210. Hosoiri, Y., Harada, Y., and Kuroda, R. (2003). Construction of a backcross progeny collection of dextral and sinistral individuals of a freshwater gastropod, *Lymnaea stagnalis. Dev. Genes Evol.* **213**, 193–198.

1211. Hosokawa, R., Urata, M., Han, J., Zehnaly, A., Bringas, P., Jr., Nonaka, K., and Chai, Y. (2007). TGF-β mediated *Msx2* expression controls occipital somites-derived caudal region of skull development. *Dev. Biol.* **310**, 140–153.

1212. Houle, M., Sylvestre, J.-R., and Lohnes, D. (2003). Retinoic acid regulates a subset of Cdx1 function in vivo. *Development* **130**, 6555–6567. [See also (2008) *Development* **135**, 2511–2520.]

1213. Houtmeyers, E., Gosselink, R., Gayan-Ramirez, G., and Decramer, M. (1999). Regulation of mucociliary clearance in health and disease. *Eur. Respir. J.* **13**, 1177–1188.

1214. Howard, B. and Ashworth, A. (2006). Signaling pathways implicated in early mammary gland morphogenesis and breast cancer. *PLoS Genet.* **2** #8, 1121–1130 (e112).

1215. Howard, I.P. and Rogers, B.J. (1996). "Binocular Vision and Stereopsis." Oxford Univ. Pr., New York.

1216. Hozumi, S., Maeda, R., Taniguchi, K., Kanai, M., Shirakabe, S., Sasamura, T., Spéder, P., Noselli, S., Aigaki, T., Murakami, R., and Matsuno, K. (2006). An unconventional myosin in *Drosophila* reverses the default handedness in visceral organs. *Nature* **440**, 798–802.

1217. Hu, D. and Helms, J.A. (1999). The role of Sonic hedgehog in normal and abnormal craniofacial morphogenesis. *Development* **126**, 4873–4884.

1218. Huang, R., Zhi, Q., Schmidt, C., Wilting, J., Brand-Saberi, B., and Christ, B. (2000). Sclerotomal origin of the ribs. *Development* **127**, 527–532.

1219. Huang, X. and Saint-Jeannet, J.-P. (2004). Induction of the neural crest and the opportunities of life on the edge. *Dev. Biol.* **275**, 1–11.

1220. Huber, B.A., Sinclair, B.J., and Schmitt, M. (2007). The evolution of asymmetric genitalia in spiders and insects. *Biol. Rev.* **82**, 647–698.

1221. Huber, L. (2000). Psychophylogenesis: innovations and limitations in the evolution of cognition. *In* "The Evolution of Cognition" (C. Heyes and L. Huber, eds.). MIT Press, Cambridge, Mass., pp. 23–41.

1222. Hublin, J.-J. (2005). Evolution of the human brain and comparative paleoanthropology. *In* "From Monkey Brain to Human Brain" (S. Dehaene, J.-R. Duhamel, M.D. Hauser, and G. Rizzolatti, eds.). M.I.T. Pr., Cambridge, Mass., pp. 57–71.

1223. Hublin, J.-J. and Coqueugniot, H. (2006). Absolute or proportional brain size: that is the question. A reply to Leigh's (2006) comments. *J. Human Evol.* **50**, 109–113.

1224. Hughes, A.L. (2007). Looking for Darwin in all the wrong places: the misguided quest for positive selection at the nucleotide sequence level. *Heredity* **99**, 364–373.

1225. Hughes, G.M. (1965). "Comparative Physiology of Vertebrate Respiration." Harvard Univ. Pr., Cambridge, Mass.

1226. Hughes, N.C., Haug, J.T., and Waloszek, D. (2008). Basal euarthropod development: a fossil-based perspective. *In* "Evolving Pathways: Key Themes in Evolutionary Developmental Biology" (A. Minelli and G. Fusco, eds.). Cambridge Univ. Pr., New York, pp. 281–298.

1227. Huizinga, J. (1949). "Homo ludens: A Study of the Play-Element in Culture." Routledge & Kegan Paul, Boston.

1228. Humes, D.J. and Simpson, J. (2006). Acute appendicitis. *BMJ* **333**, 530–534.

1229. Hunt, G.R., Corballis, M.C., and Gray, R.D. (2001). Laterality in tool manufacture by crows. *Nature* **414**, 707.

1230. Hunt, P. and Krumlauf, R. (1992). Hox codes and positional specification in vertebrate embryonic axes. *Annu. Rev. Cell Biol.* **8**, 227–256.

1231. Hunter, T. (1987). A thousand and one protein kinases. *Cell* **50**, 823–829.

1232. Hunter, T. (1995). Protein kinases and phosphatases: the yin and yang of protein phosphorylation and signaling. *Cell* **80**, 225–236.

1233. Hurford, J.R. (2007). "The Origins of Meaning: Language in the Light of Evolution." Oxford Univ. Pr., New York.

1234. Hurles, M.E., Dermitzakis, E.T., and Tyler-Smith, C. (2008). The functional impact of structural variation in humans. *Trends Genet.* **24**, 238–245.

1235. Huszar, D., Sharpe, A., Hashmi, S., Bouchard, B., Houghton, A., and Jaenisch, R. (1991). Generation of pigmented stripes in albino mice by retroviral marking of neural crest melanoblasts. *Development* **113**, 653–660.

1236. Hutchinson, G.E. (1979). Memories of the corixid water bugs. *Discovery* **14** #1, 11–19.

1237. Hutson, J.M. and Hasthorpe, S. (2005). Abnormalities of testicular descent. *Cell Tissue Res.* **322**, 155–158.

1238. Huxley, J.S. (1927). The size of living things. *In* "Man in the Modern World: An Eminent Scientist Looks at Life Today (Selected Essays from *Man Stands Alone* and *On Living in a Revolution*)" (J. Huxley, ed.). New Amer. Libr. World Lit., New York, pp. 78–94.

1239. Hwang, J., Mehrani, T., Millar, S.E., and Morasso, M.I. (2008). Dlx3 is a crucial regulator of hair follicle differentiation and cycling. *Development* **135**, 3149–3159.

1240. Iacono, N.L., Mantero, S., Chiarelli, A., Garcia, E., Mills, A.A., Morasso, M.I., Costanzo, A., Levi, G., Guerrini, L., and Merlo, G.R. (2008). Regulation of *Dlx5* and *Dlx6* gene expression by p63 is involved in EEC and SHFM congenital limb defects. *Development* **135**, 1377–1388.

1241. Ikeda, A. and Matsumoto, S. (1993). New results concerning the vascularization of primate hands. Part I: The palmar arterial arches in Cercopithecidae, Pongidae, Hominidae and other primates. *In* "Hands of Primates" (H. Preuschoft and D.J. Chivers, eds.). Springer-Verlag, New York, pp. 173–182.

1242. Ikoma, A., Steinhoff, M., Ständer, S., Yosipovitch, G., and Schmelz, M. (2006). The neurobiology of itch. *Nature Rev. Neurosci.* **7**, 535–547.

1243. Ingber, D.E. and Levin, M. (2007). What lies at the interface of regenerative medicine and developmental biology? *Development* **134**, 2541–2547.

1244. Ingham, P.W. and Placzek, M. (2006). Orchestrating ontogenesis: variations on a theme by sonic hedgehog. *Nature Rev. Genet.* **7**, 841–850.

1245. Innis, J.W., Margulies, E.H., and Kardia, S. (2002). Integrative biology and the developing limb bud. *Evol. Dev.* **4**, 378–389.

1246. Inomata, H., Haraguchi, T., and Sasai, Y. (2008). Robust stability of the embryonic axial pattern requires a secreted scaffold for Chordin degradation. *Cell* **134**, 854–865.

1247. Inoue, S. and Matsuzawa, T. (2007). Working memory of numerals in chimpanzees. *Curr. Biol.* **17**, R1004–R1005.

1248. Isaac, A., Rodriguez-Esteban, C., Ryan, A., Altabef, M., Tsukui, T., Patel, K., Tickle, C., and Izpisúa-Belmonte, J.-C. (1998). Tbx genes and limb identity in chick embryo development. *Development* **125**, 1867–1875.

1249. Isalan, M., Lemerle, C., Michalodimitrakis, K., Horn, C., Beltrao, P., Raineri, E., Garriga-Canut, M., and Serrano, L. (2008). Evolvability and hierarchy in rewired bacterial gene networks. *Nature* **452**, 840–845.

1250. Isbell, L.A. (2006). Snakes as agents of evolutionary change in primate brains. *J. Human Evol.* **51**, 1–35.

1251. Isensee, J. and Ruiz Noppinger, P. (2007). Sexually dimorphic gene expression in mammalian somatic tissue. *Gend. Med.* **4 Suppl. B**, S75–S95.

1252. Ishii, M., Tachiwana, T., Hoshino, A., Tsunekawa, N., Hiramatsu, R., Matoba, S., Kanai-Azuma, M., Kawakami, H., Kurohmaru, M., and Kanai, Y. (2007). Potency of testicular somatic environment to support spermatogenesis in XX/Sry transgenic male mice. *Development* **134**, 449–454.

1253. Ishijima, S., Sekiguchi, K., and Hiramoto, Y. (1988). Comparative study of the beat patterns of American and Asian horseshoe crab sperm: evidence for a role of the central pair complex in forming planar waveforms in flagella. *Cell Motil. Cytoskel.* **9**, 264–270.

1254. Ishimaru, Y., Komatsu, T., Kasahara, M., Katoh-Fukui, Y., Ogawa, H., Toyama, Y., Maekawa, M., Toshimori, K., Chandraratna, R.A.S., Morohashi, K.-i., and Yoshioka, H. (2008). Mechanism of asymmetric ovarian development in chick embryos. *Development* **135**, 677–685.

1255. Istrail, S., De-Leon, S.B.-T., and Davidson, E. (2007). The regulatory genome and the computer. *Dev. Biol.* **310**, 187–195.

1256. Iwasaki, S.-i. (2002). Evolution of the structure and function of the vertebrate tongue. *J. Anat.* **201**, 1–13.

1257. Iwasato, T., Katoh, H., Nishimaru, H., Ishikawa, Y., Inoue, H., Saito, Y.M., Ando, R., Iwama, M., Takahashi, R., Negishi, M., and Itohara, S. (2007). Rac-GAP α-chimerin regulates motor-circuit formation as a key mediator of ephrinB3/EphA4 forward signaling. *Cell* **130**, 742–753.

1258. Izpisúa Belmonte, J.C. (1999). How the body tells left from right. *Sci. Am.* **280** #6, 46–51.

1259. Jablonka, E. and Lamb, M.J. (1990). The evolution of heteromorphic sex chromosomes. *Biol. Rev.* **65**, 249–276.

1260. Jablonski, N.G. (2006). "Skin: A Natural History." Univ. of Calif. Pr., Berkeley.

1261. Jackson, C., McCabe, B.J., Nicol, A.U., Grout, A.S., Brown, M.W., and Horn, G. (2008). Dynamics of a memory trace: effects of sleep on consolidation. *Curr. Biol.* **18**, 393–400.

1262. Jackson, C.E., Callies, Q.C., Krull, E.A., and Mehregan, A. (1975). Hairy cutaneous malformations of palms and soles. *Arch. Dermatol.* **111**, 1146–1149.

1263. Jackson, G.M. (2006). Tourette's Syndrome. *Curr. Biol.* **16**, R443–R444.
1264. Jackson, H.G. (1913). *Eupagurus. Proc. Trans. Liverpool Biol. Soc.* **27**, 495–573.
1265. Jacob, F. (1977). Evolution and tinkering. *Science* **196**, 1161–1166.
1266. Jacob, F. (1982). "The Possible and the Actual." Univ. Wash. Pr., Seattle.
1267. Jacobs, D.K., Nakanishi, N., Yuan, D., Camara, A., Nichols, S.A., and Hartenstein, V. (2007). Evolution of sensory structures in basal metazoa. *Integr. Comp. Biol.* **47**, 712–723.
1268. Jahoda, C.A.B. (1998). Cellular and developmental aspects of androgenetic alopecia. *Exp. Dermatol.* **7**, 235–248.
1269. Jain, A.K., Prabhakar, S., and Pankanti, S. (2002). On the similarity of identical twin fingerprints. *Patt. Recog.* **35**, 2653–2663.
1270. James, W. (1893). "Psychology." Holt, New York.
1271. Jäncke, L. (2008). Music, memory and emotion. *J. Biol.* **7**, e21.
1272. Jane, S.M., Ting, S.B., and Cunningham, J.M. (2005). Epidermal impermeable barriers in mouse and fly. *Curr. Opin. Gen. Dev.* **15**, 447–453.
1273. Jangir, O.P., Modi, D., and Sharma, M. (2005). Effect of vitamin A on lens regeneration in pigs. *Indian J. Exp. Biol.* **43**, 679–685.
1274. Jangir, O.P., Shekhawat, D.V.S., Prakash, A., Swami, K.K., and Suthar, P. (2001). Homeotic regeneration of eye in amphibian tadpoles and its enhancement by vitamin A. *J. Biosci.* **26**, 577–581.
1275. Jangir, O.P., Suthar, P., Shekhawat, D.V.S., Acharya, P., Swami, K.K., and Sharma, M. (2005). The "Third Eye"—A new concept of trans-differentiation of pineal gland into median eye in amphibian tadpoles of *Bufo melanostictus. Indian J. Exp. Biol.* **43**, 671–678.
1276. Janvier, P. (2008). Squint of the fossil flatfish. *Nature* **454**, 169–170.
1277. Janzen, F.J. and Paukstis, G.L. (1991). Environmental sex determination in reptiles: ecology, evolution, and experimental design. *Q. Rev. Biol.* **66**, 149–179.
1278. Jarman, A.P. (2000). Developmental genetics: Vertebrates and insects see eye to eye. *Curr. Biol.* **10**, R857–R859.
1279. Jarman, A.P., Grell, E.H., Ackerman, L., Jan, L.Y., and Jan, Y.N. (1994). *atonal* is the proneural gene for *Drosophila* photoreceptors. *Nature* **369**, 398–400.
1280. Jarne, P. and Auld, J.R. (2006). Animals mix it up too: the distribution of self-fertilization among hermaphroditic animals. *Evolution* **60**, 1816–1824.
1281. Jeffery, C.J. (2003). Moonlighting proteins: old proteins learning new tricks. *Trends Genet.* **19**, 415–417.
1282. Jeffery, G. (1997). The albino retina: an abnormality that provides insight into normal retinal development. *Trends Neurosci.* **20**, 165–169.
1283. Jeffery, G. and Erskine, L. (2005). Variations in the architecture and development of the vertebrate optic chiasm. *Prog. Retin. Eye Res.* **24**, 721–753.
1284. Jeffery, G., Schütz, G., and Montoliu, L. (1994). Correction of abnormal retinal pathways found with albinism by introduction of a functional tyrosinase gene in transgenic mice. *Dev. Biol.* **166**, 460–464.
1285. Jeffery, W.R. (2001). Cavefish as a model system in evolutionary developmental biology. *Dev. Biol.* **231**, 1–12.
1286. Jeffery, W.R. (2006). Ascidian neural crest-like cells: phylogenetic distribution, relationship to larval complexity, and pigment cell fate. *J. Exp. Zool. (Mol. Dev. Evol.)* **306B**, 470–480.
1287. Jeffery, W.R., Strickler, A.G., and Yamamoto, Y. (2004). Migratory neural crest-like cells form body pigmentation in a urochordate embryo. *Nature* **431**, 696–699.
1288. Jen, J.C., Chan, W.-M., Bosley, T.M., Wan, J., Carr, J.R., Rüb, U., Shattuck, D., Salamon, G., Kudo, L.C., Ou, J., Lin, D.D.M., Salih, M.A.M., Kansu, T., al Dhalaan, H., al Zayed, Z., MacDonald, D.B., Stigsby, B., Plaitakis, A., Dretakis, E.K., Gottlob, I., Pieh, C., Traboulsi, E.I., Wang, Q., Wang, L., Andrews, C., Yamada, K., Demer, J.L., Karim, S., Alger, J.R., Geschwind, D.H., Deller, T., Sicotte, N.L., Nelson, S.F., Baloh, R.W., and Engle, E.C. (2004). Mutations in a human *ROBO* gene disrupt hindbrain axon pathway crossing and morphogenesis. *Science* **304**, 1509–1513.
1289. Jenner, R.A. (2006). Metazoan phylogeny. *In* "Evolution of Nervous Systems," Vol. 1 (J.H. Kaas, ed.). Elsevier, Oxford, pp. 17–40.
1290. Jenner, R.A. and Wills, M.A. (2007). The choice of model organisms in evo-devo. *Nature Rev. Genet.* **8**, 311–319.
1291. Jenny, A. and Mlodzik, M. (2006). Planar cell polarity signaling: a common mechanism for cellular polarization. *Mt. Sinai J. Med.* **73**, 738–750.
1292. Jeong, J., Li, X., McEvilly, R.J., Rosenfeld, M.G., Lufkin, T., and Rubenstein, J.L.R. (2008). Dlx genes pattern mammalian jaw primordium by regulating both lower jaw-specific and upper jaw-specific genetic programs. *Development* **135**, 2905–2916.
1293. Jeong, J., Mao, J., Tenzen, T., Kottmann, A.H., and McMahon, A.P. (2004). Hedgehog signaling in the neural crest cells regulates the patterning and growth of facial primordia. *Genes Dev.* **18**, 937–951.
1294. Jeong, S., Rokas, A., and Carroll, S.B. (2006). Regulation of body pigmentation by the Abdominal-B Hox protein and its gain and loss in *Drosophila* evolution. *Cell* **125**, 1387–1399.
1295. Jerison, H.J. (1973). "Evolution of the Brain and Intelligence." Acad. Pr., New York.
1296. Jerison, H.J. (2001). The study of primate brain evolution: where do we go from here? *In* "Evolutionary Anatomy of the Primate Cerebral Cortex" (D. Falk

and K.R. Gibson, eds.). Cambridge Univ. Pr., New York, pp. 305–337.

1297. Jermiin, L.S., Poladian, L., and Charleston, M.A. (2005). Is the "Big Bang" in animal evolution real? *Science* **310**, 1910–1911.

1298. Jernvall, J. (2000). Linking development with generation of novelty in mammalian teeth. *PNAS* **97** #6, 2641–2645.

1299. Jernvall, J. and Salazar-Cuidad, I. (2007). The economy of tinkering mammalian teeth. *In* "Tinkering: The Microevolution of Development," *Novartis Found. Symp. 284* (G. Bock and J. Goode, eds.). Wiley, Chichester, U.K., pp. 207–224.

1300. Jernvall, J. and Thesleff, I. (2000). Reiterative signaling and patterning during mammalian tooth morphogenesis. *Mechs. Dev.* **92**, 19–29.

1301. Jesson, L.K. and Barrett, S.C.H. (2002). Solving the puzzle of mirror-image flowers. *Nature* **417**, 707.

1302. Johanson, Z., Joss, J., Boisvert, C.A., Ericsson, R., Sutija, M., and Ahlberg, P.E. (2007). Fish fingers: Digit homologues in Sarcopterygian fish fins. *J. Exp. Zool. (Mol. Dev. Evol.)* **308B**, 757–768.

1303. John, B. and Miklos, G.L.G. (1988). "The Eukaryotic Genome in Development and Evolution." Allen & Unwin, London.

1304. Johnson, J.B., Burt, D.B., and DeWitt, T.J. (2008). Form, function, and fitness: pathways to survival. *Evolution* **62**, 1243–1251.

1305. Johnson, M.S. (1987). Adaptation and rules of form: chirality and shape in *Partula suturalis*. *Evolution* **41**, 672–675.

1306. Johnson, P.T.J., Lunde, K.B., Ritchie, E.G., and Launer, A.E. (1999). The effect of trematode infection on amphibian limb development and survivorship. *Science* **284**, 802–804.

1307. Jollie, M.T. (1977). Segmentation of the vertebrate head. *Amer. Zool.* **17**, 323–333.

1308. Jonasova, K. and Kozmik, Z. (2008). Eye evolution: lens and cornea as an upgrade of animal visual system. *Sems. Cell Dev. Biol.* **19**, 71–81.

1309. Jones, A.G., Arnold, S.J., and Bürger, R. (2007). The mutation matrix and the evolution of evolvability. *Evolution* **61**, 727–745.

1310. Jones, C.A. and Li, D.Y. (2007). Common cues regulate neural and vascular patterning. *Curr. Opin. Gen. Dev.* **17**, 332–336.

1311. Jones, M.P., Pierce, K.E., Jr., and Ward, D. (2007). Avian vision: a review of form and function with special consideration to birds of prey. *J. Exotic Pet Med.* **16**, 69–87.

1312. Jones, P., Chase, K., Martin, A., Davern, P., Ostrander, E.A., and Lark, K.G. (2008). Single-nucleotide-polymorphism-based association mapping of dog stereotypes. *Genetics* **179**, 1033–1044.

1313. Jones, S. (1999). "Darwin's Ghost: *The Origin of Species* Updated." Ballantine Bks., New York.

1314. Jourdain, R. (1997). "Music, the Brain, and Ecstasy: How Music Captures Our Imagination." William Morrow, New York.

1315. Judson, O. (2002). "Dr. Tatiana's Sex Advice to All Creation." Metropolitan Bks., New York.

1316. Judson, O. (2007). Hiber Nation. *Nat. Hist.* **116** #10, 16–19.

1317. Judson, O. (2008). Chromosomagnon man. *Nat. Hist.* **117** #5, 17–20.

1318. Judson, O.P. (2005). Anticlimax. *Nature* **436**, 916–917.

1319. Jugessur, A. and Murray, J.C. (2005). Orofacial clefting: recent insights into a complex trait. *Curr. Opin. Gen. Dev.* **15**, 270–278. [See also (2008) *Dev. Biol.* **321**, 273–282.]

1320. Juslin, P.N. and Sloboda, J.A. (2001). "Music and Emotion: Theory and Research." Oxford Univ. Pr., New York.

1321. Juste, J. and Ibáñez, C. (1993). An asymmetric dental formula in a mammal, the São Tomé Island fruit bat *Myonycteris brachycephala* (Mammalia: Megachiroptera). *Can. J. Zool.* **71**, 221–224.

1322. Kaas, J.H. (1993). Evolution of multiple areas and modules within neocortex. *Persp. Dev. Neurobiol.* **1**, 101–107.

1323. Kaas, J.H. (2000). Organizing principles of sensory representations. *In* "Evolutionary Developmental Biology of the Cerebral Cortex," *Novartis Found. Symp. 228* (G.R. Bock and G. Cardew, eds.). Wiley, New York, pp. 188–205.

1324. Kaas, J.H. (2006). Evolution of the neocortex. *Curr. Biol.* **16**, R910–R914.

1325. Kaas, J.H. and Collins, C.E. (2001). Evolving ideas of brain evolution. *Nature* **411**, 141–142.

1326. Kadonaga, J.T. (1998). Eukaryotic transcription: an interlaced network of transcription factors and chromatin-modifying machines. *Cell* **92**, 307–313.

1327. Kaestner, K.H. and Knöchel, W. (2000). Unified nomenclature for the winged helix/forkhead transcription factors. *Genes Dev.* **14**, 142–146.

1328. Kagan, J. (2007). "What is Emotion? History, Measures, and Meanings." Yale Univ. Pr., New Haven, Conn.

1329. Kaneshiro, K.Y. (1969). A study of the relationships of Hawaiian *Drosophila* species based on external male genitalia. *In* "Studies in Genetics," Vol. 5 (Pub. #6918) (M.R. Wheeler, ed.). Univ. Texas Pr., pp. 55–70.

1330. Kangas, A.T., Evans, A.R., Thesleff, I., and Jernvall, J. (2004). Nonindependence of mammalian dental characters. *Nature* **432**, 211–214.

1331. Kango-Singh, M., Singh, A., and Sun, Y.H. (2003). Eyeless collaborates with Hedgehog and Decapentaplegic signaling in *Drosophila* eye induction. *Dev. Biol.* **256**, 48–60.

1332. Kantaputra, P.N. and Gorlin, R.J. (1992). Double dens invaginatus of molarized maxillary central incisors, premolarization of maxillary lateral incisors, multituberculism of the mandibular incisors, canines and first premolar, and sensorineural hearing loss. *Clin. Dysmorphol.* **1**, 128–136.

1333. Karam, J.A. and Baker, L.A. (2004). True hermaphroditism. *N. Engl. J. Med.* **350** #4, 393.

1334. Kardon, G., Heanue, T.A., and Tabin, C.J. (2004). The *Pax/Six/Eya/Dach* network in development and evolution. *In* "Modularity in Development and Evolution," (G. Schlosser and G.P. Wagner, eds.). Univ. of Chicago Pr.: Chicago, Ill., pp. 59–80.
1335. Kardong, K.V. (2002). "Vertebrates: Comparative Anatomy, Function, Evolution," 3rd ed. McGraw-Hill, New York.
1336. Kardong, K.V. (2005). "An Introduction to Biological Evolution." McGraw-Hill, New York.
1337. Karmiloff-Smith, A. (2007). Williams Syndrome. *Curr. Biol.* **17**, R1035–R1036.
1338. Karner, C., Wharton, K.A., Jr., and Carroll, T.J. (2006). Planar cell polarity and vertebrate organogenesis. *Sems. Cell Dev. Biol.* **17**, 194–203.
1339. Karp, G. and Berrill, N.J. (1981). "Development.," 2nd ed. McGraw-Hill, New York.
1340. Kashalikar, S.J. (1988). An explanation for the development of decussations in the central nervous system. *Med. Hypotheses* **26**, 1–8.
1341. Kaskan, P.M. and Finlay, B.L. (2001). Encephalization and its developmental structure: how many ways can a brain get big? *In* "Evolutionary Anatomy of the Primate Cerebral Cortex" (D. Falk and K.R. Gibson, eds.). Cambridge Univ. Pr., New York, pp. 14–29.
1342. Kathirithamby, J. and Johnston, J.S. (2004). The discovery after 94 years of the elusive female of a myrmecolacid (Strepsiptera), and the cryptic species of *Caenocholax fenyesi* Pierce sensu lato. *Proc. Roy. Soc. Lond. B* **271 Suppl. 3**, S5–S8.
1343. Katz, M.J. and Grenander, U. (1982). Developmental matching and the numerical matching hypothesis for neuronal cell death. *J. Theor. Biol.* **98**, 501–517.
1344. Kauffman, S.A. (1983). Developmental constraints: internal factors in evolution. *In* "Development and Evolution," *Symp. Brit. Soc. Dev. Biol.*, Vol. 6 (B.C. Goodwin, N. Holder, and C.C. Wylie, eds.). Cambridge Univ. Pr., Cambridge, pp. 195–225.
1345. Kaufman, M.H. (2004). The embryology of conjoined twins. *Childs Nerv. Syst.* **20**, 508–525.
1346. Kavanagh, K.D., Evans, A.R., and Jernvall, J. (2007). Predicting evolutionary patterns of mammalian teeth from development. *Nature* **449**, 427–432.
1347. Kavanau, J.L. (1997). Origin and evolution of sleep: roles of vision and endothermy. *Brain Res. Bull.* **42**, 245–264.
1348. Kawamura, K., Kouki, T., Kawahara, G., and Kikuyama, S. (2002). Hypophyseal development in vertebrates from amphibians to mammals. *Gen. Comp. Endocrinol.* **126**, 130–135.
1349. Kawanishi, C.Y., Hartig, P., Bobseine, K.L., Schmid, J., Cardon, M., Massenburg, G., and Chernoff, N. (2003). Axial skeletal and hox expression domain alterations induced by retinoic acid, valproic acid, and bromoxynil during murine development. *J. Biochem. Mol. Toxicol.* **17**, 346–356.
1350. Kawauchi, H. and Sower, S.A. (2006). The dawn and evolution of hormones in the adenohypophysis. *Gen. Comp. Endocrinol.* **148**, 3–14.
1351. Kay, E.H. and Hoekstra, H.E. (2008). Rodents. *Curr. Biol.* **18**, R406–R410.
1352. Kay, J.N. and Baier, H. (2004). Out-foxing fate: molecular switches create neuronal diversity in the retina. *Neuron* **43**, 759–760.
1353. Kay, J.N., Link, B.A., and Baier, H. (2005). Staggered cell-intrinsic timing of *ath5* expression underlies the wave of ganglion cell neurogenesis in the zebrafish retina. *Development* **132**, 2573–2585.
1354. Kay, R.F., Ross, C., and Williams, B.A. (1997). Anthropoid origins. *Science* **275**, 797–804.
1355. Keddy-Hector, A.C. (1992). Mate choice in non-human primates. *Amer. Zool.* **32**, 62–70.
1356. Kee, Y., Hwang, B.J., Sternberg, P.W., and Bronner-Fraser, M. (2007). Evolutionary conservation of cell migration genes: from nematode neurons to vertebrate neural crest. *Genes Dev.* **21**, 391–396.
1357. Keene, H.J. (1991). On heterochrony in heterodonty: a review of some problems in tooth morphogenesis and evolution. *Yrbk. Phys. Anthrop.* **34**, 251–282.
1358. Keisman, E.L. and Baker, B.S. (2001). The *Drosophila* sex determination hierarchy modulates *wingless* and *decapentaplegic* signaling to deploy *dachshund* sex—specifically in the genital imaginal disc. *Development* **128**, 1643–1656.
1359. Keisman, E.L., Christiansen, A.E., and Baker, B.S. (2001). The sex determination gene *doublesex* regulates the A/P organizer to direct sex-specific patterns of growth in the *Drosophila* genital imaginal disc. *Dev. Cell* **1**, 215–225.
1360. Keith, A. (1948). "A New Theory of Human Evolution." Watts, London.
1361. Kelley, J. and Qinghua, X. (1991). Extreme sexual dimorphism in a Miocene hominoid. *Nature* **352**, 151–153.
1362. Kelly, D.A. (2002). The functional morphology of penile erection: tissue designs for increasing and maintaining stiffness. *Integ. Comp. Biol.* **42**, 216–221.
1363. Kendrick, K.M., Hinton, M.R., Atkins, K., Haupt, M.A., and Skinner, J.D. (1998). Mothers determine sexual preferences. *Nature* **395**, 229–230.
1364. Kennedy, H., Douglas, R., Knoblauch, K., and Dehay, C. (2007). Self-organization and pattern formation in primate cortical networks. *In* "Cortical Development: Genes and Genetic Abnormalities," *Novartis Found. Symp. 288* (G. Bock and J. Goode, eds.). Wiley, Hoboken, N.J., pp. 178–198.
1365. Keplinger, B.L., Guo, X., Quine, J., Feng, Y., and Cavener, D.R. (2001). Complex organization of promoter and enhancer elements regulate the tissue- and developmental stage-specific expression of the *Drosophila melanogaster Gld* gene. *Genetics* **157**, 699–716.

1366. Ker, R.F., Bennett, M.B., Bibby, S.R., Kester, R.C., and Alexander, R.M. (1987). The spring in the arch of the human foot. *Nature* **325**, 147–149.

1367. Kessel, M. (1992). Respecification of vertebral identities by retinoic acid. *Development* **115**, 487–501.

1368. Kessel, M. and Gruss, P. (1991). Homeotic transformations of murine vertebrae and concomitant alteration of *Hox* codes induced by retinoic acid. *Cell* **67**, 89–104.

1369. Khaitovich, P., Weiss, G., Lachmann, M., Hellmann, I., Enard, W., Muetzel, B., Wirkner, U., Ansorge, W., and Pääbo, S. (2004). A neutral model of transcriptome evolution. *PLoS Biol.* **2** #5, 682–689 (e132). [See also (2008) *Nature Rev. Genet.* **9**, 965–974.]

1370. Kharlamova, A.V., Trut, L.N., Carrier, D.R., Chase, K., and Lark, K.G. (2007). Genetic regulation of canine skeletal traits: trade-offs between the hind limbs and forelimbs in the fox and dog. *Integr. Comp. Biol.* **47**, 373–381.

1371. Kidd, W. (1920). "Initiative in Evolution." H. F. & G. Witherby, London.

1372. Kiefer, J.C. (2006). Emerging developmental model systems. *Dev. Dynamics* **235**, 2895–2899.

1373. Kiefer, J.C. (2007). Back to basics: *Sox* genes. *Dev. Dynamics* **236**, 2356–2366.

1374. Kiefer, J.C. (2007). Epigenetics in development. *Dev. Dynamics* **236**, 1144–1156.

1375. Kieser, J.A. and Groeneveld, H.T. (1988). Allometric relations of teeth and jaws in man. *Am. J. Phys. Anthrop.* **77**, 57–67.

1376. Kim, J.S., Jin, D.I., Lee, J.H., Son, D.S., Lee, S.H., Yi, Y.J., and Park, C.S. (2005). Effects of teat number on litter size in gilts. *Anim. Reprod. Sci.* **90**, 111–116.

1377. Kim, S., Bardwell, V.J., and Zarkower, D. (2007). Cell type-autonomous and non-autonomous requirements for *Dmrt1* in postnatal testis differentiation. *Dev. Biol.* **307**, 314–327.

1378. Kim, S., Namekawa, S.H., Niswander, L.M., Ward, J.O., Lee, J.T., Bardwell, V.J., and Zarkower, D. (2007). A mammal-specific *doublesex* homolog associates with male sex chromatin and is required for male meiosis. *PLoS Genet.* **3** #4, 559–571 (e62).

1379. Kim, S.Y., Paylor, S.W., Magnuson, T., and Schumacher, A. (2007). Juxtaposed Polycomb complexes co-regulate vertebral identity. *Development* **133**, 4957–4968.

1380. Kim, T.-D., Woo, K.-C., Cho, S., Ha, D.-C., Jang, S.K., and Kim, K.-T. (2007). Rhythmic control of AANAT translation by hnRNP Q in circadian melatonin production. *Genes Dev.* **21**, 797–810.

1381. Kim, Y. and Capel, B. (2006). Balancing the bipotential gonad between alternative organ fates: a new perspective on an old problem. *Dev. Dynamics* **235**, 2292–2300.

1382. Kim, Y., Kobayashi, A., Sekido, R., DiNapoli, L., Brennan, J., Chaboissier, M.-C., Poulat, F., Behringer, R.R., Lovell-Badge, R., and Capel, B. (2006). *Fgf9* and *Wnt4* act as antagonistic signals to regulate mammalian sex determination. *PLoS Biol.* **4** #6, 1000–1009 (e187). [See also (2008) *Curr. Opin. Gen. Dev.* **18**, 499–505.]

1383. Kimbel, W.H. (2004). Becoming bipeds. *Am. Sci.* **92**, 274–276.

1384. Kimura, J. and Deutsch, G.H. (2007). Key mechanisms of early lung development. *Pediatr. Dev. Pathol.* **10**, 335–347.

1385. Kimura, K.-i., Usui, K., and Tanimura, T. (1994). Female myoblasts can participate in the formation of a male-specific muscle in *Drosophila*. *Zool. Sci.* **11**, 247–251.

1386. King, N., Hittinger, C.T., and Carroll, S.B. (2003). Evolution of key cell signaling and adhesion protein families predates animal origins. *Science* **301**, 361–363.

1387. King, R.A., Townsend, D., Oetting, W., Summers, C.G., Olds, D.P., White, J.G., and Spritz, R.A. (1991). Temperature-sensitive tyrosinase associated with peripheral pigmentation in oculocutaneous albinism. *J. Clin. Invest.* **87**, 1046–1053.

1388. Kingsley, M.C.S. and Ramsay, M.A. (1988). The spiral in the tusk of the narwhal. *Artic* **41**, 236–238.

1389. Kinsman, S.L., Plawner, L.L., and Hahn, J.S. (2000). Holoprosencephaly: recent advances and new insights. *Curr. Opin. Neurol.* **13**, 127–132.

1390. Kirk, E.C. (2006). Visual influences on primate encephalization. *J. Human Evol.* **51**, 76–90.

1391. Kirkpatrick, M. and Price, T. (2008). In sight of speciation. *Nature* **455**, 601–602.

1392. Kirschner, M. and Gerhart, J. (1998). Evolvability. *Proc. Natl. Acad. Sci. USA* **95**, 8420–8427.

1393. Kirschner, M.W. and Gerhart, J.C. (2005). "The Plausibility of Life: Resolving Darwin's Dilemma." Yale Univ. Pr., New Haven, Conn.

1394. Klar, A.J.S. (2005). A 1927 study supports a current genetic model for inheritance of human scalp hairwhorl orientation and hand-use preference traits. *Genetics* **170**, 2027–2030.

1395. Klattig, J. and Englert, C. (2007). The Müllerian duct: recent insights into its development and regression. *Sex. Dev.* **1**, 271–278.

1396. Klein, D.C. (2006). Evolution of the vertebrate pineal gland: the AANAT hypothesis. *Chronobiol. Internat.* **23**, 5–20.

1397. Klein, O.D., Minowada, G., Peterkova, R., Kangas, A., Yu, B.D., Lesot, H., Peterka, M., Jernvall, J., and Martin, G. (2006). Sprouty genes control diastema tooth development via bidirectional antagonism of epithelial-mesenchymal FGF signaling. *Dev. Cell* **11**, 181–190.

1398. Kleiner, K. (2007). Keep your eyes on the eyes. *Sci. Am. Mind* **18** #1, 11.

1399. Kley, N.J. and Kearney, M. (2007). Adaptations for digging and burrowing. *In* "Fins into Limbs: Evolution, Development, and Transformation" (B.K. Hall, ed.). Univ. Chicago Pr., Chicago, Ill., pp. 284–309.

1400. Klingenberg, C.P. (2005). Developmental constraints, modules, and evolvability. *In* "Variation: A Central Concept in Biology" (B. Hallgrímsson and B.K. Hall, eds.). Elsevier Acad. Pr., New York, pp. 219–247.
1401. Klingenberg, C.P., McIntyre, G.S., and Zaklan, S.D. (1998). Left-right asymmetry of fly wings and the evolution of body axes. *Proc. Roy. Soc. Lond. B* **265**, 1255–1259.
1402. Klingenberg, C.P. and Nijhout, H.F. (1999). Genetics of fluctuating asymmetry: a developmental model of developmental instability. *Evolution* **53**, 358–375.
1403. Klonisch, T., Fowler, P.A., and Hombach-Klonisch, S. (2004). Molecular and genetic regulation of testis descent and external genitalia development. *Dev. Biol.* **270**, 1–18.
1404. Kmita, M., Fraudeau, N., Hérault, Y., and Duboule, D. (2002). Serial deletions and duplications suggest a mechanism for the collinearity of *Hoxd* genes in limbs. *Nature* **420**, 145–150.
1405. Knecht, A.K. and Bronner-Fraser, M. (2002). Induction of the neural crest: a multigene process. *Nature Rev. Genet.* **3**, 453–461.
1406. Knezevic, V., De Santo, R., Schughart, K., Huffstadt, U., Chiang, C., Mahon, K.A., and Mackem, S. (1997). *Hoxd-12* differentially affects preaxial and postaxial chondrogenic branches in the limb and regulates *Sonic hedgehog* in a positive feedback loop. *Development* **124**, 4523–4536.
1407. Knowles, L.L., Carstens, B.C., and Keat, M.L. (2007). Coupling genetic and ecological-niche models to examine how past population distributions contribute to divergence. *Curr. Biol.* **17**, 940–946.
1408. Knudsen, E.I. (1981). The hearing of the barn owl. *Sci. Am.* **245** #6, 113–125.
1409. Knudsen, E.I. and Knudsen, P.F. (1985). Vision guides the adjustment of auditory localization in young barn owls. *Science* **230**, 545–548.
1410. Kobayashi, H. and Kohshima, S. (2001). Unique morphology of the human eye and its adaptive meaning: comparative studies on external morphology of the primate eye. *J. Human Evol.* **40**, 419–435.
1411. Kobayashi, T., Kajiura-Kobayashi, H., Guan, G., and Nagahama, Y. (2008). Sexual dimorphic expression of *DMRT1* and *Sox9a* during gonadal differentiation and hormone-induced sex reversal in the teleost fish Nile tilapia (*Oreochromis niloticus*). *Dev. Dynamics* **237**, 297–306.
1412. Koch, C. (2008). Rendering the visible invisible. *Sci. Am. Mind* **19** #5, 18–19.
1413. Koentges, G. (2008). Evolution of anatomy and gene control: evo-devo meets systems biology. *Nature* **451**, 658–663.
1414. Koentges, G. (2008). Teeth in double trouble. *Nature* **455**, 747–748.
1415. Kohlsdorf, T. and Wagner, G.P. (2006). Evidence for the reversibility of digit loss: a phylogenetic study of limb evolution in *Bachia* (Gymnophthalmidae: Squamata). *Evolution* **60**, 1896–1912.
1416. Kokko, H. and Jennions, M.D. (2008). Parental investment, sexual selection and sex ratios. *J. Evol. Biol.* **21**, 919–948.
1417. Komisaruk, B.R., Beyer-Flores, C., and Whipple, B. (2006). "The Science of Orgasm." Johns Hopkins Univ. Pr., Baltimore, Maryland.
1418. Kondo, M., Froschauer, A., Kitano, A., Nanda, I., Hornung, U., Volff, J.-N., Asakawa, S., Mitani, H., Naruse, K., Tanaka, M., Schmid, M., Shimizu, N., Schartl, M., and Shima, A. (2002). Molecular cloning and characterization of *DMRT* genes from the medaka *Oryzias latipes* and the platyfish *Xiphophorus maculatus*. *Gene* **295**, 213–222.
1419. Kondo, T. (2006). Epigenetic alchemy for cell fate conversion. *Curr. Opin. Gen. Dev.* **16**, 502–507.
1420. Kondrashov, F.A. (2005). In search of the limits of evolution. *Nature Gen.* **37**, 9–10.
1421. Konishi, M. (1983). Night owls are good listeners. *Nat. Hist.* **92** #9, 56–69.
1422. Koopman, P. and Loffler, K.A. (2003). Sex determination: the fishy tale of Dmrt1. *Curr. Biol.* **13**, R177–R179.
1423. Kopp, A., Duncan, I., Godt, D., and Carroll, S.B. (2000). Genetic control and evolution of sexually dimorphic characters in *Drosophila*. *Nature* **408**, 553–559 (*cf.* correction in *Nature* **410**: 611).
1424. Kopp, A., Graze, R.M., Xu, S., Carroll, S.B., and Nuzhdin, S.V. (2003). Quantitative trait loci responsible for variation in sexually dimorphic traits in *Drosophila melanogaster*. *Genetics* **163**, 771–787.
1425. Kopp, A. and True, J.R. (2002). Evolution of male sexual characters in the Oriental Drosophila melanogaster species group. *Evol. Dev.* **4**, 278–291. [See also (2009) *Evol. Dev.* **11**, 205–218.]
1426. Korf, H.-W., Schomerus, C., and Stehle, J.H. (1998). "The Pineal Organ, Its Hormone Melatonin, and the Photoneuroendocrine System." *Advances in Anatomy, Embryology, and Cell Biology*, Vol. 146. Springer-Verlag, New York.
1427. Kornak, U. and Mundlos, S. (2003). Genetic disorders of the skeleton: a developmental approach. *Am. J. Hum. Genet.* **73**, 447–474.
1428. Kornberg, R.D. (1999). Eukaryotic transcriptional control. *Trends Genet.* **15**, M46–M49.
1429. Kornberg, T.B. and Guha, A. (2007). Understanding morphogen gradients: a problem of dispersion and containment. *Curr. Opin. Gen. Dev.* **17**, 264–271.
1430. Kosaki, K. and Casey, B. (1998). Genetics of human left-right axis malformations. *Sems. Cell Dev. Biol.* **9**, 89–99.
1431. Koshland, D.E., Jr. (1994). The key-lock theory and the induced fit theory. *Angew. Chem. Int. Ed. Engl.* **33**, 2375–2378.
1432. Kosukegawa, I., Yoshimoto, M., Isogai, S., Nonaka, S., and Yamashita, T. (2006). Piriformis syndrome resulting from a rare anatomic variation. *Spine* **31** #18, E664–E666.

1433. Koukoura, O., Sifakis, S., Stratoudakis, G., Mantas, N., Kaminopetros, P., and Koumantakis, E. (2006). A case report of recurrent anencephaly and literature review. *Clin. Exp. Obstet. Gynecol.* **33**, 185–189.

1434. Koyanagi, M., Kubokawa, K., Tsukamoto, H., Shichida, Y., and Terakita, A. (2005). Cephalochordate melanopsin: evolutionary linkage between invertebrate visual cells and vertebrate photosensitive retinal ganglion cells. *Curr. Biol.* **15**, 1065–1069.

1435. Kozmik, Z. (2005). Pax genes in eye development and evolution. *Curr. Opin. Gen. Dev.* **15**, 430–438.

1436. Kozu, T., Suda, K., and Toki, F. (1995). Pancreatic development and anatomical variation. *Gastrointest. Endoscopy Clinics N. Amer.* **5** #1, 1–30.

1437. Krause, J., Lalueza-Fox, C., Orlando, L., Enard, W., Green, R.E., Burbano, H.A., Hublin, J.-J., Hänni, C., Fortea, J., de la Rasilla, M., Bertranpetit, J., Rosas, A., and Pääbo, S. (2007). The derived *FOXP2* variant of modern humans was shared with Neandertals. *Curr. Biol.* **17**, 1908–1912.

1438. Krebs, J.R. (1991). The case of the curious bill. *Nature* **349**, 465.

1439. Krishnan, V. and Nestler, E.J. (2008). The molecular neurobiology of depression. *Nature* **455**, 894–902.

1440. Kristan, W.B. (2008). Neuronal decision-making circuits. *Curr. Biol.* **18**, R928–R932.

1441. Kronhamn, J., Frei, E., Daube, M., Jiao, R., Shi, Y., Noll, M., and Rasmuson-Lestander, Å. (2002). Headless flies produced by mutations in the paralogous *Pax6* genes *eyeless* and *twin of eyeless*. *Development* **129**, 1015–1026.

1442. Krubitzer, L. (1995). The organization of neocortex in mammals: are species differences really so different? *Trends Neurosci.* **18**, 408–417.

1443. Krubitzer, L. (2007). The magnificent compromise: cortical field evolution in mammals. *Neuron* **56**, 201–208.

1444. Krubitzer, L. and Kaas, J. (2005). The evolution of the neocortex in mammals: how is phenotypic diversity generated? *Curr. Opin. Neurobiol.* **15**, 444–453.

1445. Krubitzer, L.A. (2000). How does evolution build a complex brain? *In* "Evolutionary Developmental Biology of the Cerebral Cortex," *Novartis Found. Symp. 228* (G.R. Bock and G. Cardew, eds.). Wiley, New York, pp. 206–226.

1446. Krützen, M., Mann, J., Heithaus, M.R., Connor, R.C., Bejder, L., and Sherwin, W.B. (2005). Cultural transmission of tool use in bottlenose dolphins. *PNAS* **102** #25, 8939–8943.

1447. Kucera, T., Eglinger, J., Strilic, B., and Lammert, E. (2007). Vascular lumen formation from a cell biological perspective. *In* "Vascular Development" (D.J. Chadwick and J. Goode, eds.). Wiley, Chichester, U.K., pp. 46–60.

1448. Kukalová-Peck, J. (1983). Origin of the insect wing and wing articulation from the arthropodan leg. *Can. J. Zool.* **61**, 1618–1669.

1449. Kumar, A., Godwin, J.W., Gates, P.B., Garza-Garcia, A.A., and Brockes, J.P. (2007). Molecular basis for the nerve dependence of limb regeneration in an adult vertebrate. *Science* **318**, 772–777.

1450. Kumar, P., Yuan, X., Kumar, M.R., Kind, R., Li, X., and Chadha, R.K. (2007). The rapid drift of the Indian tectonic plate. *Nature* **449**, 894–897.

1451. Kumar, V., Fausto, N., and Abbas, A. (2004). "Robbins and Cotran Pathologic Basis of Disease." 7th ed. Saunders, Philadelphia.

1452. Kunhardt, P.B., Jr., Kunhardt, P.B., III, and Kunhardt, P.W. (1995). "P. T. Barnum: America's Greatest Showman." Knopf, New York.

1453. Kunwar, P.S., Siekhaus, D.E., and Lehmann, R. (2006). In vivo migration: a germ cell perspective. *Annu. Rev. Cell Dev. Biol.* **22**, 237–265.

1454. Kuratani, S. (2005). Craniofacial development and the evolution of the vertebrates: the old problems on a new background. *Zool. Sci.* **22**, 1–19.

1455. Kusakabe, R. and Kuratani, S. (2007). Evolutionary perspectives from development of mesodermal components in the lamprey. *Dev. Dynamics* **236**, 2410–2420.

1456. Kushner, H.I. (1999). "A Cursing Brain? The Histories of Tourette Syndrome." Harvard Univ. Pr., Cambridge, Mass.

1457. Lacalli, T. (1996). Dorsoventral axis inversion: a phylogenetic perspective. *BioEssays* **18**, 251–254.

1458. Lacalli, T. (2003). Body plans and simple brains. *Nature* **424**, 263–264.

1459. Lacalli, T. (2004). Light on ancient photoreceptors. *Nature* **432**, 454–455.

1460. Lacalli, T.C. (2001). New perspectives on the evolution of protochordate sensory and locomotory systems, and the origin of brains and heads. *Phil. Trans. Roy. Soc. Lond. B* **356**, 1565–1572.

1461. Lacalli, T.C. (2004). Sensory systems in amphioxus: a window on the ancestral chordate condition. *Brain Behav. Evol.* **64**, 148–162.

1462. Lacalli, T.C. (2005). Protochordate body plan and the evolutionary role of larvae: old controversies resolved? *Can. J. Zool.* **83**, 216–224.

1463. Lacalli, T.C. (2008). Basic features of the ancestral chordate brain: a protochordate perspective. *Brain Res. Bull.* **75**, 319–323.

1464. Lacalli, T.C. (2008). Mucus secretion and transport in amphioxus larvae: organization and ultrastructure of the food trapping system, and implications for head evolution. *Acta Zool. (Stockholm)* **89**, 219–230.

1465. Lachance, J. (2008). A fundamental relationship between genotype frequencies and fitnesses. *Genetics* **180**, 1087–1093.

1466. Lagnado, L. (2002). Signal amplification: Let's turn down the lights. *Curr. Biol.* **12**, R215–R217.

1467. Lahn, B.T. and Page, D.C. (1999). Four evolutionary strata on the human X chromosome. *Science* **286**, 964–967.

1468. Lai, C.S.L., Fisher, S.E., Hurst, J.A., Vargha-Khadem, F., and Monaco, A.P. (2001). A forkhead-domain gene is mutated in a severe speech and language disorder. *Nature* **413**, 519–523.
1469. Laitman, J.T. (1984). The anatomy of human speech. *Nat. Hist.* **93** #8, 20–27.
1470. Laitman, J.T. and Reidenberg, J.S. (1988). Advances in understanding the relationship between the skull base and larynx with comments on the origins of speech. *Human Evol.* **3**, 99–109.
1471. Lambert, C.C. (2005). Historical introduction, overview, and reproductive biology of the protochordates. *Can. J. Zool.* **83**, 1–7.
1472. Lambert, D. (1993). "The Ultimate Dinosaur Book." Dorling Kindersley, New York.
1473. Lambot, M.-A., Depasse, F., Noel, J.-C., and Vanderhaeghen, P. (2005). Mapping labels in the human developing visual system and the evolution of binocular vision. *J. Neurosci.* **25**, 7232–7237. [See also (2008) *Annu. Rev. Neurosci.* **31**, 295–315.]
1474. LaMendola, N.P. and Bever, T.G. (1997). Peripheral and cerebral asymmetries in the rat. *Science* **278**, 483–486.
1475. Lamont, A.M. (2005). What do monkeys' music choices mean? *Trends Cognitive Sci.* **9**, 359–361.
1476. Lancaster, J.B. (1975). "Primate Behavior and the Emergence of Human Culture." Holt, Rinehart, and Winston, New York.
1477. Lanctôt, C., Moreau, A., Chamberland, M., Tremblay, M.L., and Drouin, J. (1999). Hindlimb patterning and mandible development require the *Ptx1* gene. *Development* **126**, 1805–1810.
1478. Land, M.F. and Nilsson, D.-E. (2002). "Animal Eyes." Oxford Univ. Pr., New York.
1479. Landauer, W. (1958). On phenocopies, their developmental physiology and genetic meaning. *Am. Nat.* **92**, 201–213.
1480. Lande, R. (1976). Natural selection and random genetic drift. *Evolution* **30**, 314–334. [See also (2009) *Nature Rev. Genet.* **10**, 195–205.]
1481. Lande, R. (1981). Models of speciation by sexual selection on polygenic traits. *Proc. Natl. Acad. Sci. USA* **78**, 3721–3725.
1482. Lander, A., King, T., and Brown, N.A. (1998). Left-right development: mammalian phenotypes and conceptual models. *Sems. Cell Dev. Biol.* **9**, 35–41.
1483. Lander, A.D. (2007). Morpheus unbound: Reimagining the morphogen gradient. *Cell* **128**, 245–256.
1484. Landry, C.R., Lemos, B., Rifkin, S.A., Dickinson, W.J., and Hartl, D.L. (2007). Genetic properties influencing the evolvability of gene expression. *Science* **317**, 118–121.
1485. Lane, N. (2007). Reading the book of death. *Nature* **448**, 122–125.
1486. Lang, D., Brown, C.B., and Epstein, J.A. (2004). Neural crest formation and craniofacial development. *In* "Inborn Errors of Development: The Molecular Basis of Clinical Disorders of Morphogenesis," *Oxford Monographs on Medical Genetics, No. 49* (C.J. Epstein, R.P. Erickson, and A. Wynshaw-Boris, eds.). Oxford Univ. Pr, New York, pp. 67–74.
1487. Lang, R.A. (2004). Pathways regulating lens induction in the mouse. *Int. J. Dev. Biol.* **48**, 783–791.
1488. Langdon, J.H. (2005). "The Human Strategy: An Evolutionary Perspective on Human Anatomy." Oxford Univ. Pr., New York.
1489. Lange, K.E. (2002). Wolf to woof: The evolution of dogs. *Nat. Geogr.* **201**#1, 2–11.
1490. Langley, L.L., Telford, I.R., and Christensen, J.B. (1974). "Dynamic Anatomy & Physiology," 4th ed. McGraw-Hill, New York.
1491. Lark, K.G., Chase, K., and Sutter, N.B. (2006). Genetic architecture of the dog: sexual size dimorphism and functional morphology. *Trends Genet.* **22**, 537–544.
1492. Lauder, G.V. (1982). Historical biology and the problem of design. *J. Theor. Biol.* **97**, 57–67.
1493. Lauder, G.V. (1991). Biomechanics and evolution: integrating physical and historical biology in the study of complex systems. *In* "Biomechanics in Evolution" (J.M.V. Rayner and R.J. Wootton, eds.). Cambridge Univ. Pr., New York, pp. 1–19.
1494. Laurent-Vannier, A., Fadda, G., Laigle, P., Dusser, A., and Leroy-Malherbe, V. (1999). Syndrome de Foix-Chavany-Marie d'origine traumatique chez l'enfant. *Rev. Neurol. (Paris)* **155**, 387–390.
1495. Lavado, A. and Montoliu, L. (2006). New animal models to study the role of tyrosinase in normal retinal development. *Frontiers Biosci.* **11**, 743–752.
1496. LaVelle, M. (1995). Natural selection and developmental sexual variation in the human pelvis. *Am. J. Phys. Anthrop.* **98**, 59–72.
1497. Lawrence, P.A. (2004). Last hideout of the unknown? *Nature* **429**, 247.
1498. Lawson, R. (1979). The comparative anatomy of the coelom and of the digestive and respiratory systems. *In* "Hyman's Comparative Vertebrate Anatomy," 3rd ed. (M.H. Wake, ed.). Univ. Chicago Pr, Chicago, Ill., pp. 378–447.
1499. Layton, W.M., Jr. (1976). Random determination of a developmental process: reversal of normal visceral asymmetry in the mouse. *J. Hered.* **67**, 336–338.
1500. Lazar, M.A. (2008). How now, brown fat? *Science* **321**, 1048–1049.
1501. Le Douarin, N.M., Creuzet, S., Couly, G., and Dupin, E. (2004). Neural crest plasticity and its limits. *Development* **131**, 4637–4650.
1502. Le Poole, C. and Boissy, R.E. (1997). Vitiligo. *Semin. Cutan. Med. Surg.* **16**, 3–14.
1503. Lebrecht, S. and Badre, D. (2008). Emotional regulation, or: How I learned to stop worrying and love the nucleus accumbens. *Neuron* **59**, 841–843.
1504. Ledford, H. (2008). Disputed definitions. *Nature* **455**, 1023–1028.

1505. Lee, A.C., Kamalam, A., Adams, S.M., and Jobling, M.A. (2004). Molecular evidence for absence of Y-linkage of the Hairy Ears trait. *Eur. J. Hum. Genet.* **12**, 1077–1079.

1506. Lee, H.J. and Graham, J.B. (2002). Their game is mud. *Nat. Hist.* **111** #7, 42–47.

1507. Lee, K.A., King, R.A., and Summers, C.G. (2001). Stereopsis in patients with albinism: clinical evidence. *J. AAPOS* **5**, 98–104.

1508. Lee, R.B. and DeVore, I., eds. (1968). "Man the Hunter." Aldine Pub. Co., Chicago, Ill.

1509. Lee, S.-H., Bédard, O., Buchtová, M., Fu, K., and Richman, J.M. (2004). A new origin for the maxillary jaw. *Dev. Biol.* **276**, 207–224.

1510. Lee, S.-J. (2007). Sprinting without myostatin: a genetic determinant of athletic prowess. *Trends Genet.* **23**, 475–477.

1511. Lee, W.-C. and Davies, J.A. (2007). Epithelial branching: the power of self-loathing. *BioEssays* **29**, 205–207.

1512. Lehoczky, J.A. and Innis, J.W. (2008). BAC transgenic analysis reveals enhancers sufficient for *Hoxa13* and neighborhood gene expression in mouse embryonic distal limbs and genital bud. *Evol. Dev.* **10**, 421–432.

1513. Leigh, S.R. and Shea, B.T. (1995). Ontogeny and the evolution of adult body size dimorphism in apes. *Am. J. Primatol.* **36**, 37–60.

1514. Leitzell, K. (2008). Just a smile. *Sci. Am. Mind* **19** #2, 8.

1515. Lejeune, J. and Turpin, R. (1961). Détection chromosomique d'une mosaïque artificielle humaine. *Comptes Rendus l'Acad. Sci. Série III.* **252**, 3148–3150.

1516. Lemaire, P., Smith, W.C., and Nishida, H. (2008). Ascidians and the plasticity of the chordate developmental program. *Curr. Biol.* **18**, R620–R631.

1517. LeMay, M. (1976). Morphological cerebral asymmetries of modern man, fossil man, and nonhuman primate. *In* "Origins and Evolution of Language and Speech," *Ann. N. Y. Acad. Sci.*, Vol. 280 (S.R. Harnad, H.D. Steklis, and J. Lancaster, eds.). N.Y. Acad. Sci., New York, pp. 349–366.

1518. Lemon, B. and Tjian, R. (2000). Orchestrated response: a symphony of transcription factors for gene control. *Genes Dev.* **14**, 2551–2569.

1519. Lemons, D. and McGinnis, W. (2006). Genomic evolution of Hox gene clusters. *Science* **313**, 1918–1922.

1520. Lenhoff, H.M., Wang, P.P., Greenberg, F., and Bellugi, U. (1997). Williams Syndrome and the brain. *Sci. Am.* **277** #6, 68–73.

1521. Lens, L., Van Dongen, S., Kark, S., and Matthysen, E. (2002). Fluctuating asymmetry as an indicator of fitness: can we bridge the gap between studies? *Biol. Rev.* **77**, 27–38.

1522. Leonard, W.R. (2002). Food for thought: Dietary change was a driving force in human evolution. *Sci. Am.* **287** #6, 106–115.

1523. Leroi, A.M. (2000). The scale independence of evolution. *Evol. Dev.* **2**, 67–77.

1524. Leroi, A.M. (2003). "Mutants: On Genetic Variety and the Human Body." Viking Pr., New York. [See also Blumberg, M.S. (2009) "Freaks of Nature." Oxford Univ. Pr., New York.]

1525. Lettice, L.A., Hill, A.E., Devenney, P.S., and Hill, R.E. (2008). Point mutations in a distant sonic hedgehog *cis*-regulator generate a variable regulatory output responsible for preaxial polydactyly. *Hum. Mol. Genet.* **17**, 978–985.

1526. Letzkus, P., Ribi, W.A., Wood, J.T., Zhu, H., Zhang, S.-W., and Srinivasan, M.V. (2006). Lateralization of olfaction in the honeybee *Apis mellifera*. *Curr. Biol.* **16**, 1471–1476.

1527. Leucht, P., Minear, S., Ten Berge, D., Nusse, R., and Helms, J.A. (2008). Translating insights from development into regenerative medicine: the function of Wnts in bone biology. *Sems Cell Dev. Biol.* **19**, 434–443.

1528. Levin, M. and Palmer, A.R. (2007). Left-right patterning from the inside out: widespread evidence for intracellular control. *BioEssays* **29**, 271–287. [See also (2008) Special issue of *Dev. Dynamics* 237 #12.]

1529. Levin, M., Roberts, D.J., Holmes, L.B., and Tabin, C. (1996). Laterality defects in conjoined twins. *Nature* **384**, 321.

1530. Levine, M. and Tjian, R. (2003). Transcription regulation and animal diversity. *Nature* **424**, 147–151.

1531. Levinton, J.S. (1986). Developmental constraints and evolutionary saltations: a discussion and critique. *In* "Genetics, Development, and Evolution," *17th Stadler Genetics Symp.*, (J.P. Gustafson, G.L. Stebbins, and F.J. Ayala, eds.). Plenum, New York, pp. 253–288.

1532. Levitin, D. (2007). "This Is Your Brain on Music: Understanding a Human Obsession." Penguin, New York.

1533. Levitt, P. and Eagleson, K.L. (2000). Regionalization of the cerebral cortex: developmental mechanisms and models. *In* "Evolutionary Developmental Biology of the Cerebral Cortex," *Novartis Found. Symp. 228* (G.R. Bock and G. Cardew, eds.). Wiley, New York, pp. 173–187.

1534. Lewin, R. (1993). "The Origin of Modern Humans." Sci. Am. Libr., New York.

1535. Lewis, E.B. (1994). Homeosis: the first 100 years. *Trends Genet.* **10**, 341–343.

1536. Lewis, J., Slack, J.M.W., and Wolpert, L. (1977). Thresholds in development. *J. Theor. Biol.* **65**, 579–590.

1537. Lewis, J.H. and Wolpert, L. (1976). The principle of non-equivalence in development. *J. Theor. Biol.* **62**, 479–490.

1538. Li, G., Wang, J., Rossiter, S.J., Jones, G., and Zhang, S. (2007). Accelerated *FoxP2* evolution in echolocating bats. *PLoS ONE* **9**, e900.

1539. Li, X., Wärri, A., Mäkelä, S., Ahonen, T., Streng, T., Santti, R., and Poutanen, M. (2002). Mammary gland development in transgenic male mice expressing human P450 aromatase. *Endocrinology* **143**, 4074–4083.

1540. Li, Y., Gordon, J., Manley, N.R., Litingtung, Y., and Chiang, C. (2008). Bmp4 is required for tracheal formation: a novel mouse model for tracheal agenesis. *Dev. Biol.* **322**, 145–155.

1541. Li-Kroeger, D., Witt, L.M., Grimes, H.L., Cook, T.A., and Gebelein, B. (2008). Hox and Senseless antagonism functions as a molecular switch to regulate EGF secretion in the *Drosophila* PNS. *Dev. Cell* **15**, 298–308.

1542. Lichanska, A.M. and Waters, M.J. (2007). How growth hormone controls growth, obesity and sexual dimorphism. *Trends Genet.* **24**, 41–47.

1543. Lickliter, R. and Honeycutt, H. (2003). Developmental dynamics: toward a biologically plausible evolutionary psychology. *Psych. Bull.* **129**, 819–835.

1544. Lie, H.C., Rhodes, G., and Simmons, L.W. (2008). Genetic diversity revealed in human faces. *Evolution* **62**, 2473–2486.

1545. Lieberman, D.E. and Hall, B.K. (2007). The evolutionary developmental biology of tinkering: an introduction to the challenge. *In* "Tinkering: The Microevolution of Development," *Novartis Found. Symp. 284* (G. Bock and J. Goode, eds.). Wiley, Chichester, U.K., pp. 1–19.

1546. Lieberman, P. (1991). "Uniquely Human: The Evolution of Speech, Thought, and Selfless Behavior." Harvard Univ. Pr., Cambridge, Mass.

1547. Lieberman, P. (1992). Human speech and language. *In* "The Cambridge Encyclopedia of Human Evolution" (S. Bunney, S. Jones, R. Martin, and D. Pilbeam, eds.). Cambridge Univ. Pr., New York, pp. 134–137.

1548. Liem, K.F. (1973). Evolutionary strategies and morphological innovations: cichlid pharyngeal jaws. *Syst. Zool.* **22**, 425–441.

1549. Liem, K.F. (1988). Form and function of lungs: the evolution of air breathing mechanisms. *Amer. Zool.* **28**, 739–759.

1550. Lin, B., Wang, S.W., and Masland, R.H. (2004). Retinal ganglion cell type, size, and spacing can be specified independent of homotypic dendritic contacts. *Neuron* **43**, 475–485.

1551. Lin, C.H., Liu, J.H., Osterburg, J.W., and Nicol, J.D. (1982). Fingerprint comparison. I. Similarity of fingerprints. *J. Forensic Sci.* **27**, 290–304.

1552. Lin, C.R., Kioussi, C., O'Connell, S., Briata, P., Szeto, D., Liu, F., Izpisúa-Belmonte, J.C., and Rosenfeld, M.G. (1999). Pitx2 regulates lung asymmetry, cardiac positioning and pituitary and tooth morphogenesis. *Nature* **401**, 279–282.

1553. Lin, D., Huang, Y., He, F., Gu, S., Zhang, G., Chen, Y., and Zhang, Y. (2007). Expression survey of genes critical for tooth development in the human embryonic tooth germ. *Dev. Dynamics* **236**, 1307–1312.

1554. Lin, S.-Y. and Burdine, R.D. (2005). Brain asymmetry: switching from left to right. *Curr. Biol.* **15**, R343–R345.

1555. Linden, D.J. (2007). "The Accidental Mind." Harvard Univ. Pr., Cambridge, Mass.

1556. Lipset, D. (1980). "Gregory Bateson: The Legacy of a Scientist." Prentice-Hall, Englewood Cliffs, N.J.

1557. Litingtung, Y., Dahn, R.D., Li, Y., Fallon, J.F., and Chiang, C. (2002). *Shh* and *Gli3* are dispensable for limb skeleton formation but regulate digit number and identity. *Nature* **418**, 979–983.

1558. Little, A.C., Jones, B.C., Waitt, C., Tiddeman, B.P., Feinberg, D.R., Perrett, D.I., Apicella, C.L., and Marlowe, F.W. (2008). Symmetry is related to sexual dimorphism in faces: data across cultures and species. *PLoS ONE* 3 #5, e2106.

1559. Liu, C., Liu, W., Lu, M.-F., Brown, N.A., and Martin, J.F. (2001). Regulation of left–right asymmetry by thresholds of Pitx2c activity. *Development* **128**, 2039–2048.

1560. Liu, K.J., Arron, J.R., Stankunas, K., Crabtree, G.R., and Longaker, M.T. (2007). Chemical rescue of cleft palate and midline defects in conditional GSK-3β mice. *Nature* **446**, 79–82.

1561. Liu, W., Selever, J., Lu, M.-F., and Martin, J.F. (2003). Genetic dissection of *Pitx2* in craniofacial development uncovers new functions in branchial arch morphogenesis, late aspects of tooth morphogenesis and cell migration. *Development* **130**, 6375–6385.

1562. Liversedge, S.P., Rayner, K., White, S.J., Findlay, J.M., and McSorley, E. (2006). Binocular coordination of the eyes during reading. *Curr. Biol.* **16**, 1726–1729.

1563. Llewellyn, M.E., Barretto, R.P.J., Delp, S.L., and Schnitzer, M.J. (2008). Minimally invasive high-speed imaging of sarcomere contractile dynamics in mice and humans. *Nature* **454**, 784–788.

1564. Lnenicka, G.A., Blundon, J.A., and Govind, C.K. (1988). Early experience influences the development of bilateral asymmetry in a lobster motoneuron. *Dev. Biol.* **129**, 84–90.

1565. Lo, L., Dormand, E., Greenwood, A., and Anderson, D.J. (2002). Comparison of the generic neuronal differentiation and neuron subtype specification functions of mammalian *achaete-scute* and *atonal* homologs in cultured neural progenitor cells. *Development* **129**, 1553–1567.

1566. Logan, M. (2003). Finger or toe: the molecular basis of limb identity. *Development* **130**, 6401–6410.

1567. Logan, M. and Tabin, C.J. (1999). Role of Pitx1 upstream of Tbx4 in specification of hindlimb identity. *Science* **283**, 1736–1739.

1568. Logan, M.A. and Vetter, M.L. (2004). Do-it-yourself tiling: dendritic growth in the absence of homotypic contacts. *Neuron* **43**, 439–440.

1569. Lohmueller, K.E., Indap, A.R., Schmidt, S., Boyko, A.R., Hernandez, R.D., Hubisz, M.J., Sninsky, J.J., White, T.J., Sunyaev, S.R., Nielsen, R., Clark, A.G., and Bustamante, C.D. (2008). Proportionally more deleterious genetic variation in European than in African populations. *Nature* **451**, 994–997.

1570. Lohnes, D. (2003). The Cdx1 homeodomain protein: an integrator of posterior signaling in the mouse. *BioEssays* **25**, 971–980.

1571. Long, H., Sabatier, C., Ma, L., Plump, A., Yuan, W., Ornitz, D.M., Tamada, A., Murakami, F., Goodman, C.S., and Tessier-Lavigne, M. (2004). Conserved roles for Slit and Robo proteins in midline commissural axon guidance. *Neuron* **42**, 213–223.

1572. Long, M.E. (1987). What is this thing called sleep? *Nat. Geogr.* **172** #6, 786–821.

1573. Lopatin, A.V. and Averianov, A.O. (2006). An Aegialodontid upper molar and the evolution of mammal dentition. *Science* **313**, 1092.

1574. López-Gracia, M.L. and Ros, M.A. (2007). "Left–Right Asymmetry in Vertebrate Development." *Advances in Anatomy, Embryology, and Cell Biology*, Vol. 188. Springer-Verlag, New York.

1575. Loughry, W.J., Prodöhl, P.A., McDonough, C.M., and Avise, J.C. (1998). Polyembryony in armadillos. *Am. Sci.* **86**, 274–279.

1576. Louis, M., Holm, L., Sánchez, L., and Kaufman, M. (2003). A theoretical model for the regulation of *Sex-lethal*, a gene that controls sex determination and dosage compensation in *Drosophila melanogaster*. *Genetics* **165**, 1355–1384.

1577. Lours, C. and Dietrich, S. (2005). The dissociation of the Fgf-feedback loop controls the limbless state of the neck. *Development* **132**, 5553–5564.

1578. Lovejoy, C.O. (1981). The origin of man. *Science* **211**, 341–350.

1579. Lovejoy, C.O. (1988). Evolution of human walking. *Sci. Am.* **259** #5, 118–125.

1580. Lovejoy, C.O. (1993). Modeling human origins: are we sexy because we're smart, or smart because we're sexy? *In* "The Origin and Evolution of Humans and Humanness" (D.T. Rasmussen, ed.). Jones & Bartlett, Boston, pp. 1–28.

1581. Lovejoy, C.O., Heiple, K.G., and Burstein, A.H. (1973). The gait of *Australopithecus*. *Am J. Phys. Anthrop.* **38**, 757–779.

1582. Lovejoy, C.O., McCollum, M.A., Reno, P.L., and Rosenman, B.A. (2003). Developmental biology and human evolution. *Annu. Rev. Anthrop.* **32**, 85–109.

1583. Lowe, C.J., Terasaki, M., Wu, M., Freeman, R.M., Jr., Runft, L., Kwan, K., Haigo, S., Aronowicz, J., Lander, E., Gruber, C., Smith, M., Kirschner, M., and Gerhart, J. (2006). Dorsoventral patterning in hemichordates: insights into early chordate evolution. *PLoS Biology* **4** #9, 1603–1619 (e291).

1584. Lowe, C.J., Wu, M., Salic, A., Evans, L., Lander, E., Stange-Thomann, N., Gruber, C.E., Gerhart, J., and Kirschner, M. (2003). Anteroposterior patterning in hemichordates and the origins of the chordate nervous system. *Cell* **113**, 853–865.

1585. Lu, M.-F., Pressman, C., Dyer, R., Johnson, R.L., and Martin, J.F. (1999). Function of Rieger syndrome gene in left–right asymmetry and craniofacial development. *Nature* **401**, 276–278.

1586. Lubarsky, B. and Krasnow, M.A. (2003). Tube morphogenesis: making and shaping biological tubes. *Cell* **112**, 19–28.

1587. Lucas, P.W. (2004). "How Teeth Work." Cambridge Univ. Pr., New York.

1588. Lucas, R.J., Douglas, R.H., and Foster, R.G. (2001). Characterization of an ocular photopigment capable of driving pupillary constriction in mice. *Nat. Neurosci.* **4**, 621–626.

1589. Lucchesi, J.C. (1983). Curt Stern: 1902–1981. *Genetics* **103**, 1–4.

1590. Luckett, W.P. (1993). An ontogenetic assessment of dental homologies in therian mammals. *In* "Mammalian Phylogeny: Mesozoic Differentiation, Multituberculates, Monotremes, Early Therians, and Marsupials" (F.S. Szalay, M.J. Novacek, and M.C. McKenna, eds.). Springer-Verlag, New York, pp. 182–204.

1591. Ludwig, M.Z., Palsson, A., Alekseeva, E., Bergman, C.M., Nathan, J., and Kreitman, M. (2005). Functional evolution of a *cis*-regulatory module. *PLoS Biol.* **3** #4, 588–598 (e93).

1592. Lufkin, T., Mark, M., Hart, C.P., Dollé, P., LeMeur, M., and Chambon, P. (1992). Homeotic transformation of the occipital bones of the skull by ectopic expression of a homeobox gene. *Nature* **359**, 835–841.

1593. Lummaa, V., Haukioja, E., Lemmetyinen, R., and Pikkola, M. (1998). Natural selection on human twinning. *Nature* **394**, 533–534.

1594. Lumsden, D. (2002). Crossing the symbolic threshold: a critical review of Terrence Deacon's *The Symbolic Species*. *Philos. Psych.* **15**, 155–171.

1595. Luo, J. and Elledge, S.J. (2008). Deconstructing oncogenesis. *Nature* **453**, 995–996.

1596. Luo, L., Callaway, E.M., and Svoboda, K. (2008). Genetic dissection of neural circuits. *Neuron* **57**, 634–660.

1597. Luo, Z.-X. (2007). Transformation and diversification in early mammal evolution. *Nature* **450**, 1011–1019.

1598. Lupski, J.R. (2007). An evolution revolution provides further revelation. *BioEssays* **29**, 1182–1184.

1599. Lutz, N., Meyrat, B.J., Guignard, J.P., and Hohlfeld, J. (2004). Mermaid syndrome: virtually no hope for survival. *Pediatr. Surg. Int.* **20**, 559–561.

1600. Lutz, P.L. and Bentley, T.B. (1985). Respiratory physiology of diving in the sea turtle. *Copeia* 1985, 671–679.

1601. Lutz, T. (1999). "Crying: The Natural and Cultural History of Tears." W. Norton, New York.

1602. Lyamin, O., Pryaslova, J., Lance, V., and Siegel, J. (2005). Continuous activity in cetaceans after birth. *Nature* **435**, 1177.

1603. Lynch, J.A., Brent, A.E., Leaf, D.S., Pultz, M.A., and Desplan, C. (2006). Localized maternal *orthodenticle* patterns anterior and posterior in the long germ wasp *Nasonia*. *Nature* **439**, 728–732.

1604. Lynch, K.W. and Maniatis, T. (1996). Assembly of specific SR protein complexes on distinct regulatory elements of the *Drosophila doublesex* splicing enhancer. *Genes Dev.* **10**, 2089–2101.
1605. Lynch, M. (2007). The evolution of genetic networks by non-adaptive processes. *Nature Rev. Genet.* **8**, 803–813.
1606. Lynch, M. (2007). The frailty of adaptive hypotheses for the origins of organismal complexity. *PNAS* **104 Suppl. 1**, 8597–8604.
1607. Lynch, M. (2008). The cellular, developmental and population-genetic determinants of mutation-rate evolution. *Genetics* **180**, 933–943.
1608. Lynch, M. and Conery, J.S. (2000). The evolutionary fate and consequences of duplicate genes. *Science* **290**, 1151–1155.
1609. Lynch, V.J. (2008). Clitoral and penile size variability are not significantly different: lack of evidence for the byproduct theory of the female orgasm. *Evol. Dev.* **10**, 396–397.
1610. Lynch, V.J. and Wagner, G.P. (2005). The birth of the uterus. *Nat. Hist.* **114** #10, 36–41.
1611. Lynch, V.J. and Wagner, G.P. (2008). Resurrecting the role of transcription factor change in developmental evolution. *Evolution* **62**, 2131–2154.
1612. Lynch, W. (2007). "Owls of the United States and Canada: A Complete Guide to Their Biology and Behavior." Johns Hopkins Univ. Pr., Baltimore, Md.
1613. Ma, N.S. and Geffner, M.E. (2008). Gynecomastia in prepubertal and pubertal men. *Curr. Opin. Pediatr.* **20**, 465–470.
1614. MacArthur, J.W. and Ford, N. (1937). "A Biological Study of the Dionne Quintuplets: An Identical Set." Univ. Toronto Pr., Toronto, Canada.
1615. Mace, G. (1992). Differences between the sexes. *In* "The Cambridge Encyclopedia of Human Evolution" (S. Bunney, S. Jones, R. Martin, and D. Pilbeam, eds.). Cambridge Univ. Pr.: New York, pp. 52–55.
1616. Machin, G.A. (1993). Conjoined twins: implications for blastogenesis. *Birth Defects Orig. Article Ser.* **29** #1, 141–179.
1617. Macías, A., Romero, N.M., Martín, F., Suárez, L., Rosa, A.L., and Morata, G. (2004). PVF1/PVR signaling and apoptosis promotes the rotation and dorsal closure of the *Drosophila* male terminalia. Int. J. *Dev. Biol.* **48**, 1087–1094.
1618. Macías-Flores, M.A., García-Cruz, D., Rivera, H., Escobar-Luján, M., Melendrez-Vega, A., Rivas-Campos, D., Rodríguez-Collazo, F., Moreno-Arellano, I., and Cantú, J.M. (1984). A new form of hypertrichosis inherited as an X-linked dominant trait. *Hum. Genet.* **66**, 66–70.
1619. Mackay, R.S. and Liaw, H.M. (1981). Dolphin vocalization mechanisms. *Science* **212**, 676–678.
1620. Maclean, K. and Dunwoodie, S.L. (2004). Breaking symmetry: a clinical overview of left-right patterning. *Clin. Genet.* **65**, 441–457.
1621. Maden, M. (1999). Heads or tails? Retinoic acid will decide. *BioEssays* **21**, 809–812.
1622. Maden, M., Gribbin, M.C., and Summerbell, D. (1983). Axial organisation in developing and regenerating vertebrate limbs. *In* "Development and Evolution," *Symp. Brit. Soc. Dev. Biol.*, Vol. **6** (B.C. Goodwin, N. Holder, and C.C. Wylie, eds.). Cambridge Univ. Pr., Cambridge, pp. 381–397.
1623. Maderson, P.F.A. (1972). When? Why? and How?: Some speculations on the evolution of the vertebrate integument. *Amer. Zool.* **12**, 159–171.
1624. Maderson, P.F.A., ed. (1987). "Developmental and Evolutionary Aspects of the Neural Crest." Wiley, New York.
1625. Maeda, R., Hozumi, S., Taniguchi, K., Sasamura, T., Murakami, R., and Matsuno, K. (2007). Roles of *single-minded* in the left-right asymmetric development of the *Drosophila* embryonic gut. *Mechs. Dev.* **124**, 204–217.
1626. Mahadevan, N.R., Horton, A.C., and Gibson-Brown, J.J. (2004). Developmental expression of the amphioxus *Tbx1/10* gene illuminates the evolution of vertebrate branchial arches and sclerotome. *Dev. Genes Evol.* **214**, 559–566.
1627. Mainguy, G., In der Rieden, P.M.J., Berezikov, E., Woltering, J.M., Plasterk, R.H.A., and Durston, A.J. (2003). A position-dependent organisation of retinoid response elements is conserved in the vertebrate *Hox* clusters. *Trends Genet.* **19**, 476–479.
1628. Majerus, M. (2008). What to tell Darwin about Darwinism. *Trends Ecol. Evol.* **23**, 357–358.
1629. Makagon, M.M., Funayama, E.S., and Owren, M.J. (2008). An acoustic analysis of laughter produced by congenitally deaf and normally hearing college students. *J. Acoust. Soc. Am.* **124**, 472–483.
1630. Malaschichev, Y.B. and Wasserug, R.J. (2004). Left and right in the amphibian world: which way to develop and where to turn? *BioEssays* **26**, 512–522.
1631. Malinowski, W. and Wierzba, W. (1998). Twin reversed arterial perfusion syndrome. *Acta Genet. Med. Gemellol. (Roma)* **47**, 75–87.
1632. Mällo, T., Matrov, D., Herm, L., Kõiv, K., Eller, M., Rinken, A., and Harro, J. (2007). Tickling-induced 50-kHz ultrasonic vocalization is individually stable and predicts behaviour in tests of anxiety and depression in rats. *Behav. Brain Res.* **184**, 57–71.
1633. Mangale, V.S., Hiorkawa, K.E., Satyaki, P.R.V., Gokulchandran, N., Chikbire, S., Subramanian, L., Shetty, A.S., Martynoga, B., Paul, J., Mai, M.V., Li, Y., Flanagan, L.A., Tole, S., and Monuki, E.S. (2008). Lhx2 selector activity specifies cortical identity and suppresses hippocampal organizer fate. *Science* **319**, 304–309.
1634. Mania, J.N. (2002). Structure, function and evolution of the gas exchangers: comparative perspectives. *J. Anat.* **201**, 281–304.

1635. Mann, F., Harris, W.A., and Holt, C.E. (2004). New views on retinal axon development: a navigation guide. *Int. J. Dev. Biol.* **48**, 957–964.

1636. Mann, R.S. and Carroll, S.B. (2002). Molecular mechanisms of selector gene function and evolution. *Curr. Opin. Gen. Dev.* **12**, 592–600.

1637. Manni, L., Agnoletto, A., Zaniolo, G., and Burighel, P. (2005). Stomodeal and neurohypophysial placodes in *Ciona intestinalis*: insights into the origin of the pituitary gland. *J. Exp. Zool. (Mol. Dev. Evol.)* **304B**, 324–339.

1638. Manouvrier-Hanu, S., Holder-Espinasse, M., and Lyonnet, S. (1999). Genetics of limb anomalies in humans. *Trends Genet.* **15**, 409–417.

1639. Manseau, L., Baradaran, A., Brower, D., Budhu, A., Elefant, F., Phan, H., Philp, A.V., Yang, M., Glover, D., Kaiser, K., Palter, K., and Selleck, S. (1997). GAL4 enhancer traps expressed in the embryo, larval brain, imaginal discs, and ovary of *Drosophila*. *Dev. Dynamics* **209**, 310–322.

1640. Manzi, M. and Coomes, O.T. (2002). Cormorant fishing in southwestern China: a traditional fishery under siege. *Geogr. Rev.* **92**, 597–603.

1641. Marcil, A., Dumontier, É., Chamberland, M., Camper, S.A., and Drouin, J. (2003). *Pitx1* and *Pitx2* are required for development of hindlimb buds. *Development* **130**, 45–55.

1642. Marcucio, R.S., Cordero, D.R., Hu, D., and Helms, J.A. (2005). Molecular interactions coordinating the development of the forebrain and face. *Dev. Biol.* **284**, 48–61.

1643. Marcus, G. (2008). "Kludge: The Haphazard Construction of the Human Mind." Houghton Mifflin, New York.

1644. Maretto, S., Müller, P.-S., Aricescu, A.R., Cho, K.W.Y., Bikoff, E.K., and Robertson, E.J. (2008). Ventral closure, headfold fusion and definitive endoderm migration defects in mouse embryos lacking the fibronectin leucine-rich transmembrane protein FLRT3. *Dev. Biol.* **318**, 184–193.

1645. Maricich, S.M. and Zoghbi, H.Y. (2006). Getting back to basics. *Cell* **126**, 11–15.

1646. Marieb, E.N. (1996). "Human Anatomy and Physiology Laboratory Manual," 5th ed. Benjamin Cummings, Menlo Park, Calif.

1647. Marin, M.L., Tobias, M.L., and Kelley, D.B. (1990). Hormone-sensitive stages in the sexual differentiation of laryngeal muscle fiber number in *Xenopus laevis*. *Development* **110**, 703–712.

1648. Marino, C., Penedo, M.G., Penas, M., Carreira, M.J., and Gonzalez, F. (2006). Personal authentication using digital retinal images. *Pattern Anal. Applic.* **9**, 21–33.

1649. Marino, L. (2004). Dolphin cognition. *Curr. Biol.* **14**, R910–R911.

1650. Marino, L. (2005). Big brains do matter in new environments. *PNAS* **102** #15, 5306–5307.

1651. Marino, L. (2008). Beautiful minds—for how long? *PLoS Biol.* **6** #7, 1377–1378 (e189).

1652. Marino, L., Connor, R.C., Fordyce, R.E., Herman, L.M., Hof, P.R., Lefebvre, L., Lusseau, D., McCowan, B., Nimchinsky, E.A., Pack, A.A., Rendell, L., Reidenberg, J.S., Reiss, D., Uhen, M.D., Van der Gucht, E., and Whitehead, H. (2007). Cetaceans have complex brains for complex cognition. *PLoS Biol.* **5** #5, 966–972.

1653. Marino, L., McShea, D.W., and Uhen, M.D. (2004). Origin and evolution of large brains in toothed whales. *Anat. Rec.* **281A**, 1247–1255.

1654. Mark, R. (1996). Architecture and evolution. *Am. Sci.* **84**, 383–389.

1655. Markert, C.L. and Petters, R.M. (1978). Manufactured hexaparental mice show that adults are derived from three embryonic cells. *Science* **202**, 56–58.

1656. Marlétaz, F., Holland, L.Z., Laudet, V., and Schubert, M. (2006). Retinoic acid signaling and the evolution of chordates. *Int. J. Biol. Sci.* **2**, 38–47.

1657. Marmor, M.F., Choi, S.S., Zawadzki, R.J., and Werner, J.S. (2008). Visual insignificance of the foveal pit. *Arch. Ophthalmol.* **126**, 907–913.

1658. Marquardt, T., Ashery-Padan, R., Andrejewski, N., Scardigli, R., Guillemot, F., and Gruss, P. (2001). Pax6 is required for the multipotent state of retinal progenitor cells. *Cell* **105**, 43–55.

1659. Marshall, C.R., Raff, E.C., and Raff, R.A. (1994). Dollo's law and the death and resurrection of genes. *Proc. Natl. Acad. Sci. USA* **91**, 12283–12287.

1660. Marshall, L., Helgadóttir, H., Mölle, M., and Born, J. (2006). Boosting slow oscillations during sleep potentiates memory. *Nature* **444**, 610–613.

1661. Marshall, W.H. (1959). Plato's myth of Aristophanes and Shelley's Panthea. *Classical J.* **55** #3, 121–123.

1662. Marte, B. (2006). Red-eye redirected. *Nature* **443**, 928.

1663. Martin, G., Rojas, L.M., Ramírez, Y., and McNeil, R. (2004). The eyes of oilbirds (*Steatornis caripensis*): pushing at the limits of sensitivity. *Naturwissenschaften* **91**, 26–29.

1664. Martin, G.N. and Gray, C.D. (1996). The effects of audience laughter on men's and women's responses to humor. *J. Soc. Psychol.* **136**, 221–231.

1665. Martin, G.R. (1986). Shortcomings of an eagle's eye. *Nature* **319**, 357.

1666. Martin, G.R. (1998). The roles of FGFs in the early development of vertebrate limbs. *Genes Dev.* **12**, 1571–1586.

1667. Martin, L. (1991). Teeth, sex and species. *Nature* **352**, 111–112.

1668. Martin, P. and Parkhurst, S.M. (2004). Parallels between tissue repair and embryo morphogenesis. *Development* **131**, 3021–3034.

1669. Martin, R.E., Pine, R.H., and DeBlase, A.F. (2001). "A Manual of Mammology.," 3rd ed. McGraw-Hill, New York.

1670. Martin, T. and Luo, Z.-X. (2005). Homoplasy in the mammalian ear. *Science* **307**, 861–862.

1671. Martín-Loeches, M. (2006). On the uniqueness of humankind: is language working memory the final piece that made us human? *J. Human Evol.* **50**, 226–229.
1672. Martindale, M.Q. and Henry, J.Q. (1998). The development of radial and biradial symmetry: the evolution of bilaterality. *Amer. Zool.* **38**, 672–684.
1673. Martinez Arias, A. and Stewart, A. (2002). "Molecular Principles of Animal Development." Oxford Univ. Pr., New York.
1674. Martini, F.H., Ober, W.C., Garrison, C.W., Welch, K., Hutchings, R.T., and Ireland, K. (2004). "Fundamentals of Anatomy & Physiology," 6th ed. Benjamin Cummings, San Francisco, Calif.
1675. Marx, J. (2004). How cells endure low oxygen. *Science* **303**, 1454–1456.
1676. Marzke, M.W. and Marzke, R.F. (2000). Evolution of the human hand: approaches to acquiring, analysing and interpreting the anatomical evidence. *J. Anat.* **197**, 121–140.
1677. Masand, P., Popli, A.P., and Weilburg, J.B. (1995). Sleepwalking. *Am. Fam. Physician* **51**, 649–654.
1678. Mashanov, V.S., Zueva, O.R., Heinzeller, T., Aschauer, B., and Dolmatov, I.Y. (2007). Developmental origin of the adult nervous system in a holothurian: an attempt to unravel the enigma of neurogenesis in echinioderms. *Evol. Dev.* **9**, 244–256.
1679. Masland, R.H. (1986). The functional architecture of the retina. *Sci. Am.* **255** #6, 102–111.
1680. Mason, F. (1928). "Creation by Evolution." Macmillan, New York.
1681. Mason, S.F. (1991). "Chemical Evolution: Origin of the Elements, Molecules, and Living Systems." Oxford Univ. Pr., New York.
1682. Mason, S.F. (1991). Origins of the handedness of biological molecules. *In* "Biological Asymmetry and Handedness," *Ciba Found. Symp. 162,* (G.R. Bock and J. Marsh, eds.). Wiley, New York, pp. 3–15.
1683. May, R. and Thurner, J. (1957). The cause of the predominantly sinistral occurrence of thrombosis of the pelvic veins. *Angiology* **8**, 419–427.
1684. May, R.M., Lawton, J.H., and Stork, N.E. (1995). Assessing extinction rates. *In* "Extinction Rates" (J.H. Lawton and R.M. May, eds.). Oxford Univ. Pr., New York, pp. 1–24.
1685. Maynard Smith, J. (1978). "The Evolution of Sex." Cambridge Univ. Pr., New York.
1686. Maynard Smith, J. (1998). "Shaping Life: Genes, Embryos and Evolution." Yale Univ. Pr., New Haven.
1687. Maynard Smith, J., Burian, R., Kauffman, S., Alberch, P., Campbell, J., Goodwin, B., Lande, R., Raup, D., and Wolpert, L. (1985). Developmental constraints and evolution. *Q. Rev. Biol.* **60**, 265–287.
1688. Maynard Smith, J. and Sondhi, K.C. (1960). The genetics of a pattern. *Genetics* **45**, 1039–1050.
1689. Maynard Smith, J. and Szathmáry, E. (1995). "The Major Transitions in Evolution." W. H. Freeman, New York.
1690. Mayr, E. (1976). "Evolution and the Diversity of Life." Harvard Univ. Pr., Cambridge, Mass.
1691. Mayr, E. (1991). "One Long Argument: Charles Darwin and the Genesis of Modern Evolutionary Thought." Harvard Univ. Pr., Cambridge, Mass.
1692. McAvoy, J.W. (1980). Induction of the eye lens. *Differentiation* **17**, 137–149.
1693. McCabe, K.L., Gunther, E.C., and Reh, T.A. (1999). The development of the pattern of retinal ganglion cells in the chick retina: mechanisms that control differentiation. *Development* **126**, 5713–5724.
1694. McCauley, D.W. and Bronner-Fraser, M. (2006). Importance of SoxE in neural crest development and the evolution of the pharynx. *Nature* **441**, 750–752.
1695. McCollum, M. and Sharpe, P.T. (2001). Evolution and development of teeth. *J. Anat.* **199**, 153–159.
1696. McCollum, M.A. and Sharpe, P.T. (2001). Developmental genetics and early hominid craniodental evolution. *BioEssays* **23**, 481–493.
1697. McCollum, M.A., Sherwood, C.C., Vinyard, C.J., Lovejoy, C.O., and Schachat, F. (2006). Of muscle-bound crania and human brain evolution: the story behind the MYH16 headlines. *J. Human Evol.* **50**, 232–236.
1698. McCrory, P. and Bell, S. (1999). Nerve entrapment syndromes as a cause of pain in the hip, groin and buttock. *Sports Med.* **27**, 261–274.
1699. McCune, A.R. and Carlson, R.L. (2004). Twenty ways to lose your bladder: common natural mutants in zebrafish and widespread convergence of swim bladder loss among teleost fishes. *Evol. Dev.* **6**, 246–259.
1700. McDermott, J. (2008). The evolution of music. *Nature* **453**, 287–288.
1701. McDermott, J. and Hauser, M.D. (2007). Nonhuman primates prefer slow tempos but dislike music overall. *Cognition* **104**, 654–668.
1702. McGhee, G.R., Jr. (2007). "The Geometry of Evolution: Adaptive Landscapes and Theoretical Morphospaces." Cambridge Univ. Pr., New York.
1703. McGinnis, W. (1994). A century of homeosis, a decade of homeoboxes. *Genetics* **137**, 607–611.
1704. McGlinn, E. and Tabin, C.J. (2006). Mechanistic insight into how Shh patterns the vertebrate limb. *Curr. Opin. Gen. Dev.* **16**, 426–432.
1705. McGrath, J., Somlo, S., Makova, S., Tian, X., and Brueckner, M. (2003). Two populations of node monocilia initiate left-right asymmetry in the mouse. *Cell* **114**, 61–73.
1706. McHenry, H.M. (2006). Tempo and mode in human evolution. *In* "The Human Evolution Source Book," 2nd ed. *Advances in Human Evolution Series* (R.L. Ciochon and J.G. Fleagle, eds.). Pearson Prentice Hall: Upper Saddle River, N.J., pp. 108–115.
1707. McIntyre, D.C., Rakshit, S., Yallowitz, A.R., Loken, L., Jeannotte, L., Capecchi, M.R., and Wellik, D.M. (2007). Hox patterning of the vertebrate rib cage. *Development* **134**, 2981–2989.

1708. McKee, J.K. (1999). The autocatalytic nature of hominid evolution in African Plio-Pleistocene environments. *In* "African Biogeography, Climate Change, and Human Evolution" (T.G. Bromage and F. Schrenk, eds.). Oxford Univ. Pr., New York, pp. 57–67.

1709. McKinney, M.L., ed. (1988). "Heterochrony in Evolution: A Multidisciplinary Approach." Plenum Pr., New York.

1710. McKinney, M.L. and McNamara, K.J. (1991). "Heterochrony: The Evolution of Ontogeny." Plenum Pr., New York.

1711. McKusick, V.A. (1998). "Mendelian Inheritance in Man: A Catalog of Human Genes and Genetic Disorders," 12th ed. Johns Hopkins Univ. Pr., Baltimore, Md.

1712. McLaughlin, P.A. (1974). The hermit crabs (Crustacea Decapoda, Paguridea) of Northwestern North America. *Zool. Verh.* **130**, 1–397.

1713. McManus, C. (2002). "Right Hand, Left Hand: The Origins of Asymmetry in Brains, Bodies, Atoms and Cultures." Harvard Univ. Pr., Cambridge, Mass.

1714. McManus, C. (2005). Reversed bodies, reversed brains, and (some) reversed behaviors: of zebrafish and men. *Dev. Cell* **8**, 796–797.

1715. McManus, I.C. (1991). The inheritance of left-handedness. *In* "Biological Asymmetry and Handedness," *Ciba Found. Symp. 162* (G.R. Bock and J. Marsh, eds.). Wiley, New York, pp. 251–281.

1716. McNamara, J.M., Barta, Z., Fromhage, L., and Houston, A.I. (2008). The coevolution of choosiness and cooperation. *Nature* **451**, 189–192.

1717. McNamara, K.J. (2002). Sequential hypermorphosis: stretching ontogeny to the limit. *In* "Human Evolution through Developmental Change" (N. Minugh-Purvis and K.J. McNamara, eds.). Johns Hopkins Univ. Pr., Baltimore, Md, pp. 102–121.

1718. Mead, J.G. (1975). Anatomy of the external nasal passages and facial complex in the Delphinidae (Mammalia: Cetacea). *Smithsonian Contrib. Zool.* **207**, 72 pp.

1719. Mead, L.S. and Arnold, S.J. (2004). Quantitative genetic models of sexual selection. *Trends Ecol. Evol.* **19**, 264–271.

1720. Medawar, P.B. (1957). The imperfections of man. *In* "The Uniqueness of the Individual" (P.B. Medawar, ed.). Basic Books: New York, pp. 122–133.

1721. Medawar, P.B. (1957). "The Uniqueness of the Individual." Basic Bks. New York.

1722. Medvedev, Z.A. (1990). An attempt at a rational classification of theories of ageing. *Biol. Rev.* **65**, 375–398.

1723. Meinertzhagen, I.A. (1991). Evolution of the cellular organization of the arthropod compound eye and optic lobe. *In* "Evolution of the Eye and Visual System" (J.R. Cronly-Dillon and R.L. Gregory, eds.). CRC Pr., Boston., pp. 341–363.

1724. Meinhardt, H. (1982). "Models of Biological Pattern Formation." Acad. Pr., New York.

1725. Mellon, D., Jr. (1978). Limb morphology and function are transformed by contralateral nerve section in snapping shrimps. *Nature* **272**, 246–248.

1726. Melyan, Z., Tarttelin, E.E., Bellingham, J., Lucas, R.J., and Hankins, M.W. (2005). Addition of human melanopsin renders mammalian cells photoresponsive. *Nature* **433**, 741–745.

1727. Menaker, M. and Zimmerman, N. (1976). Role of the pineal in the circadian system of birds. *Amer. Zool.* **16**, 45–55.

1728. Menke, D.B., Guenther, C., and Kingsley, D.M. (2008). Dual hindlimb control elements in the Tbx4 gene and region-specific control of bone size in vertebrate limbs. *Development* **135**, 2543–2553.

1729. Merika, M. and Thanos, D. (2001). Enhanceosomes. Curr. Opin. Gen. Dev. **11**, 205–208.

1730. Merino, R., Macias, D., Gañan, Y., Rodriguez-Leon, J., Economides, A.N., Rodriguez-Estaban, C., Izpisua-Belmonte, J.C., and Hurle, J.M. (1999). Control of digit formation by activin signalling. *Development* **126**, 2161–2170.

1731. Mesecar, A.D. and Koshland, D.E., Jr. (2000). A new model for protein stereospecificity. *Nature* **403**, 614–615.

1732. Messenger, J.B. (1991). Photoreception and vision in molluscs. *In* "Evolution of the Eye and Visual System" (J.R. Cronly-Dillon and R.L. Gregory, eds.). CRC Pr., Boston., pp. 364–397.

1733. Metzger, R.J., Klein, O.D., Martin, G.R., and Krasnow, M.A. (2008). The branching programme of mouse lung development. *Nature* **453**, 745–750.

1734. Metzger, R.J. and Krasnow, M.A. (1999). Genetic control of branching morphogenesis. *Science* **284**, 1635–1639. [See also (2008) *Science* **322**, 1506–1509.]

1735. Meulemans, D. and Bronner-Fraser, M. (2005). Central role of gene cooption in neural crest evolution. *J. Exp. Zool. (Mol. Dev. Evol.)* **304B**, 298–303.

1736. Meyer, M., Baumann, S., Wildgruber, D., and Alter, K. (2007). How the brain laughs. Comparative evidence from behavioral, electrophysiological and neuroimaging studies in human and monkey. *Behav. Brain Res.* **182**, 245–260.

1737. Micevych, P.E., Coquelin, A., and Arnold, A.P. (1986). Immunohistochemical distribution of Substance P, serotonin, and methionine enkephalin in sexually dimorphic nuclei of the rat lumbar spinal cord. *J. Comp. Neurol.* **248**, 235–244.

1738. Migeon, B.R. (2007). "Females are Mosaics: X Inactivation and Sex Differences in Disease." Oxford Univ. Pr., New York.

1739. Mignot, E. (2008). Why we sleep: the temporal organization of recovery. *PLoS Biol.* **6** #4, 661–669 (e106).

1740. Mikkola, M.L. (2007). Genetic basis of skin appendage development. *Sems. Cell Dev. Biol.* **18**, 225–236.

1741. Miklos, G.L.G. and Rubin, G.M. (1996). The role of the Genome Project in determining gene function: insights from model organisms. *Cell* **86**, 521–529.
1742. Miklosi, A. (2007). "Dog Behaviour, Evolution and Cognition." Oxford Univ. Pr., New York.
1743. Miletich, I. and Sharpe, P.T. (2004). Neural crest contribution to mammalian tooth formation. *Birth Defects Res. (Part C)* **72**, 200–212.
1744. Milius, S. (2001). Face the music. *Nat. Hist.* **110** #10, 48–57.
1745. Miller, A. (1941). Position of adult testes in *Drosophila melanogaster* Meigen. *Proc. Natl. Acad. Sci.* **27**, 35–41.
1746. Miller, A. (1950). The internal anatomy and histology of the imago of *Drosophila melanogaster. In* "Biology of *Drosophila*" (M. Demerec, ed.). Hafner: New York, pp. 420–534.
1747. Miller, G. (2007). Grasping for clues to the biology of itch. *Science* **318**, 188–189.
1748. Miller, G. (2007). Hunting for meaning after midnight. *Science* **315**, 1360–1363.
1749. Miller, G. (2008). Berkeley hyenas face an uncertain future. *Science* **319**, 722–723.
1750. Miller, G.S., Jr. (1931). Human hair and primate patterning. *Smithsonian Misc. Coll. (Publ. No. 3130)* **85** #10, 1-13 (plus 5 plates).
1751. Miller, K. and Doman, J.M.R. (**1996**). Together forever. *Life* **April 1996**, 44–56.
1752. Miller, K.R. (1994). Life's grand design. *Technol. Rev.* **97** #2, 24–32.
1753. Miller, S.W., Hayward, D.C., Bunch, T.A., Miller, D.J., Ball, E.E., Bardwell, V.J., Zarkower, D., and Brower, D.L. (2003). A DM domain protein from a coral, Acropora millepora, homologous to proteins important for sex determination. *Evol. Dev.* **5**, 251–258.
1754. Millien, G., Beane, J., Lenburg, M., Tsao, P.-N., Lu, J., Spira, A., and Ramirez, M.I. (2008). Characterization of the mid-foregut transcriptome identifies genes regulated during lung bud induction. *Gene Expr. Patt.* **8**, 124–139.
1755. Millien, V. (2006). Morphological evolution is accelerated among island mammals. *PLoS Biol.* **4** #10, 1863–1868 (e321; correction in v. 4, #11, p. 2165).
1756. Milton, K. (1993). Diet and primate evolution. *Sci. Am.* **269** #2, 86–93.
1757. Milton, R.C.d., Milton, S.C.F., and Kent, S.B.H. (1992). Total chemical synthesis of a D-enzyme: the enantiomers of HIV-1 protease show demonstration of reciprocal chiral substrate specificity. *Science* **256**, 1445–1448.
1758. Mima, T., Ohuchi, H., Noji, S., and Mikawa, T. (1995). FGF can induce outgrowth of somatic mesoderm both inside and outside of limb-forming regions. *Dev. Biol.* **167**, 617–620.
1759. Minelli, A. (2000). Limbs and tail as evolutionarily diverging duplicates of the main body axis. *Evol. Dev.* **2**, 157–165.
1760. Minelli, A. (2002). Homology, limbs, and genitalia. *Evol. Dev.* **4**, 127–132.
1761. Minelli, A. (2003). The origin and evolution of appendages. *Int. J. Dev. Biol.* **47**, 573–581.
1762. Minelli, A. and Fusco, G. (2005). Conserved versus innovative features in animal body organization. *J. Exp. Zool. (Mol. Dev. Evol.)* **304B**, 520–525.
1763. Minelli, A. and Fusco, G., eds. (2008). "Evolving Pathways: Key Themes in Evolutionary Developmental Biology." Cambridge Univ. Pr., New York.
1764. Minetti, A.E. (2001). Walking on other planets. *Nature* **409**, 467–469.
1765. Minguillon, C., Del Buono, J., and Logan, M.P. (2005). *Tbx5* and *Tbx4* are not sufficient to determine limb-specific morphologies but have common roles in initiating limb outgrowth. *Dev. Cell* **8**, 75–84.
1766. Minsuk, S.B. and Raff, R.A. (2002). Pattern formation in a pentameral animal: induction of early adult rudiment development in sea urchins. *Dev. Biol.* **247**, 335–350. [See also (2009) *Dev. Genes Evol.* **219**, 89–101.]
1767. Misro, A.K. and Radhika, V. (2008). A case of congenital absence/rudimentary vermiform appendix. *Bombay Hosp. J.* **50**, 293–294.
1768. Mitchell, B., Jacobs, R., Li, J., Chien, S., and Kintner, C. (2007). A positive feedback mechanism governs the polarity and motion of motile cilia. *Nature* **447**, 97–101.
1769. Mitsiadis, T.A., Caton, J., De Bari, C., and Bluteau, G. (2008). The large functional spectrum of the heparin-binding cytokines MK and HB-GAM in continuously growing organs: the rodent incisor as a model. *Dev. Biol.* **320**, 256–266.
1770. Mitsiadis, T.A. and Smith, M.M. (2006). How do genes make teeth to order through development? *J. Exp. Zool. (Mol. Dev. Evol.)* **306B**, 177–182.
1771. Mittenthal, J.E. (1981). The rule of normal neighbors: A hypothesis for morphogenetic pattern regulation. *Dev. Biol.* **88**, 15–26.
1772. Mitteroecker, P. and Bookstein, F. (2008). The evolutionary role of modularity and integration in the hominid cranium. *Evolution* **62**, 943–958.
1773. Mittwoch, U. (2000). Genetics of sex determination: exceptions that prove the rule. *Molec. Genet. Metab.* **71**, 405–410.
1774. Mittwoch, U. (2001). Genetics of mammalian sex determination: some unloved exceptions. *J. Exp. Zool.* **290**, 484–489.
1775. Mittwoch, U. (2006). Sex is a threshold dichotomy mimicking a single gene effect. *Trends Genet.* **22**, 96–100.
1776. Mizunami, M. (1994). Processing of contrast signals in the insect ocellar system. *Zool. Sci.* **11**, 175–190.
1777. Mizunami, M. (1995). Functional diversity of neural organization in insect ocellar systems. *Vision Res.* **35**, 443–452.

1778. Mizutani, C.M. and Bier, E. (2008). EvoD/Vo: the origins of BMP signalling in the neuroectoderm. *Nature Rev. Genet.* **9**, 663–677.

1779. Mizutani, C.M., Meyer, N., Roelink, H., and Bier, E. (2006). Threshold-dependent BMP-mediated repression: a model for a conserved mechanism that patterns the neuroectoderm. *PLoS Biol.* **4** #10, 1777–1788 (e313).

1780. Mo, R., Kim, J.H., Zhang, J., Chiang, C., Hui, C.C., and Kim, P.C.W. (2001). Anorectal malformations caused by defects in sonic hedgehog signaling. *Am. J. Pathol.* **159**, 765–774.

1781. Mobbs, D., Greicius, M.D., Abdel-Azim, E., Menon, V., and Reiss, A.L. (2003). Humor modulates the mesolimbic reward centers. *Neuron* **40**, 1041–1048.

1782. Moczek, A.P. (2008). On the origins of novelty in development and evolution. *BioEssays* **30**, 432–447.

1783. Mofid, A., Alinaghi, S.A.S., Zandieh, S., and Yazdani, T. (2007). Hirsutism. *Int. J. Clin. Pract.* **62**, 433–443.

1784. Mogilner, A. and Edelsteinkeshet, L. (1995). Selecting a common direction. 1. How orientational order can arise from simple contact responses between interacting cells. *J. Math. Biol.* **33**, 619–660.

1785. Mohit, P., Makhijani, K., Madhavi, M.B., Bharathi, V., Lal, A., Sirdesai, G., Reddy, V.R., Ramesh, P., Kannan, R., Dhawan, J., and Shashidhara, L.S. (2006). Modulation of AP and DV signaling pathways by the homeotic gene *Ultrabithorax* during haltere development in Drosophila. *Dev. Biol.* **291**, 356–367.

1786. Mohun, J., ed. (2006). "Human." Dorling Kindersley, New York.

1787. Moller, A.P. and Swaddle, J.P. (1997). "Asymmetry, Developmental Stability, and Evolution." Oxford Univ. Pr., Oxford.

1788. Molloy, P.P. and Gage, M.J.G. (2006). Evolution: Vertebrate reproductive strategies get mixed up. *Curr. Biol.* **16**, R876–R879.

1789. Molnár, Z., Hoerder-Suabedissen, A., Wang, W.Z., DeProto, J., Davies, K., Lee, S., Jacobs, E.C., Campagnoni, A.T., Paulsen, O., Piñon, M.C., and Cheung, A.F.P. (2007). Genes involved in the formation of the earliest cortical circuits. *In* "Cortical Development: Genes and Genetic Abnormalities," *Novartis Found. Symp. 288* (G. Bock and J. Goode, eds.). Wiley, Hoboken, N.J., pp. 212–229.

1790. Mon-Williams, M. and Tresilian, J.R. (2001). A simple rule of thumb for elegant prehension. *Curr. Biol.* **11**, 1058–1061.

1791. Monkhorst, K., Jonkers, I., Rentmeester, E., Grosveld, F., and Gribnau, J. (2008). X inactivation counting and choice is a stochastic process: Evidence for involvement of an X-linked activator. *Cell* **132**, 410–421.

1792. Monod, J. (1974). On chance and necessity. *In* "Studies in the Philosophy of Biology: Reduction and Related Problems" (F.J. Ayala and T. Dobzhansky, eds.). Univ. Calif. Pr., Berkeley, pp. 357–375.

1793. Montagna, W. (1972). The skin of nonhuman primates. *Amer. Zool.* **12**, 109–124.

1794. Montavon, T., Le Garrec, J.-F., Kerszberg, M., and Duboule, D. (2008). Modeling Hox gene regulation in digits: reverse collinearity and the molecular origin of thumbness. *Genes Dev.* **22**, 346–359.

1795. Monteiro, L.R., Bonato, V., and dos Reis, S.F. (2005). Evolutionary integration and morphological diversification in complex morphological structures: mandible shape divergence in spiny rats (Rodentia, Echimyidae). *Evol. Dev.* **7**, 429–439.

1796. Montero, J.A. and Hurlé, J.M. (2007). Deconstructing digit chondrogenesis. *BioEssays* **29**, 725–737.

1797. Montero, J.A., Lorda-Diez, C.I., Gañan, Y., Macias, D., and Hurle, J.M. (2008). Activin/TGFβ and BMP crosstalk determines digit chrondrogenesis. *Dev. Biol.* **321**, 343–356.

1798. Moorad, J.A. and Linksvayer, T.A. (2008). Levels of selection on threshold characters. *Genetics* **179**, 899–905.

1799. Moore, J.A. (1987). Science as a way of knowing–developmental biology. *Amer. Zool.* **27**, 415–573.

1800. Moore, K.B., Schneider, M.L., and Vetter, M.L. (2002). Posttranslational mechanisms control the timing of bHLH function and regulate retinal cell fate. *Neuron* **34**, 183–195.

1801. Moore, K.L. and Persaud, T.V.N. (2008). "The Developing Human: Clinically Oriented Embryology," 8th ed. Saunders, Philadelphia.

1802. Moran, N.A. (2002). Microbial minimalism: genome reduction in bacterial pathogens. *Cell* **108**, 583–586.

1803. Morell, V. (1991). A hand on the bird—and one on the bush. *Science* **254**, 33–34.

1804. Morell, V. (2008). Minds of their own: Animals are smarter than you think. *Nat. Geogr.* **213** #3, 36–61.

1805. Moreno, T.A. and Kintner, C. (2004). Regulation of segmental patterning by retinoic acid signaling during Xenopus somitogenesis. *Dev. Cell* **6**, 205–218.

1806. Morey, D.F. (1994). The early evolution of the domestic dog. *Am. Sci.* **82**, 336–347.

1807. Morgan, M.A., Thurnau, G.R., and Fishburne, J.I. (1986). The fetal-pelvic index as an indicator of fetal-pelvic disproportion: a preliminary report. *Am. J. Obstet. Gynecol.* **155**, 608–613.

1808. Morgan, M.J. (1992). On the evolutionary origin of right handedness. *Curr. Biol.* **2**, 15–17.

1809. Morgan, T. (2007). Turner syndrome: diagnosis and management. *Am. Fam. Physician* **76**, 405–410.

1810. Morin-Kensicki, E.M., Melancon, E., and Eisen, J.S. (2002). Segmental relationship between somites and vertebral column in zebrafish. *Development* **129**, 3851–3860.

1811. Morrison, P. (1979). Termites and telescopes. *NOVA/BBC TV Program Transcript: 1979 Jacob Bronowski Memorial Lecture*, 1–17.

1812. Morrison, P. and Morrison, P. (1999). Walk, run—and skip. *Sci. Am.* **280** #3, 111–113.

1813. Morton, D.J. (1935). "The Human Foot." Columbia Univ. Pr., Morningside Heights, N.Y.
1814. Moses, A.M., Pollard, D.A., Nix, D.A., Iyer, V.N., Li, X.-Y., Biggin, M.D., and Eisen, M.B. (2006). Large-scale turnover of functional transcription factor binding sites in Drosophila. *PLoS Comput. Biol.* **2** #10, 1219–1231 (e130).
1815. Moses, K. (2006). Fly eyes get the whole picture. *Nature* **443**, 638–639.
1816. Mosher, D.S., Quignon, P., Bustamante, C.D., Sutter, N.B., Mellersh, C.S., Parker, H.G., and Ostrander, E.A. (2007). A mutation in the myostatin gene increases muscle mass and enhances racing performance in heterozygote dogs. *PLoS Genetics* **3** #5, 779–786 (e79).
1817. Moussian, B. and Uv, A.E. (2005). An ancient control of epithelial barrier formation and wound healing. *BioEssays* **27**, 987–990.
1818. Mrackova, M., Nicolas, M., Hobza, R., Negrutiu, I., Monéger, F., Widmer, A., Vyskot, B., and Janousek, B. (2008). Independent origin of sex chromosomes in two species of the genus Silene. *Genetics* **179**, 1129–1133.
1819. Mu, X. and Klein, W.H. (2004). A gene regulatory hierarchy for retinal ganglion cell specification and differentiation. *Sems. Cell Dev. Biol.* **15**, 115–123.
1820. Mueller, S.V., Jackson, G.M., Dhalla, R., Datsopoulos, S., and Hollis, C.P. (2006). Enhanced cognitive control in young people with Tourette's Syndrome. *Curr. Biol.* **16**, 570–573.
1821. Müller, G.B. (1991). Experimental strategies in evolutionary embryology. *Amer. Zool.* **31**, 605–615.
1822. Müller, G.B. (2007). Evo-devo: extending the evolutionary synthesis. *Nature Rev. Genet.* **8**, 943–949.
1823. Müller, G.B. (2008). Evo-devo as a discipline. *In* "Evolving Pathways: Key Themes in Evolutionary Developmental Biology" (A. Minelli and G. Fusco, eds.). Cambridge Univ. Pr., New York, pp. 5–30.
1824. Müller, G.B. and Newman, S.A. (2005). The innovation triad: an EvoDevo agenda. *J. Exp. Zool. (Mol. Dev. Evol.)* **304B**, 487–503.
1825. Muller, H.J. (1939). Reversibility in evolution considered from the standpoint of genetics. *Biol. Rev.* **14**, 261–280.
1826. Muller, J.K., Prather, D.R., and Nascone-Yoder, N.M. (2003). Left-right asymmetric morphogenesis in the Xenopus digestive system. *Dev. Dynamics* **228**, 672–682.
1827. Muller, M.N., Thompson, M.E., and Wrangham, R.W. (2006). Male chimpanzees prefer mating with old females. *Curr. Biol.* **16**, 2234–2238.
1828. Müller, W.E.G. (2001). How was metazoan threshold crossed? The hypothetical Urmetazoa. *Comp. Biochem. Physiol. Part A* **129**, 433–460.
1829. Muneoka, K., Han, M., and Gardiner, D.M. (2008). Regrowing human limbs. *Sci. Am.* **298** #4, 56–63.
1830. Muñoz-Chápuli, R., Carmona, R., Guadix, J.A., Macías, D., and Pérez-Pomares, J.M. (2005). The origin of the endothelial cells: an evo-devo approach for the invertebrate/vertebrate transition of the circulatory system. *Evol. Dev.* **7**, 351–358.
1831. Muragaki, Y., Mundlos, S., Upton, J., and Olsen, B.R. (1996). Altered growth and branching patterns in synpolydactyly caused by mutations in HOXD13. *Science* **272**, 548–551.
1832. Murdock, C. and Wibbels, T. (2006). *Dmrt1* expression in response to estrogen treatment in a reptile with temperature-dependent sex determination. *J. Exp. Zool. (Mol. Dev. Evol.)* **306B**, 134–139.
1833. Murisier, F., Guichard, S., and Beermann, F. (2007). Distinct distal regulatory elements control tyrosinase expression in melanocytes and the retinal pigment epithelium. *Dev. Biol.* **303**, 838–847.
1834. Murphy, J.B., Gutiérrez-Alonso, G., Nance, R.D., Fernández-Suárez, J., Keppie, J.D., Quesada, C., Strachan, R.A., and Dostal, J. (2008). Tectonic plates come apart at the seams. *Am. Sci.* **96**, 129–137.
1835. Murphy, N. and Brown, W.S. (2007). "Did My Neurons Make Me Do It? Philosophical and Neurobiological Perspectives on Moral Responsibility and Free Will." Oxford Univ. Pr., New York.
1836. Murray, J.D. (1989). "Mathematical Biology." Springer-Verlag, Berlin.
1837. Murray, J.M. (1991). Structure of flagellar microtubules. *Int. Rev. Cytol.* **125**, 47–93.
1838. Muster, A.J., Idriss, R.F., and Backer, C.L. (2001). The left-sided aortic arch in humans, viewed as the end-result of natural selection during vertebrate evolution. *Cardiol. Young* **11**, 111–122.
1839. Mustonen, T., Pispa, J., Mikkola, M.L., Pummila, M., Kangas, A.T., Pakkasjärvi, L., Jaatinen, R., and Thesleff, I. (2003). Stimulation of ectodermal organ development by Ectodysplasin-A1. *Dev. Biol.* **259**, 123–136.
1840. Myers, S., Freeman, C., Auton, A., Donnelly, P., and McVean, G. (2008). A common sequence motif associated with recombination hot spots and genome instability in humans. *Nature Genet.* **40**, 1124–1129.
1841. Mysorekar, V.V., Rao, S.G., and Sundari, N. (2007). Sirenomelia: a case report. *Indian J. Pathol. Microbiol.* **50**, 359–361.
1842. Nagashima, H., Kuraku, S., Uchida, K., Ohya, Y.K., Narita, Y., and Kuratani, S. (2007). On the carapacial ridge in turtle embryos: its developmental origin, function and the chelonian body plan. *Development* **134**, 2219–2226.
1843. Nagy, J.D., Victor, E.M., and Cropper, J.H. (2007). Why don't all whales have cancer? A novel hypothesis resolving Peto's paradox. *Integr. Comp. Biol.* **47**, 317–328.
1844. Naiche, L.A. and Papaioannou, V.E. (2007). *Tbx4* is not required for hindlimb identity or post-bud hindlimb outgrowth. *Development* **134**, 93–103.
1845. Naitoh, T. and Wassersug, R. (1996). Why are toads right-handed? *Nature* **380**, 30–31.
1846. Nakamura, T., Mine, N., Nakaguchi, E., Mochizuki, A., Yamamoto, M., Yashiro, K., Meno, C., and

Hamada, H. (2006). Generation of robust left-right asymmetry in the mouse embryo requires a self-enhancement and lateral-inhibition system. *Dev. Cell* **11**, 495–504.

1847. Nakanishi, K. (1991). Why 11-cis-retinal? *Amer. Zool.* **31**, 479–489.
1848. Napier, J. (1993). "Hands," revised ed. Princeton Univ. Pr., Princeton, N.J.
1849. Napier, J.R. (1976). The human hand. *Carolina Biol. Readers* #61, 16 pp.
1850. Narita, Y. and Kuratani, S. (2005). Evolution of the vertebral formulae in mammals: a perspective on developmental constraints. *J. Exp. Zool. (Mol. Dev. Evol.)* **304B**, 91–106.
1851. Nash, W.G. (1976). Patterns of pigmentation color states regulated by the y locus in Drosophila melanogaster. *Dev. Biol.* **48**, 336–343.
1852. Nash, W.G. and Yarkin, R.J. (1974). Genetic regulation and pattern formation: a study of the *yellow* locus in *Drosophila melanogaster. Genet. Res., Camb.* **24**, 19–26.
1853. Nath, R.K. and Paizi, M. (2007). Scapular deformity in obstetric plexus palsy: a new finding. *Surg. Radiol. Anat.* **29**, 133–140.
1854. Naugler, W.E., Sakurai, T., Kim, S., Maeda, S., Kim, K.H., Elsharkawy, A.M., and Karin, M. (2007). Gender disparity in liver cancer due to sex differences in MyD88-dependent IL-6 production. *Science* **317**, 121–124.
1855. Nawshad, A. (2008). Palatal seam disintegration: to die or not to die? That is no longer the question. *Dev. Dynamics* **237**, 2643–2656.
1856. Nayak, S., Goree, J., and Schedl, T. (2005). fog-2 and the evolution of self-fertile hermaphroditism in Caenorhabditis. *PLoS Biol.* 3 #1, 57–71 (e6).
1857. Needham, J. (1933). On the dissociability of the fundamental processes in ontogenesis. *Biol. Rev.* **8**, 180–223.
1858. Neel, J.V. (1983). Curt Stern: 1902–1981. *Annu. Rev. Genet.* **17**, 1–10.
1859. Neel, J.V. (1987). Curt Stern: August 30, 1902–October 23, 1981. *Biogr. Memoirs Nat. Acad. Sci. U. S.* **56**, 442–473.
1860. Nehls, M., Pfeifer, D., Schorpp, M., Hedrich, H., and Boehm, T. (1994). New member of the winged-helix protein family disrupted in mouse and rat nude mutations. *Nature* **372**, 103–107.
1861. Nelsen, E.M., Frankel, J., and Jenkins, L.M. (1989). Non-genic inheritance of cellular handedness. *Development* **105**, 447–456.
1862. Nelson, C. (2004). Selector genes and the genetic control of developmental modules. *In* "Modularity in Development and Evolution" (G. Schlosser and G.P. Wagner, eds.). Univ. Chicago Pr., Chicago, Ill.,
1863. Nelson, C.M., VanDuijn, M.M., Inman, J.L., Fletcher, D.A., and Bissell, M.J. (2006). Tissue geometry determines sites of mammary branching morphogenesis in organotypic cultures. *Science* **314**, 298–300.
1864. Nemeschkal, H.L. (1999). Morphometric correlation patterns of adult birds (Fringillidae: Passeriformes and Columbiformes) mirror the expression of developmental control genes. *Evolution* **53**, 899–918.
1865. Nern, A., Zhu, Y., and Zipursky, S.L. (2008). Local N-cadherin interactions mediate distinct steps in the targeting of lamina neurons. *Neuron* **58**, 34–41.
1866. Netter, F.H. (2003). "Atlas of Human Anatomy," 3rd ed. Icon Learning Systems, Teterboro, N.J.
1867. Neubüser, A., Peters, H., Balling, R., and Martin, G.R. (1997). Antagonistic interactions between FGF and BMP signaling pathways: a mechanism for positioning the sites of tooth formation. *Cell* **90**, 247–255.
1868. Neumann, C.J. and Nuesslein-Volhard, C. (2000). Patterning of the zebrafish retina by a wave of Sonic hedgehog activity. *Science* **289**, 2137–2139.
1869. Neuweiler, G. (2000). "The Biology of Bats." Oxford Univ. Pr., New York.
1870. Neville, A.C. (1970). Cuticle ultrastructure in relation to the whole insect. *In* "Insect Ultrastructure," *5th Symp. Roy. Entomol. Soc. Lond.* (A.C. Neville, ed.). Blackwell: Oxford, pp. 17–39.
1871. Neville, A.C. (1976). "Animal Asymmetry." *The Institute of Biology's Studies in Biology No. 67,* Edward Arnold, London.
1872. Newman, S.A., Dahn, R.D., and Fallon, J.F. (2000). Fingering digit identity. *Science* **290**, 275–277.
1873. Newton, A. (1896). "A Dictionary of Birds." Adam & Charles Black, London.
1874. Ng, C.S. and Kopp, A. (2008). Sex combs are important for male mating success in *Drosophila melanogaster. Behav. Genet.* **38**, 195–201.
1875. Nicastro, D., Schwartz, C., Pierson, J., Gaudette, R., Porter, M.E., and McIntosh, J.R. (2006). The molecular architecture of axonemes revealed by cryoelectron tomography. *Science* **313**, 944–948.
1876. Nicklen, P. (2007). Arctic ivory: hunting the narwhal. *Nat. Geogr.* **212** #2, 110–129.
1877. Nicolau, M.C., Akaârir, M., Gamundí, A., González, J., and Rial, R.V. (2000). Why we sleep: the evolutionary pathway to the mammalian sleep. *Prog. Neurobiol.* **62**, 379–406.
1878. Niederreither, K. and Dollé, P. (2008). Retinoic acid in development: towards an integrated view. *Nature Rev. Genet.* **9**, 541–553.
1879. Nielsen, C. (2008). Six major steps in animal evolution: are we derived sponge larvae? *Evol. Dev.* **10**, 241–257.
1880. Nielsen, T.A. and Stenstrom, P. (2005). What are the memory sources of dreaming? *Nature* **437**, 1286–1289.
1881. Nieschalk, U. and Demes, B. (1993). Biomechanical determinants of reduction of the second ray in Lorisinae. *In* "Hands of Primates" (H. Preuschoft and D.J. Chivers, eds.). Springer-Verlag: New York, pp. 225–234.

1882. Niimura, Y. and Nei, M. (2007). Extensive gains and losses of olfactory receptor genes in mammalian evolution. *PLoS ONE* **8**, e708.

1883. Nilsson, D.-E. (2004). Eye evolution: a question of genetic promiscuity. *Curr. Opin. Neurobiol.* **14**, 407–414.

1884. Nilsson, D.-E. (2005). Photoreceptor evolution: Ancient siblings serve different tasks. *Curr. Biol.* **15**, R94–R96.

1885. Nilsson, G.E. (1999). The cost of a brain. *Nat. Hist.* **108** #10, 66–72.

1886. Nisar, P.J. and Scholefield, J.H. (2003). Managing haemorrhoids. *BMJ* **327**, 847–851.

1887. Nishimura, T., Mikami, A., Suzuki, J., and Matsuzawa, T. (2006). Descent of the hyoid in chimpanzees: evolution of face flattening and speech. *J. Human Evol.* **51**, 244–254.

1888. Nissim, S. and Tabin, C. (2004). Development of the limbs. *In* "Inborn Errors of Development: The Molecular Basis of Clinical Disorders of Morphogenesis," *Oxford Monographs on Medical Genetics, No. 49* (C.J. Epstein, R.P. Erickson, and A. Wynshaw-Boris, eds.). Oxford Univ. Pr., New York, pp. 148–167.

1889. Niswander, L. (2003). Pattern formation: old models out on a limb. *Nature Rev. Genet.* **4**, 133–143.

1890. Nitecki, M.H. (1990). The plurality of evolutionary innovations. *In* "Evolutionary Innovations" (M.H. Nitecki, ed.). Univ. Chicago Pr., Chicago, IL, pp. 3–18.

1891. Nixon, M. and Young, J.Z. (2003). "The Brains and Lives of Cephalopods." Oxford Univ. Pr., New York.

1892. Nojima, D., Linck, R.W., and Egelman, E.H. (1995). At least one of the protofilaments in flagellar microtubules is not composed of tubulin. *Curr. Biol.* **5**, 158–167.

1893. Norberg, R.Å. (1977). Occurrence and independent evolution of bilateral ear asymmetry in owls and implications on owl taxonomy. *Phil. Trans. Roy. Soc. Lond. B* **280**, 375–408.

1894. Norberg, U.M. (1985). Flying, gliding, and soaring. *In* "Functional Vertebrate Morphology" (M. Hildebrand, D.M. Bramble, K.F. Liem, and D.B. Wake, eds.). Harvard Univ. Pr., Cambridge, Mass., pp. 129–158.

1895. Norden, J.J. and Constantine-Paton, M. (1994). Dynamics of retinotectal synaptogenesis in normal and 3-eyed frogs: evidence for the postsynaptic regulation of synapse number. *J. Comp. Neurol.* **348**, 461–479.

1896. Nordström, K., Wallén, R., Seymour, J., and Nilsson, D. (2003). A simple visual system without neurons in jellyfish larvae. *Proc. Roy. Soc. Lond. B* **270**, 2349–2354.

1897. Norman, M.D., Paul, D., Finn, J., and Tregenza, T. (2002). First encounter with a live male blanket octopus: the world's most sexually size-dimorphic large animal. *New Zealand J. Marine Freshwater Res.* **36**, 733–736.

1898. Norris, D. (2005). Breaking the left-right axis: do nodal parcels pass a signal to the left? *BioEssays* **27**, 991–994.

1899. Northcutt, R.G. (2002). Understanding vertebrate brain evolution. *Integr. Comp. Biol.* **42**, 743–756.

1900. Northcutt, R.G. (2005). The New Head Hypothesis revisited. *J. Exp. Zool. (Mol. Dev. Evol.)* **304B**, 274–297.

1901. Northoff, G. and Panksepp, J. (2008). The trans-species concept of self and the subcortical–cortical midline system. *Trends Cogn. Sci.* **12**, 259–264.

1902. Nowicki, J.L. and Burke, A.C. (2000). Hox genes and morphological identity: axial versus lateral patterning in the vertebrate mesoderm. *Development* **127**, 4265–4275.

1903. Noyes, M.B., Christensen, R.G., Wakabayashi, A., Stormo, G.D., Brodsky, M.H., and Wolfe, S.A. (2008). Analysis of homeodomain specificities allows the family-wide prediction of preferred recognition sites. *Cell* **133**, 1277–1289.

1904. O'Brien, S.J., Nash, W.G., Wildt, D.E., Bush, M.E., and Benveniste, R.E. (1985). A molecular solution to the riddle of the giant panda's phylogeny. *Nature* **317**, 140–144.

1905. O'Higgins, P. and Elton, S. (2007). Walking on trees. *Science* **316**, 1292–1294.

1906. O'Kane, C.J. (2003). Modelling human diseases in Drosophila and Caenorhabditis. *Sems. Cell Dev. Biol.* **14**, 3–10.

1907. O'Leary, D.D.M., Chou, S.-J., and Sahara, S. (2007). Area patterning of the mammalian cortex. *Neuron* **56**, 252–269.

1908. O'Leary, D.D.M. and Sahara, S. (2008). Genetic regulation of arealization of the neocortex. *Curr. Opin. Neurobiol.* **18**, 90–100.

1909. Oard, D.W. (2008). Unlocking the potential of the spoken word. *Science* **321**, 1787–1788.

1910. Odelberg, S.J. (2002). Inducing cellular dedifferentiation: a potential method for enhancing endogenous regeneration in mammals. *Sems. Cell Dev. Biol.* **13**, 335–343.

1911. Oetting, W.S. (2000). The tyrosinase gene and oculocutaneous albinism Type 1 (OCA1): A model for understanding the molecular biology of melanin formation. *Pigment Cell Res.* **13**, 320–325.

1912. Oetting, W.S., Fryer, J.P., Shriram, S., and King, R.A. (2003). Oculocutaneous albinism Type 1: the last 100 years. *Pigment Cell Res.* **16**, 307–311.

1913. Oftedal, O.T. (2002). The mammary gland and its origin during synapsid evolution. *J. Mammary Gland Biol. Neoplasia* **7**, 225–252.

1914. Oftedal, O.T. (2002). The origin of lactation as a water source for parchment-shelled eggs. *J. Mammary Gland Biol. Neoplasia* **7**, 253–266.

1915. Oginuma, M., Niwa, Y., Chapman, D.L., and Saga, Y. (2008). Mesp2 and Tbx6 cooperatively create periodic patterns coupled with the clock machinery during mouse somitogenesis. *Development* **135**, 2555–2562.

1916. Ogura, A., Ikeo, K., and Gojobori, T. (2005). Estimation of ancestral gene set of bilaterian animals and its implication to dynamic change of gene content in bilaterian evolution. *Gene* **345**, 65–71.

1917. Ohazama, A. and Sharpe, P.T. (2004). Development of epidermal appendages: teeth and hair. *In* "Inborn Errors of Development: The Molecular Basis of Clinical Disorders of Morphogenesis," *Oxford Monographs on Medical Genetics, No. 49* (C.J. Epstein, R.P. Erickson, and A. Wynshaw-Boris, eds.). Oxford Univ. Pr., New York, pp. 199–209.

1918. Ohlsson, R., Kanduri, C., Whitehead, J., Pfeifer, S., Lobanenkov, V., and Feinberg, A.P. (2003). Epigenetic variability and the evolution of human cancer. *Adv. Cancer Res.* **88**, 145–168.

1919. Ohno, S. (1972). Gene duplication, mutation load, and mammalian genetic regulatory systems. *J. Med. Genet.* **9**, 254–263.

1920. Ohnuma, S.-i., Hopper, S., Wang, K.C., Philpott, A., and Harris, W.A. (2002). Co-ordinating retinal histogenesis: early cell-cycle exit enhances early cell fate determination in the Xenopus retina. *Development* **129**, 2435–2446.

1921. Ohuchi, H., Nakagawa, T., Yamauchi, M., Ohata, Y., Yoshioka, H., Kuwana, T., Mima, T., Nohno, T., and Noji, S. (1995). An additional limb can be induced from the flank of the chick embryo by FGF4. *Biochem. Biophys. Res. Commun.* **209**, 809–816.

1922. Ohuchi, H., Takeuchi, J., Yoshioka, H., Ishimaru, Y., Ogura, K., Takahashi, N., Ogura, T., and Noji, S. (1998). Correlation of wing-leg identity in ectopic FGF-induced chimeric limbs with the differential expression of chick *Tbx5* and *Tbx4*. *Development* **125**, 51–60.

1923. Okada, Y., Nonaka, S., Tanaka, Y., Saijoh, Y., Hamada, H., and Hirokawa, N. (1999). Abnormal nodal flow precedes situs inversus in iv and inv mice. *Molec. Cell* **4**, 459–468.

1924. Okada, Y., Takeda, S., Tanaka, Y., Izpisúa Belmonte, J.-C., and Hirokawa, N. (2005). Mechanism of nodal flow: a conserved symmetry breaking event in left-right axis determination. *Cell* **121**, 633–644.

1925. Okano, T., Yoshizawa, T., and Fukada, Y. (1994). Pinopsin is a chicken pineal photoreceptive molecule. *Nature* **372**, 94–97.

1926. Okanoya, K. (2007). Language evolution and an emergent property. *Curr. Opin. Neurobiol.* **17**, 271–276.

1927. Oldham, M.C., Horvath, S., and Geschwind, D.H. (2006). Conservation and evolution of gene coexpression networks in human and chimpanzee brains. *PNAS* **103** #47, 17973–17978.

1928. Olivera-Martinez, I., Coltey, M., Dhouailly, D., and Pourquié, O. (2000). Mediolateral somitic origin of ribs and dermis determined by quail-chick chimeras. *Development* **127**, 4611–4617.

1929. Olivier, E., Davare, M., Andres, M., and Fadiga, L. (2007). Precision grasping in humans: from motor control to cognition. *Curr. Opin. Neurobiol.* **17**, 644–648.

1930. Olsen, S.R. and Wilson, R.I. (2008). Cracking neural circuits in a tiny brain: new approaches for understanding the neural circuitry of *Drosophila*. *Trends Neurosci.* **31**, 512–520.

1931. Olshansky, S.J., Carnes, B.A., and Butler, R.N. (2001). If humans were built to last. *Sci. Am.* **284** #3, 50–55.

1932. Olshansky, S.J., Carnes, B.A., and Cassel, C. (1990). In search of Methuselah: estimating the upper limits to human longevity. *Science* **250**, 634–640.

1933. Olson, E.N. (2006). Gene regulatory networks in the evolution and development of the heart. *Science* **313**, 1922–1927.

1934. Olson, K.R. (1994). Circulatory anatomy in bimodally breathing fish. *Amer. Zool.* **34**, 280–288.

1935. Olsson, A. and Ochsner, K.N. (2008). The role of social cognition in emotion. *Trends Cogn. Sci.* **12**, 65–71.

1936. Olsson, L., Ericsson, R., and Cerny, R. (2005). Vertebrate head development: segmentation, novelties, and homology. *Theory Biosci.* **124**, 145–163.

1937. Omoto, C.K. and Kung, C. (1980). Rotation and twist of the central-pair microtubules in the cilia of *Paramecium*. *J. Cell Biol.* **87**, 33–46.

1938. Oostra, R.-J., Hennekam, R.C.M., de Rooij, L., and Moorman, A.F.M. (2005). Malformations of the axial skeleton in *Museum Vrolik* I: Homeotic transformations and numerical anomalies. *Am. J. Med. Genet.* **134A**, 268–281.

1939. Opitz, J.M. (1996). Limb anomalies from evolutionary, developmental, and genetic perspectives. *Birth Defects: Orig. Article Ser.* **30** #1, 35–77.

1940. Oppenheim, R.W. (1991). Cell death during development of the nervous system. *Annu. Rev. Neurosci.* **14**, 453–501.

1941. Oppenheimer, J.M. (1974). Asymmetry revisited. *Amer. Zool.* **14**, 867–879.

1942. Orgogozo, V., Muro, N.M., and Stern, D.L. (2007). Variation in fiber number of a male-specific muscle between *Drosophila* species: a genetic and developmental analysis. *Evol. Dev.* **9**, 368–377.

1943. Ortíz-Barrientos, D. and Noor, M.A.F. (2005). Evidence for a one-allele assortative mating locus. *Science* **310**, 1467.

1944. Osborn, J.W. (1984). From reptile to mammal: evolutionary considerations of the dentition with emphasis on tooth attachment. *Symp. Zool. Soc. Lond.* **52**, 549–574.

1945. Osman, A., Muller, K.M., Syre, P., and Russ, B. (2005). Paradoxical lateralization of brain potentials during imagined foot movements. *Brain Res. Cogn. Brain Res.* **24**, 727–731.
1946. Osorio, D. (1991). Patterns of function and evolution in the arthropod optic lobe. *In* "Evolution of the Eye and Visual System" (J.R. Cronly-Dillon and R.L. Gregory, eds.). CRC Pr., Boston, pp. 203–228.
1947. Osorio, D. (2007). *Spam* and the evolution of the fly's eye. *BioEssays* **29**, 111–115.
1948. Osorio, D. and Bacon, J.P. (1994). A good eye for arthropod evolution. *BioEssays* **16**, 419–424.
1949. Oster, G. and Alberch, P. (1982). Evolution and bifurcation of developmental programs. *Evolution* **36**, 444–459.
1950. Oster, G.F., Shubin, N., Murray, J.D., and Alberch, P. (1988). Evolution and morphogenetic rules: the shape of the vertebrate limb in ontogeny and phylogeny. *Evolution* **42**, 862–884.
1951. Ostrander, E.A. (2007). Genetics and the shape of dogs. *Am. Sci.* **95**, 406–413.
1952. Ota, K.G., Kuraku, S., and Kuratani, S. (2007). Hagfish embryology with reference to the evolution of the neural crest. *Nature* **446**, 672–675.
1953. Ota, S., Zhou, Z.-Q., Keene, D.R., Knoepfler, P., and Hurlin, P.J. (2007). Activities of N-Myc in the developing limb link control of skeletal size with digit separation. *Development* **134**, 1583–1592.
1954. Otiang'a-Owiti, G.E., Oduor-Okelo, D., Kamau, G.K., Makori, N., and Hendrickx, A.G. (1997). Morphology of a six-legged goat with duplication of the intestinal, lower urinary, and genital tracts. *Anat. Rec.* **247**, 432–438.
1955. Otto, S.P., Servedio, M.R., and Nuismer, S.L. (2008). Frequency-dependent selection and the evolution of assortative mating. *Genetics* **179**, 2091–2112.
1956. Ottolenghi, C., Uda, M., Crisponi, L., Omari, S., Cao, A., Forabosco, A., and Schlessinger, D. (2006). Determination and stability of sex. *BioEssays* **29**, 15–25.
1957. Õunap, K., Uibo, O., Zordania, R., Kiho, L., Ilus, T., Õiglane-Shlik, E., and Bartsch, O. (2004). Three patients with 9p deletions including DMRT1 and DMRT2: a girl with XY complement, bilateral ovotestes, and extreme growth retardation, and two XX females with normal pubertal development. *Am. J. Med. Genet.* **130A**, 415–423.
1958. Ouweneel, W.J. (1976). Developmental genetics of homoeosis. *Adv. Genet.* **18**, 179–248.
1959. Oviedo, N.J. and Levin, M. (2007). Gap junctions provide new links in left–right patterning. *Cell* **129**, 645–647.
1960. Özbudak, E.M. and Pourquié, O. (2008). The vertebrate segmentation clock: the tip of the iceberg. *Curr. Opin. Gen. Dev.* **18**, 317–323.
1961. Ozeki, H., Kurihara, Y., Tonami, K., Watatani, S., and Kurihara, H. (2004). Endothelin-1 regulates the dorsoventral branchial arch patterning in mice. *Mechs. Dev.* **121**, 387–395.
1962. Packer, C. (1998). Why menopause? *Nat. Hist.* **107** #6, 24–26. [See also (2007) *Curr. Biol.* **17**, 2150-2156.]
1963. Packer, C., Tatar, M., and Collins, A. (1998). Reproductive cessation in female mammals. *Nature* **392**, 807–811.
1964. Padian, K. (2008). Darwin's enduring legacy. *Nature* **451**, 632–634.
1965. Padmanabhan, R. (2006). Etiology, pathogenesis and prevention of neural tube defects. *Congenit. Anom. (Kyoto)* **46**, 55–67.
1966. Pagán-Westphal, S.M. and Tabin, C.J. (1998). The transfer of left–right positional information during chick embryogenesis. *Cell* **93**, 25–35.
1967. Page, D.C. (2004). On low expectations exceeded; or, the genomic salvation of the Y chromosome. *Am. J. Hum. Genet.* **74**, 399–402.
1968. Pagel, M. (2007). What is the latest theory of why humans lost their body hair? *Sci. Am.* **297** #3, 124.
1969. Paintal, H.S. and Kuschner, W.G. (2007). Aspiration syndromes: 10 clinical pearls every physician should know. *Int. J. Clin. Pract.* **61**, 846–852.
1970. Pajni-Underwood, S., Wilson, C.P., Elder, C., Mishina, Y., and Lewandoski, M. (2007). BMP signals control limb bud interdigital programmed cell death by regulating FGF signaling. *Development* **134**, 2359–2368.
1971. Palczewski, K. (2006). G protein-coupled receptor rhodopsin. *Annu. Rev. Biochem.* **75**, 743–767.
1972. Palestis, B. (2006). Yawn. *Am. Sci.* **94**, 101.
1973. Pallavi, S.K., Kannan, R., and Shashidhara, L.S. (2006). Negative regulation of Egfr/Ras pathway by *Ultrabithorax* during haltere development in Drosophila. *Dev. Biol.* **296**, 340–352.
1974. Palmer, A.R. (1996). From symmetry to asymmetry: phylogenetic patterns of asymmetry variation in animals and their evolutionary significance. *Proc. Natl. Acad. Sci. USA* **93**, 14279–14286.
1975. Palmer, A.R. (1996). Waltzing with asymmetry. *BioScience* **46**, 518–532.
1976. Palmer, A.R. (2004). Symmetry breaking and the evolution of development. *Science* **306**, 828–833.
1977. Palmer, A.R. (2005). Antisymmetry. *In* "Variation: A Central Concept in Biology" (B. Hallgrímsson and B.K. Hall, eds.). Elsevier Acad. Pr., New York, pp. 359–397.
1978. Palmer, A.R. and Lewontin, R.C. (2004). Selection for asymmetry. *Science* **306**, 812–813.
1979. Palmer, A.R. and Strobeck, C. (1986). Fluctuating asymmetry: measurement, analysis, patterns. *Annu. Rev. Ecol. Syst.* **17**, 391–421.
1980. Palmer, A.R., Strobeck, C., and Chippindale, A.K. (1993). Bilateral variation and the evolutionary origin of macroscopic asymmetries. *Genetica* **89**, 201–218.
1981. Panda, S., Nayak, S.K., Campo, B., Walker, J.R., Hogenesch, J.B., and Jegla, T. (2005). Illumination of the melanopsin signaling pathway. *Science* **307**, 600–604.

1982. Panksepp, J. (2005). Beyond a joke: From animal laughter to human joy? *Science* **308**, 62–63.
1983. Panksepp, J. (2007). Neuroevolutionary sources of laughter and social joy: modeling primal human laughter in laboratory rats. *Behav. Brain Res.* **182**, 231–244.
1984. Panksepp, J. and Burgdorf, J. (2000). 50-kHz chirping (laughter?) in response to conditioned and unconditioned tickle-induced reward in rats: effects of social housing and genetic variables. *Behav. Brain Res.* **115**, 25–38.
1985. Panman, L., Galli, A., Lagarde, N., Michos, O., Soete, G., Zuniga, A., and Zeller, R. (2006). Differential regulation of gene expression in the digit forming area of the mouse limb bud by SHH and gremlin 1/FGF-mediated epithelial-mesenchymal signalling. *Development* **133**, 3419–3428.
1986. Pantin, C.F.A. (1951). Organic design. *Advmt. Sci. Lond.* **8**, 138–150.
1987. Papastavrou, V. (1993). "Whale." Knopf, New York.
1988. Paria, B.C., Ma, W.-g., Tan, J., Raja, S., Das, S.K., Dey, S.K., and Hogan, B.L.M. (2001). Cellular and molecular responses of the uterus to embryo implantation can be elicited by locally applied growth factors. *PNAS* **98** #3, 1047–1052.
1989. Parichy, D.M. (2005). Variation and developmental biology: prospects for the future. *In* "Variation: A Central Concept in Biology" (B. Hallgrímsson and B.K. Hall, eds.). Elsevier Acad. Pr., New York, pp. 475–498.
1990. Paris, M., Escriva, H., Schubert, M., Brunet, F., Brtko, J., Ciesielski, F., Roecklin, D., Vivat-Hannah, V., Jamin, E.L., Cravedi, J.-P., Scanlan, T.S., Renaud, J.-P., Holland, N.D., and Laudet, V. (2008). Amphioxus postembryonic development reveals the homology of chordate metamorphosis. Curr. Biol. **18**, 825–830.
1991. Parisi, P., Gatti, M., Prinzi, G., and Caperna, G. (1983). Familial incidence of twinning. Nature **304**, 626–628.
1992. Park, B.K., Sperber, S.M., Choudhury, A., Ghanem, N., Hatch, G.T., Sharpe, P.T., Thomas, B.L., and Ekker, M. (2004). Intergenic enhancers with distinct activities regulate Dlx gene expression in the mesenchyme of the branchial arches. *Dev. Biol.* **268**, 532–545.
1993. Park, J.H., Scheerer, P., Hofmann, K.P., and Choe, H.-W. (2008). Crystal structure of the ligand-free G-protein-coupled receptor opsin. *Nature* **454**, 183–187.
1994. Parker, G.A. (1970). Sperm competition and its evolutionary consequences in the insects. *Biol. Rev.* **45**, 525–567.
1995. Pasteur, L. (1850). Recherches sur les propriétés spécifiques des deux acides qui composent l'acide racémique. *Ann. Chim. Phys.* **28**, 56–99.
1996. Patikoglou, G. and Burley, S.K. (1997). Eukaryotic transcription factor-DNA complexes. *Annu. Rev. Biophys. Biomol. Struct.* **26**, 289–325.
1997. Pattatucci, A.M., Otteson, D.C., and Kaufman, T.C. (1991). A functional and structural analysis of the *Sex combs reduced locus* of *Drosophila melanogaster. Genetics* **129**, 423–441.
1998. Patten, B.M. (1958). "Foundations of Embryology." McGraw-Hill, New York.
1999. Patterson, J.S. and Klingenberg, C.P. (2007). Developmental buffering: how many genes? *Evol. Dev.* **9**, 525–526.
2000. Patterson, K.D., Drysdale, T.A., and Krieg, P.A. (2000). Embryonic origins of spleen asymmetry. *Development* **127**, 167–175.
2001. Paul, J.S., Fwu-Shan, S., and Luft, A.R. (2006). Early adaptations in somatosensory cortex after focal ischemic injury to motor cortex. *Exp. Brain Res.* **168**, 178–185.
2002. Pawlowski, B. (1999). Loss of oestrus and concealed ovulation in human evolution: The case against the Sexual Selection Hypothesis. *Curr. Anthrop.* **40**, 257–275.
2003. Peaker, M. (1994). Male suckling. *Nature* **371**, 292.
2004. Pearce, J. and Govind, C.K. (1987). Spontaneous generation of bilateral symmetry in the paired claws and closer muscles of adult snapping shrimps. *Development* **100**, 57–63.
2005. Pearl, R. (1913). On the correlation between the number of mammae of the dam and size of litter in mammals. I. Interracial correlation. *Proc. Soc. Exp. Biol. Med.* **11**, 27–30.
2006. Pearl, R. (1913). On the correlation between the number of mammae of the dam and size of litter in mammals. II. Intraracial correlation in swine. *Proc. Soc. Exp. Biol. Med.* **11**, 31–32.
2007. Pearman, P.B., Guisan, A., Broennimann, O., and Randin, C.F. (2008). Niche dynamics in space and time. *Trends Ecol. Evol.* **23**, 149–158.
2008. Pearse, R.V., II, Scherz, P.J., Campbell, J.K., and Tabin, C.J. (2007). A cellular lineage analysis of the chick limb bud. *Dev. Biol.* **310**, 388–400.
2009. Pearson, H. (2007). The roots of accomplishment. *Nature* **446**, 20–21.
2010. Pearson, J.C., Lemons, D., and McGinnis, W. (2005). Modulating *Hox* gene functions during animal body patterning. *Nature Rev. Genet.* **6**, 893–904.
2011. Peirson, S. and Foster, R.G. (2006). Melanopsin: another way of signaling light. *Neuron* **49**, 331–339.
2012. Penn, J.K.M. and Schedl, P. (2007). The master switch gene Sex-lethal promotes female development by negatively regulating the N-signaling pathway. *Dev. Cell* **12**, 275–286.
2013. Pennisi, E. (2002). Evo-devo devotees eye ocular origins and more. *Science* **296**, 1010–1011.
2014. Pennisi, E. (2004). The primate bite: brawn versus brain? *Science* **303**, 1957.
2015. Pennisi, E. (2006). Mining the molecules that made our mind. *Science* **313**, 1908–1911.
2016. Pennock, R.T. (2007). Evolution—once more, with feeling. *Am. Sci.* **95**, 528–531.

2017. Pérez-Barbería, F.J., Shultz, S., and Dunbar, R.I.M. (2007). Evidence for coevolution of sociality and relative brain size in three orders of mammals. *Evolution* **61**, 2811–2821.
2018. Perrimon, N. and McMahon, A.P. (1999). Negative feedback mechanisms and their roles during pattern formation. *Cell* **97**, 13–16.
2019. Perry, S.F. and Sander, M. (2004). Reconstructing the evolution of the respiratory apparatus in tetrapods. *Respir. Physiol. Neurobiol.* **144**, 125–139.
2020. Peterkova, R., Lesot, H., and Peterka, M. (2006). Phylogenetic memory of developing mammalian dentition. *J. Exp. Zool. (Mol. Dev. Evol.)* **306B**, 234–250.
2021. Peters, H. and Balling, R. (1999). Teeth: Where and how to make them. *Trends Genet.* **15**, 59–65.
2022. Peters, M. (1991). Laterality and motor control. *In* "Biological Asymmetry and Handedness," *Ciba Found. Symp. 162* (G.R. Bock and J. Marsh, eds.). Wiley, New York, pp. 300–311.
2023. Peterson, K.J., McPeek, M.A., and Evans, D.A.D. (2005). Tempo and mode of early animal evolution: inferences from rocks, Hox, and molecular clocks. *Paleobiology* **31**, 36–55.
2024. Pettigrew, J.B. (1908). "Design in Nature" Vol. 1. Longmans, Green, & Co., London.
2025. Pettigrew, J.D. (1991). Evolution of binocular vision. *In* "Evolution of the Eye and Visual System" (J.R. Cronly-Dillon and R.L. Gregory, eds.). CRC Pr., Boston, pp. 271–283.
2026. Pettigrew, J.D., Collin, S.P., and Ott, M. (1999). Convergence of specialised behaviour, eye movements and visual optics in the sandlance (Teleostei) and the chameleon (Reptilia). *Curr. Biol.* **9**, 421–424.
2027. Peyer, B. (1968). "Comparative Odontology." Univ. of Chicago Pr., Chicago, Ill.
2028. Piatigorsky, J. (1993). Puzzle of crystallin diversity in eye lenses. *Dev. Dynamics* **196**, 267–272.
2029. Piatigorsky, J. (2003). Gene sharing, lens crystallins and speculations on an eye/ear evolutionary relationship. *Integr. Comp. Biol.* **43**, 492–499.
2030. Pichaud, F. and Desplan, C. (2002). Pax genes and eye organogenesis. *Curr. Opin. Gen. Dev.* **12**, 430–434.
2031. Pichaud, F., Treisman, J., and Desplan, C. (2001). Reinventing a common strategy for patterning the eye. *Cell* **105**, 9–12.
2032. Pick, T.P. and Howden, R., eds. (1977). "Gray's Anatomy." Bounty Books, New York.
2033. Pieau, C. (1996). Temperature variation and sex determination in reptiles. *BioEssays* **18**, 19–26.
2034. Pietsch, T.W. (2005). Dimorphism, parasitism, and sex revisited: modes of reproduction among deep-sea ceratioid anglerfishes (Teleostei: Lophiiformes). *Ichthyol. Res.* **52**, 207–236.
2035. Pigliucci, M. (2008). Adaptive landscapes, phenotypic space, and the power of metaphors. *Q. Rev. Biol.* **83**, 283–287.
2036. Pinker, S. (1995). Facts about human language relevant to its evolution. *In* "Origins of the Human Brain", (J.-P. Changeux and J. Chavaillon, eds.). Clarendon Pr.: Oxford, pp. 262–283.
2037. Pinker, S. (1997). "How the Mind Works." W.W. Norton, New York.
2038. Pinker, S. and Bloom, P. (1990). Natural language and natural selection. *Behav. Brain Sci.* **13**, 707–784.
2039. Pires-daSilva, A. (2007). Evolution of the control of sexual identity in nematodes. *Sems. Cell Dev. Biol.* **18**, 362–370.
2040. Pires-daSilva, A. and Sommer, R.J. (2003). The evolution of signalling pathways in animal development. *Nature Rev. Genet.* **4**, 39–49.
2041. Pitsouli, C. and Perrimon, N. (2008). Our fly cousins' gut. *Nature* **454**, 592–593.
2042. Pittman, A.J., Law, M.-Y., and Chien, C.-B. (2008). Pathfinding in a large vertebrate axon tract: isotypic interactions guide retinotectal axons at multiple choice points. *Development* **135**, 2865–2871.
2043. Pizzari, T. (2006). Evolution: the paradox of sperm leviathans. *Curr. Biol.* **16**, R462–R464.
2044. Plachetzki, D.C. and Oakley, T.H. (2007). Key transitions during the evolution of animal phototransduction: novelty, "tree-thinking," co-option, and co-duplication. *Integr. Comp. Biol.* **47**, 759–769.
2045. Plavcan, J.M. (2001). Sexual dimorphism in primate evolution. *Yrbk. Phys. Anthrop.* **44**, 25–53.
2046. Plavcan, J.M. (2002). Taxonomic variation in the patterns of craniofacial dimorphism in primates. *J. Human Evol.* **42**, 579–608.
2047. Plavcan, J.M. and Daegling, D.J. (2006). Interspecific and intraspecific relationships between tooth size and jaw size in primates. *J. Human Evol.* **51**, 171–184.
2048. Plavcan, J.M., Lockwood, C.A., Kimbel, W.H., Lague, M.R., and Harmon, E.H. (2005). Sexual dimorphism in *Australopithecus afarensis* revisited: How strong is the case for a human-like pattern of dimorphism? *J. Human Evol.* **48**, 313–320.
2049. Plikus, M.V., Zeichner-David, M., Mayer, J.-A., Reyna, J., Bringas, P., Thewissen, J.G.M., Snead, M.L., Chai, Y., and Chuong, C.-M. (2005). Morphoregulation of teeth: modulating the number, size, shape and differentiation by tuning BMP activity. *Evol. Dev.* **7**, 440–457.
2050. Plomin, R., Owen, M.J., and McGuffin, P. (1994). The genetic basis of complex human behaviors. *Science* **264**, 1733–1739.
2051. Plön, S. and Bernard, R. (2007). Anatomy with particular reference to the female. *In* "Reproductive Biology and Phylogeny of Cetacea" (D.L. Miller, ed.). Science Pub., Enfield, N.H., pp. 147–169.
2052. Podlasek, C., Houston, J., McKenna, K.E., and McVary, K.T. (2002). Posterior Hox gene expression in developing genitalia. *Evol. Dev.* **4**, 142–163.
2053. Podulka, S., Rohrbaugh, R.W., Jr., and Bonney, R., eds. (2004). "Handbook of Bird Biology," 2nd ed. Cornell Lab Ornith., Ithaca, N.Y.

2054. Poelwijk, F.J., Kiviet, D.J., Weinreich, D.M., and Tans, S.J. (2007). Empirical fitness landscapes reveal accessible evolutionary paths. *Nature* **445**, 383–386.
2055. Polak, M., Starmer, W.T., and Wolf, L.L. (2004). Sexual selection for size and symmetry in a diversifying secondary sexual character in *Drosophila bipectinata* Duda (Diptera: Drosophilidae). *Evolution* **58**, 597–607.
2056. Polanco, J.C. and Koopman, P. (2007). Sry and the hesitant beginnings of male development. *Dev. Biol.* **302**, 13–24.
2057. Policansky, D. (1982). The asymmetry of flounders. *Sci. Am.* **246** #5, 116–122.
2058. Polizzi, A., Pavone, P., Ciancio, E., La Rosa, C., Sorge, G., and Ruggieri, M. (2005). Hypertrichosis cubiti (hairy elbow syndrome): a clue to a malformation syndrome. *J. Ped. Endocr. Metab.* **18**, 1019–1025.
2059. Polly, P.D. (2007). Development with a bite. *Nature* **449**, 413–415.
2060. Polly, P.D. (2007). Limbs in mammalian evolution. *In* "Fins Into Limbs: Evolution, Development, and Transformation" (B.K. Hall, ed.). Univ. Chicago Pr., Chicago, Ill., pp. 245–268.
2061. Polyak, S. (1957). "The Vertebrate Visual System." Univ. Chicago Pr., Chicago, Ill.
2062. Pombal, M.A., Carmona, R., Megías, M., Ruiz, A., Pérez-Pomares, J., and Muñoz-Chápuli, R. (2008). Epicardial development in lamprey supports an evolutionary origin of the vertebrate epicardium from an ancestral pronephric external glomerulus. *Evol. Dev.* **10**, 210–216.
2063. Pomerantz, J. (2001). Does the appendix serve a purpose in any animal? *Sci. Am.* **285** #5, 96.
2064. Pomiankowski, A., Nöthiger, R., and Wilkins, A. (2004). The evolution of the *Drosophila* sex-determination pathway. *Genetics* **166**, 1761–1773.
2065. Ponce, C.R. and Born, R.T. (2008). Stereopsis. *Curr. Biol.* **18**, R845–R850.
2066. Ponce de León, M., Golovanova, L., Doronichev, V., Romanova, G., Akazawa, T., Kondo, O., Ishida, H., and Zollikofer, C.P.E. (2008). Neandertal brain size at birth provides insights into the evolution of human life history. *PNAS* **105** #37, 13764–13768.
2067. Porter, M.L. and Crandall, K.A. (2003). Lost along the way: the significance of evolution in reverse. *Trends Ecol. Evol.* **18**, 541–547.
2068. Portner, M. (2008). The orgasmic mind. *Sci. Am. Mind* **19** #2, 66–71.
2069. Postlethwait, J.H. and Schneiderman, H.A. (1973). Pattern formation in imaginal discs of *Drosophila melanogaster* after irradiation of embryos and young larvae. *Dev. Biol.* **32**, 345–360.
2070. Potrzebowski, L., Vinckenbosch, N., Marques, A.C., Chalmel, F., Jégou, B., and Kaessmann, H. (2008). Chromosomal gene movements reflect the recent origin and biology of therian sex chromosomes. *PLoS Biol.* **6** #4, 709–716 (e80).
2071. Potter, M. and McLennan, J. (1992). Kiwi's egg size and moa. *Nature* **358**, 548.
2072. Pradel, J. and White, R.A.H. (1998). From selectors to realizators. *Int. J. Dev. Biol.* **42**, 417–421.
2073. Premack, D. (1976). "Intelligence in Ape and Man." Wiley, New York.
2074. Presson, J. and Jenner, J. (2008). "Biology: Dimensions of Life." McGraw-Hill, New York.
2075. Preuschoft, H. (2004). Mechanisms for the acquisition of habitual bipedality: are there biomechanical reasons for the acquisition of upright bipedal posture? *J. Anat.* **204**, 363–384.
2076. Preuschoft, H. and Chivers, D.J., eds. (1993). "Hands of Primates." Springer-Verlag, New York.
2077. Preuschoft, H., Godinot, M., Beard, C., Nieschalk, U., and Jouffroy, F.K. (1993). Biomechanical considerations to explain important morphological characters of primate hands. *In* "Hands of Primates" (H. Preuschoft and D.J. Chivers, eds.). Springer-Verlag, New York, pp. 245–256.
2078. Preuss, T.M. (2001). The discovery of cerebral diversity: an unwelcome scientific revolution. *In* "Evolutionary Anatomy of the Primate Cerebral Cortex" (D. Falk and K.R. Gibson, eds.). Cambridge Univ. Pr., New York, pp. 138–164.
2079. Price, T. (2008). "Speciation in Birds." Roberts & Co., Greenwood Village, Colo.
2080. Price, T.A.R., Bretman, A.J., Avent, T.D., Snook, R.R., Hurst, G.D.D., and Wedell, N. (2008). Sex ratio distorter reduces sperm competitive ability in an insect. *Evolution* **62**, 1644–1652.
2081. Price, T.A.R. and Hosken, D.J. (2007). Evolution: Good males are bad females. *Curr. Biol.* **17**, R168–R170.
2082. Proctor, N.S. and Lynch, P.J. (1993). "Manual of Ornithology: Avian Structure and Function." Yale Univ. Pr., New Haven, Conn.
2083. Promislow, D.E.L. and Harvey, P.H. (1990). Living fast and dying young: a comparative analysis of life-history variation among mammals. *J. Zool. Lond.* **220**, 417–437.
2084. Pröschel, M., Zhang, Z., and Parsch, J. (2006). Widespread adaptive evolution of *Drosophila* genes with sex-biased expression. *Genetics* **174**, 893–900.
2085. Prothero, D.R. and Schoch, R.M. (2002). "Horns, Tusks, and Flippers: The Evolution of Hoofed Mammals." Johns Hopkins Univ. Pr., Baltimore, Md.
2086. Provencio, I., Rodriguez, I.R., Jiang, G., Hayes, W.P., Moreira, E.F., and Rollag, M.D. (2000). A novel human opsin in the inner retina. *J. Neurosci.* **20**, 600–605.
2087. Provencio, I., Rollag, M.D., and Castrucci, A.M. (2002). Photoreceptive net in the mammalian retina. *Nature* **415**, 493.
2088. Provine, R.R. (1996). Laughter. *Am. Sci.* **84**, 38–45.
2089. Provine, R.R. (2000). The laughing species. *Nat. Hist.* **109** #10, 72–78.
2090. Provine, R.R. (2005). Yawning. *Am. Sci.* **93**, 532–539.

2091. Provine, R.R. (2006). Velocity and direction in neurobehavioral evolution: the centripetal prospective. *Behav. Brain Sci.* **29**, 21–22.
2092. Prpic, N.-M. (2008). Arthropod appendages: a prime example for the evolution of morphological diversity and innovation. *In* "Evolving Pathways: Key Themes in Evolutionary Developmental Biology" (A. Minelli and G. Fusco, eds.). Cambridge Univ. Pr., New York, pp. 381–398.
2093. Prud'homme, B., Gompel, N., and Carroll, S.B. (2007). Emerging principles of regulatory evolution. *PNAS* **104 Suppl. 1**, 8605–8612.
2094. Prud'homme, B., Gompel, N., Rokas, A., Kassner, V.A., Williams, T.M., Yeh, S.-D., True, J.R., and Carroll, S.B. (2006). Repeated morphological evolution through *cis*-regulatory changes in a pleiotropic gene. *Nature* **440**, 1050–1053.
2095. Prusinkiewicz, P., Erasmus, Y., Lane, B., Harder, L.D., and Coen, E. (2007). Evolution and development of inflorescence architectures. *Science* **316**, 1452–1456.
2096. Pueyo, J.I. and Couso, J.P. (2005). Parallels between the proximal-distal development of vertebrate and arthropod appendages: homology without an ancestor? *Curr. Opin. Gen. Dev.* **15**, 439–446.
2097. Purcell, P., Oliver, G., Mardon, G., Donner, A.L., and Maas, R.L. (2005). Pax6-dependence of Six3, Eya1 and Dach1 expression during lens and nasal placode induction. *Gene Expr. Patt.* **6**, 110–118.
2098. Purves, D., Augustine, G.J., Fitzpatrick, D., Hall, W.C., LaMantia, A.-S., McNamara, J.O., and White, L.E., eds. (2008). "Neuroscience," 4th ed. Sinauer, Sunderland, Mass.
2099. Purves, D., Williams, S.M., and Lotto, R.B. (2000). The relevance of visual perception to cortical evolution and development. *In* "Evolutionary Developmental Biology of the Cerebral Cortex," *Novartis Found. Symp. 228* (G.R. Bock and G. Cardew, eds.). Wiley, New York, pp. 240–258.
2100. Putnam, N.H., Butts, T., Ferrier, D.E.K., Furlong, R.F., Hellsten, U., Kawashima, T., Robinson-Rechavi, M., Shoguchi, E., Terry, A., Yu, J.-K., Benito-Gutiérrez, È., Dubchak, I., Garcia-Fernàndez, J., Gibson-Brown, J.J., Grigoriev, I.V., Horton, A.C., de Jong, P.J., Jurka, J., Kapitonov, V.V., Kohara, Y., Kuroki, Y., Lindquist, E., Lucas, S., Osoegawa, K., Pennacchio, L.A., Salamov, A.A., Satou, Y., Sauka-Spengler, T., Schmutz, J., Shin-I, T., Toyoda, A., Bronner-Fraser, M., Fujiyama, A., Holland, L.Z., Holland, P.W.H., Satoh, N., and Rokhsar, D.S. (2008). The amphioxus genome and the evolution of the chordate karyotype. *Nature* **453**, 1064–1071.
2101. Qiu, D., Cheng, S.-M., Wozniak, L., McSweeney, M., Perrone, E., and Levin, M. (2005). Localization and loss-of-function implicates ciliary proteins in early, cytoplasmic roles in left-right asymmetry. *Dev. Dynamics* **234**, 176–189.
2102. Qiu, X., Kumbalasiri, T., Carlson, S.M., Wong, K.Y., Krishna, V., Provencio, I., and Berson, D.M. (2005). Induction of photosensitivity by heterologous expression of melanopsin. *Nature* **433**, 745–749.
2103. Quantock, A.J. and Young, R.D. (2008). Development of the corneal stroma, and the collagen-proteoglycan associations that help define its structure and function. *Dev. Dyn.* **237**, 2607–2621.
2104. Quaranta, A., Siniscalchi, M., and Vallortigara, G. (2007). Asymmetric tail-wagging responses by dogs to different emotional stimuli. *Curr. Biol.* **17**, R199–R201.
2105. Queller, D.C. (1995). The spaniels of St. Marx and the Panglossian paradox: a critique of a rhetorical programme. *Q. Rev. Biol.* **70**, 485–489.
2106. Quinn, C.C. and Wadsworth, W.G. (2007). Axon guidance: Ephrins at WRK on the midline. *Curr. Biol.* **16**, R954–R955.
2107. Quintin, S., Gally, C., and Labouesse, M. (2008). Epithelial morphogenesis in embryos: asymmetries, motors and brakes. *Trends Genet.* **24**, 221–230.
2108. Rabosky, D.L. and Lovette, I.J. (2008). Explosive evolutionary radiations: decreasing speciation or increasing extinction through time? *Evolution* **62**, 1866–1875.
2109. Raff, E.C. and Raff, R.A. (2000). Dissociability, modularity, evolvability. *Evol. Dev.* **2**, 235–237.
2110. Raff, R.A. (2000). Evo-devo: the evolution of a new discipline. *Nature Rev. Gen.* **1**, 74–79.
2111. Raff, R.A. and Sly, B.J. (2000). Modularity and dissociation in the evolution of gene expression territories in development. *Evol. Dev.* **2**, 102–113.
2112. Ragir, S. (2001). Changes in perinatal conditions selected for neonatal immaturity. *Behav. Brain Sci.* **24** #2, 291–292.
2113. Rakic, P. and Kornack, D.R. (2001). Neocortical expansion and elaboration during primate evolution: a view from neuroembryology. *In* "Evolutionary Anatomy of the Primate Cerebral Cortex" (D. Falk and K.R. Gibson, eds.). Cambridge Univ. Pr., New York, pp. 30–56. [See also (2009) *Sci. Am.* **300** #2, 66–71.]
2114. Ralls, K. (1976). Mammals in which females are larger than males. *Q. Rev. Biol.* **51**, 245–276.
2115. Ralph, C.L. (1983). Evolution of pineal control of endocrine function in lower vertebrates. *Amer. Zool.* **23**, 597–605.
2116. Ramachandran, V.S. (1998). The neurology and evolution of humor, laughter, and smiling: the false alarm theory. *Med. Hypotheses* **51**, 351–354.
2117. Ramachandran, V.S. and Gregory, R.L. (1991). Perceptual filling in of artificially induced scotomas in human vision. *Nature* **350**, 699–702.
2118. Ramachandran, V.S. and Rogers-Ramachandran, D. (2005). Mind the gap. *Sci. Am. Mind* **16** #1, 98–100.
2119. Ramsdell, A.F. (2005). Left-right asymmetry and congenital cardiac defects: getting to the heart of the matter in vertebrate left-right axis determination. *Dev. Biol.* **288**, 1–20.

2120. Ramus, F. (2006). Genes, brain, and cognition: a roadmap for the cognitive scientist. *Cognition* **101**, 247–269.
2121. Ranade, S.S., Yang-Zhou, D., Kong, S.W., McDonald, E.C., Cook, T.A., and Pignoni, F. (2008). Analysis of the Otd-dependent transcriptome supports the evolutionary conservation of CRX/OTX/OTD functions in flies and vertebrates. *Dev. Biol.* **315**, 521–534.
2122. Randall, D.J., Burggren, W.W., Farrell, A.P., and Haswell, M.S. (1981). "The Evolution of Air Breathing in Vertebrates." Cambridge Univ. Pr., New York.
2123. Randall, V.A. (2007). Hormonal regulation of hair follicles exhibits a biological paradox. *Sems. Cell Dev. Biol.* **18**, 274–285.
2124. Randsholt, N.B. and Santamaria, P. (2008). How *Drosophila* change their combs: the Hox gene *Sex combs reduced* and sex comb variation among *Sophophora* species. *Evol. Dev.* **10**, 121–133.
2125. Rao, J. (2004). Postcards from space: Lest we forget... *Nat. Hist.* **113** #7, 18.
2126. Rasch, B. and Born, J. (2007). Maintaining memories by reactivation. *Curr. Opin. Neurobiol.* **17**, 698–703.
2127. Raschka, S. and Raschka, C. (2008). Zur Beziehung zwischen Körperbau und Appendixlänge. *Anthrop. Anz.* **66**, 67–72.
2128. Rasmussen, B., Fletcher, I.R., Brocks, J.J., and Kilburn, M.R. (2008). Reassessing the first appearance of eukaryotes and cyanobacteria. *Nature* **455**, 1101–1104.
2129. Rasskin-Gutman, D. and Esteve-Altava, B. (2008). The multiple directions of evolutionary change. *BioEssays* **30**, 521–525.
2130. Rastegar, S., Hess, I., Dickmeis, T., Nicod, J.C., Ertzer, R., Hadzhiev, Y., Thies, W.-G., Scherer, G., and Strähle, U. (2008). The words of the regulatory code are arranged in a variable manner in highly conserved enhancers. *Dev. Biol.* **318**, 366–377.
2131. Rauhut, O.W.M., Remes, K., Fechner, R., Cladera, G., and Puerta, P. (2005). Discovery of a short-necked sauropod dinosaur from the late Jurassic period of Patagonia. *Nature* **435**, 670–672.
2132. Raup, D.M. (1962). Computer as aid in describing form in gastropod shells. *Science* **138**, 150–152.
2133. Raup, D.M. (1966). Geometric analysis of shell coiling: general problems. *J. Paleontol.* **40**, 1178–1190.
2134. Ravin, J.G. and Hodge, G.P. (1969). Hypertrichosis portrayed in art. *JAMA* **207**, 533–535.
2135. Ravosa, M.J. and Savakova, D.G. (2004). Euprimate origins: the eyes have it. *J. Human Evol.* **46**, 357–364.
2136. Raya, Á. and Izpisúa Belmonte, J.C. (2006). Left-right asymmetry in the vertebrate embryo: from early information to higher-level integration. *Nature Rev. Genet.* **7**, 283–293.
2137. Rayner, J.M.V. and Wootton, R.J., eds. (1991). "Biomechanics in Evolution." Cambridge Univ. Pr., New York.
2138. Raz, N., Striem, E., Pundak, G., Orlov, T., and Zohary, E. (2007). Superior serial memory in the blind: a case of cognitive compensatory adjustment. *Curr. Biol.* **17**, 1129–1133.
2139. Ready, D.F., Hanson, T.E., and Benzer, S. (1976). Development of the *Drosophila* retina, a neurocrystalline lattice. *Dev. Biol.* **53**, 217–240.
2140. Reck-Peterson, S.L., Yildiz, A., Carter, A.P., Gennerich, A., Zhang, N., and Vale, R.D. (2006). Single-molecule analysis of dynein processivity and stepping behavior. *Cell* **126**, 335–348.
2141. Redies, C. and Puelles, L. (2001). Modularity in vertebrate brain development and evolution. *BioEssays* **23**, 1100–1111.
2142. Reebs, S. (2003). Really sinister. *Nat. Hist.* **112** #9, 16.
2143. Reebs, S. (2007). No left turn. *Nat. Hist.* **116** #5, 12.
2144. Reed, K.E. (2006). Early hominid evolution and ecological change through the African Plio-Pleistocene. *In* "The Human Evolution Source Book," 2nd ed. *Advances in Human Evolution Series*, (R.L. Ciochon and J.G. Fleagle, eds.). Pearson Prentice Hall, Upper Saddle River, N.J., pp. 197–218.
2145. Reese, B.E. and Tan, S.-S. (1998). Clonal boundary analysis in the developing retina using X-inactivation transgenic mosaic mice. *Sems. Cell Dev. Biol.* **9**, 285–292.
2146. Reeves, R.R. and Mitchell, E. (1981). The whale behind the tusk. *Nat. Hist.* **90** #8, 50–57.
2147. Reid, R.G.B. (2007). "Biological Emergences: Evolution by Natural Experiment." M.I.T. Pr., Cambridge, Mass.
2148. Reidenberg, J.S. and Laitman, J.T. (1987). Position of the larynx in Odontoceti (toothed whales). *Anat. Rec.* **218**, 98–106.
2149. Reijntjes, S., Blentic, A., Gale, E., and Maden, M. (2005). The control of morphogen signalling: regulation of the synthesis and catabolism of retinoic acid in the developing embryo. *Dev. Biol.* **285**, 224–237.
2150. Reinberger, S. (2008). Tempering tantrums. *Sci. Am. Mind* **19** #5, 72–77.
2151. Reinius, B., Saetre, P., Leonard, J.A., Blekhman, R., Merino-Martinez, R., Gilad, Y., and Jazin, E. (2008). An evolutionarily conserved sexual signature in the primate brain. *PLoS Genet.* **4** #6, e1000100.
2152. Reiss, D. and Marino, L. (2001). Mirror self-recognition in the bottlenose dolphin: a case of cognitive convergence. *PNAS* **9** #10, 5397–5942.
2153. Reiss, K.Z. (2001). Using phylogenies to study convergence: the case of the ant-eating mammals. *Amer. Zool.* **41**, 507–525.
2154. Reisz, R.R. (2006). Origin of dental occlusion in tetrapods: signal for terrestrial vertebrate evolution? *J. Exp. Zool. (Mol. Dev. Evol.)* **306B**, 261–277.
2155. Rembold, M., Loosli, F., Adams, R.J., and Wittbrodt, J. (2006). Individual cell migration serves as the

driving force for optic vesicle evagination. *Science* **313**, 1130–1134.
2156. Remulla, A. and Guilleminault, C. (2004). Somnambulism (sleepwalking). *Expert Opin. Pharmacother.* **5**, 2069–2074.
2157. Renfree, M.B., Pask, A.J., and Shaw, G. (2001). Sex down under: the differentiation of sexual dimorphisms during marsupial development. *Reprod. Fertil. Dev.* **13**, 679–690.
2158. Renfree, M.B., Robinson, E.S., Short, R.V., and Vandeberg, J.L. (1990). Mammary glands in male marsupials: I. Primordia in neonatal opossums *Didelphis virginiana* and *Monodelphis domestica. Development* **110**, 385–390.
2159. Renfree, M.B. and Shaw, G. (2001). Germ cells, gonads and sex reversal in marsupials. *Int. J. Dev. Biol.* **45**, 557–567.
2160. Renfree, M.B., Wilson, J.D., and Shaw, G. (2002). The hormonal control of sexual development. *In* "The Genetics and Biology of Sex Determination," *Novartis Found. Symp. 244* (D. Chadwick and J. Goode, eds.). Wiley, New York, pp. 136–156.
2161. Reno, P.L., McCollum, M.A., Cohn, M.J., Meindl, R.S., Hamrick, M., and Lovejoy, C.O. (2008). Patterns of correlation and covariation of anthropoid distal forelimb segments correspond to Hoxd expression territories. *J. Exp. Zool. (Mol. Dev. Evol.)* **310B**, 240–258.
2162. Reyer, H.-U. (2008). Mating with the wrong species can be right. *Trends Ecol. Evol.* **23**, 289–292.
2163. Reymond, L. (1985). Spatial visual acuity of the eagle *Aquila audax*: a behavioural, optical and anatomical investigation. *Vision Res.* **25**, 1477–1491.
2164. Rhen, T. and Crews, D. (2008). Why are there two sexes? *In* "Sex Differences in the Brain: From Genes to Behavior" (J.B. Becker, K.J. Berkley, N. Geary, E. Hampson, J.P. Herman, and E.A. Young, eds.). Oxford Univ. Pr., New York, pp. 3–14.
2165. Rhodes, G. (2006). The evolutionary psychology of facial beauty. *Annu. Rev. Psychol.* **57**, 199–226.
2166. Rice, S.H. (2002). The role of heterochrony in primate brain evolution. *In* "Human Evolution Through Developmental Change" (N. Minugh-Purvis and K.J. McNamara, eds.). Johns Hopkins Univ. Pr., Baltimore, Md., pp. 154–170.
2167. Rice, W.R. and Friberg, U. (2008). Functionally degenerate—Y not so? *Science* **319**, 42–43.
2168. Rich, T.H., Hopson, J.A., Musser, A.M., Flannery, T.F., and Vickers-Rich, P. (2005). Independent origins of middle ear bones in monotremes and therians. *Science* **307**, 910–914.
2169. Richardson, R.C. (2007). "Evolutionary Psychology as Maladapted Psychology." MIT Pr., Cambridge, Mass.
2170. Richerson, P.J. and Boyd, R. (2004). "Not by Genes Alone: How Culture Transformed Human Evolution." Chicago Univ. Pr., Chicago, Ill.
2171. Richman, J.M., Buchtová, M., and Boughner, J.C. (2006). Comparative ontogeny and phylogeny of the upper jaw skeleton in amniotes. *Dev. Dynamics* **235**, 1230–1243.
2172. Richman, J.M. and Lee, S.-H. (2003). About face: signals and genes controlling jaw patterning and identity in vertebrates. *BioEssays* **25**, 554–568.
2173. Ridderinkhof, K.R., Ullsperger, M., Crone, E.A., and Nieuwenhuis, S. (2004). The role of the medial frontal cortex in cognitive control. *Science* **306**, 443–447.
2174. Riddle, R.D., Ensini, M., Nelson, C., Tsuchida, T., Jessell, T.M., and Tabin, C. (1995). Induction of the LIM homeobox gene Lmx1 by WNT7a establishes dorsoventral pattern in the vertebrate limb. *Cell* **83**, 631–640.
2175. Rideout, E.J., Billeter, J.-C., and Goodwin, S.F. (2007). The sex-determination genes *fruitless* and *doublesex* specify a neural substrate required for courtship song. *Curr. Biol.* **17**, 1473–1478.
2176. Ridley, M. (1993). "The Red Queen: Sex and the Evolution of Human Nature." Harper Perennial, New York.
2177. Riede, T., Bronson, E., Hatzikirou, H., and Zuberbühler, K. (2006). Multiple discontinuities in nonhuman vocal tracts—A response to Lieberman (2006). *J. Human Evol.* **50**, 222–225.
2178. Riedl, R. (1978). "Order in Living Organisms: A Systems Analysis of Evolution." Wiley, New York.
2179. Rieppel, O. (2001). Turtles as hopeful monsters. *BioEssays* **23**, 987–991.
2180. Rijli, F.M., Mark, M., Lakkaraju, S., Dierich, A., Dollé, P., and Chambon, P. (1993). A homeotic transformation is generated in the rostral branchial region of the head by disruption of *Hoxa-2*, which acts as a selector gene. *Cell* **75**, 1333–1349.
2181. Rilling, J.K. (2006). Human and nonhuman primate brains: Are they allometrically scaled versions of the same design? *Evol. Anthrop.* **15**, 65–77.
2182. Rilling, J.K. and Insel, T.R. (1999). The primate neocortex in comparative perspective using magnetic resonance imaging. *J. Human Evol.* **37**, 191–223.
2183. Rincón-Limas, D.E., Lu, C.-H., Canal, I., Calleja, M., Rodríguez-Esteban, C., Izpisúa-Belmonte, J.C., and Botas, J. (1999). Conservation of the expression and function of *apterous* orthologs in *Drosophila* and mammals. *Proc. Natl. Acad. Sci. USA* **96**, 2165–2170.
2184. Rinn, J.L., Bondre, C., Gladstone, H.B., Brown, P.O., and Chang, H.Y. (2006). Anatomic demarcation by positional variation in fibroblast gene expression programs. *PLoS Genet.* **2** #7, 1084–1096 (e119).
2185. Rinn, J.L., Rozowsky, J.S., Laurenzi, I.J., Petersen, P.H., Zou, K., Zhong, W., Gerstein, M., and Snyder, M. (2004). Major molecular differences between mammalian sexes are involved in drug metabolism and renal function. *Dev. Cell* **6**, 791–800.
2186. Rinn, J.L. and Snyder, M. (2005). Sexual dimorphism in mammalian gene expression. *Trends Genet.* **21**, 298–305.

2187. Risau, W. (1977). Mechanisms of angiogenesis. *Nature* **386**, 671–674.
2188. Riyahi, K. and Shimeld, S.M. (2007). Chordate *βγ-crystallins* and the evolutionary developmental biology of the vertebrate lens. *Comp. Biochem. Physiol. B Biochem. Mol. Biol.* **147**, 347–357.
2189. Rizzino, A. (2007). A challenge for regenerative medicine: proper genetic programming, not cellular mimicry. *Dev. Dynamics* **236**, 3199–3207.
2190. Rizzolatti, G. and Fabbri-Destro, M. (2008). The mirror system and its role in social cognition. *Curr. Opin. Neurobiol.* **18**, 179–184.
2191. Rizzolatti, G. and Sinigaglia, C. (2008). "Mirrors in the Brain: How Our Minds Share Actions and Emotions." Oxford Univ. Pr., New York.
2192. Rizzoti, K. and Lovell-Badge, R. (2005). Early development of the pituitary gland: induction and shaping of Rathke's pouch. *Rev. Endocr. Metab. Disorders* **6**, 161–172.
2193. Robert, B. and Lallemand, Y. (2006). Anteroposterior patterning in the limb and digit specification: contributions of mouse genetics. *Dev. Dynamics* **235**, 2337–2352.
2194. Robert, D. and Göpfert, M.C. (2002). Novel schemes for hearing and orientation in insects. *Curr. Opin. Neurobiol.* **12**, 715–720.
2195. Robert, J.S., Hall, B.K., and Olson, W.M. (2001). Bridging the gap between developmental systems theory and evolutionary developmental biology. *BioEssays* **23**, 954–962.
2196. Robertson, R. (1993). Snail handedness. *Res. Explor.* **9** #1, 104–119.
2197. Robinson, E.S., Renfree, M.B., Short, R.V., and Vandeberg, J.L. (1991). Mammary glands in male marsupials: 2. Development of teat primordia in *Didelphis virginiana* and *Monodelphis domestica*. *Reprod. Fertil. Dev.* **3**, 295–301.
2198. Robinson, G.W. (2007). Cooperation of signalling pathways in embryonic mammary gland development. *Nature Rev. Genet.* **8**, 963–972.
2199. Robinson, G.W., Hennighausen, L., and Johnson, P.F. (2000). Side-branching in the mammary gland: the progesterone-Wnt connection. *Genes Dev.* **14**, 889–894.
2200. Robinson, R. (2008). For mammals, loss of yolk and gain of milk went hand in hand. *PLoS Biol.* **6** #3, 421 (e77).
2201. Robledo, R.F. and Lufkin, T. (2006). *Dlx5* and *Dlx6* homeobox genes are required for specification of the mammalian vestibular apparatus. *Genesis* **44**, 425–437.
2202. Roca, C. and Adams, R.H. (2007). Regulation of vascular morphogenesis by Notch signaling. *Genes Dev.* **21**, 2511–2524.
2203. Roch, F. and Akam, M. (2000). *Ultrabithorax* and the control of cell morphology in *Drosophila* halteres. *Development* **127**, 97–107.
2204. Rockman, M.V. and Stern, D.L. (2008). Tinker where the tinkering's good. *Trends Genet.* **24**, 317–319.
2205. Rodriguez-Esteban, C., Tsukui, T., Yonei, S., Magallon, J., Tamura, K., and Izpisua Belmonte, J.C. (1999). The T-box genes *Tbx4* and *Tbx5* regulate limb outgrowth and identity. *Nature* **398**, 814–818.
2206. Rodríguez-Trelles, F., Tarrío, R., and Ayala, F.J. (2005). Is ectopic expression caused by deregulatory mutations or due to gene-regulation leaks with evolutionary potential? *BioEssays* **27**, 592–601.
2207. Roff, D.A. (2005). Variation and life-history traits. *In* "Variation: A Central Concept in Biology" (B. Hallgrímsson and B.K. Hall, eds.). Elsevier Acad. Pr., New York, pp. 333–357.
2208. Rogers, A.R. and Mukherjee, A. (1992). Quantitative genetics of sexual dimorphism in human body size. *Evolution* **46**, 226–234.
2209. Rokas, A. and Carroll, S.B. (2006). Bushes in the tree of life. *PLoS* **4** #11, 1899–1904 (e352).
2210. Rokas, A., Krüger, D., and Carroll, S.B. (2005). Animal evolution and the molecular signature of radiations compressed in time. *Science* **310**, 1933–1938.
2211. Rollo, C.D. (1995). "Phenotypes: Their Epigenetics, Ecology and Evolution." Chapman & Hall, New York.
2212. Rollo, C.D. (2007). Speculations on the evolutionary ecology of *Homo sapiens* with special reference to body size, allometry and survivorship. *In* "Human Body Size and the Laws of Scaling: Physiological, Performance, Growth, Longevity and Ecological Ramifications" (T. Samaras, ed.). Nova Sci. Pub., New York, pp. 261–299.
2213. Rome, L.C. (1997). Testing a muscle's design. *Am. Sci.* **85**, 356–363.
2214. Romer, A.S. (1956). "Osteology of the Reptiles." Univ. Chicago Pr., Chicago, Ill.
2215. Rorick, A.M., Mei, W., Liette, N.L., Phiel, C., El-Hodiri, H.M., and Yang, J. (2007). PP2A:B56ϵ is required for eye induction and eye field separation. *Dev. Biol.* **302**, 477–493.
2216. Ros, M.A., Dahn, R.D., Fernandez-Teran, M., Rashka, K., Caruccio, N.C., Hasso, S.M., Bitgood, J.J., Lancman, J.J., and Fallon, J.F. (2003). The chick *oligozeugodactyly* (*ozd*) mutant lacks sonic hedgehog function in the limb. *Development* **130**, 527–537.
2217. Ros, M.A., Piedra, M.E., Fallon, J.F., and Hurle, J.M. (1997). Morphogenetic potential of the chick leg interdigital mesoderm when diverted from the cell death program. *Dev. Dynamics* **208**, 406–419.
2218. Rose, M.R. (2007). End of the line. *Q. Rev. Biol.* **82**, 395–400.
2219. Rose, M.R., Rauser, C.L., Benford, G., Matos, M., and Mueller, L.D. (2007). Hamilton's forces of natural selection after forty years. *Evolution* **61**, 1265–1276.
2220. Rosenberg, K.R. (1992). The evolution of modern human childbirth. *Yrbk. Phys. Anthrop.* **35**, 89–124.

2221. Rosenfeld, R.G. (2004). Gender differences in height: an evolutionary perspective. *J. Pediatr. Endocrinol. Metab.* **17 Suppl. 4**, 1267–1271.
2222. Rosenthal, G.G. (2007). Spatiotemporal dimensions of visual signals in animal communication. *Annu. Rev. Ecol. Evol. Syst.* **38**, 155–178.
2223. Ross, C.F. and Kirk, E.C. (2007). Evolution of eye size and shape in primates. *J. Human Evol.* **52**, 294–313.
2224. Ross, D.A. and Sasaki, C.T. (1994). Acute laryngeal obstruction. *Lancet* **344**, 1743–1748.
2225. Ross, S.A., McCaffery, P.J., Drager, U.C., and De Luca, L.M. (2000). Retinoids in embryonal development. *Physiol. Rev.* **80**, 1021–1054.
2226. Rossant, J. and Hirashima, M. (2003). Vascular development and patterning: making the right choices. *Curr. Opin. Gen. Dev.* **13**, 408–412.
2227. Rossant, J. and Howard, L. (2002). Signaling pathways in vascular development. *Annu. Rev. Cell Dev. Biol.* **18**, 541–573.
2228. Roth, G. and Dicke, U. (2005). Evolution of the brain and intelligence. *Trends Cogn. Sci.* **9**, 250–257.
2229. Roth, M.B. and Nystul, T. (2005). Buying time in suspended animation. *Sci. Am.* **29** #6, 48–55.
2230. Roth, V.L. (2005). Variation and versatility in macroevolution. *In* "Variation: A Central Concept in Biology" (B. Hallgrímsson and B.K. Hall, eds.). Elsevier Acad. Pr., New York, pp. 455–474.
2231. Roth, V.L. and Mercer, J.M. (2000). Morphometrics in development and evolution. *Amer. Zool.* **40**, 801–810.
2232. Rothchild, I. (2003). The yolkless egg and the evolution of eutherian viviparity. *Biol. Reprod.* **68**, 337–357.
2233. Rothenberg, D. (2008). "Thousand Mile Song: Whale Music in a Sea of Sound." Basic Bks. New York.
2234. Rowe, A. (1994). Legs or tails: retinoids and homeosis in frogs. *BioEssays* **16**, 53–54.
2235. Rozen, D.E., Habets, M.G.J.L., Handel, A., and de Visser, J.A.G.M. (2008). Heterogeneous adaptive trajectories of small populations on complex fitness landscapes. *PLoS ONE* **3** #3, e1715.
2236. Rozmus-Wrzesinska, M. and Pawlowski, B. (2005). Men's ratings of female attractiveness are influenced more by changes in female waist size compared with changes in hip size. *Biol. Psychol.* **68**, 299–308.
2237. Rubenstein, J.L.R. (2000). Intrinsic and extrinsic control of cortical development. *In* "Evolutionary Developmental Biology of the Cerebral Cortex," *Novartis Found. Symp. 228*, (G.R. Bock and G. Cardew, eds.). Wiley, New York, pp. 67–82.
2238. Rudel, D. and Sommer, R.J. (2003). The evolution of developmental mechanisms. *Dev. Biol.* **264**, 15–37.
2239. Ruest, L.-B., Xiang, X., Lim, K.-C., Levi, G., and Clouthier, D.E. (2004). Endothelin-A receptor-dependent and -independent signaling pathways in establishing mandibular identity. *Development* **131**, 4413–4423.
2240. Ruiz, M.F., Eirín-López, J.M., Stefani, R.N., Perondini, A.L.P., Selivon, D., and Sánchez, L. (2007). The gene *doublesex* of *Anastrepha* fruit flies (Diptera, Tephritidae) and its evolution in insects. *Dev. Genes Evol.* **217**, 725–731.
2241. Ruiz-Gonzalez, M.X. and Marin, I. (2004). New insights into the evolutionary history of type 1 rhodopsins. *J. Mol. Evol.* **58**, 348–358.
2242. Runyon, J.B. and Hurley, R.L. (2004). A new genus of long-legged flies displaying remarkable wing directional asymmetry. *Proc. Roy. Soc. Lond. B (Suppl.)* **271**, S114–S116.
2243. Rusciano, D. and Burger, M.M. (1992). Why do cancer cells metastasize into particular organs? *BioEssays* **14**, 185–194.
2244. Rusting, R.L. (2001). Hair: Why it grows, why it stops. *Sci. Am.* **284** #6, 70–79.
2245. Ruvinsky, I. and Gibson-Brown, J.J. (2000). Genetic and developmental bases of serial homology in vertebrate limb evolution. *Development* **127**, 5233–5244.
2246. Ryan, A.K., Blumberg, B., Rodriguez-Esteban, C., Yonei-Tamura, S., Tamura, K., Tsukui, T., de la Peña, J., Sabbagh, W., Greenwald, J., Choe, S., Norris, D.P., Robertson, E.J., Evans, R.M., Rosenfeld, M.G., and Izpisúa-Belmonte, J.C. (1998). Pitx2 determines left-right asymmetry of internal organs in vertebrates. *Nature* **394**, 545–551.
2247. Ryan, J.F., Mazza, M.E., Pang, K., Matus, D.Q., Baxevanis, A.D., Martindale, M.Q., and Finnerty, J.R. (2007). Pre-bilaterian origins of the Hox cluster and the Hox code: Evidence from the sea anemone, *Nematostella vectensis. PLoS ONE* **2** #1, e153.
2248. Rybczynski, N. (2004). Optimized for chewing. *Science* **306**, 2045.
2249. Sacks, O. (2007). "Musicophilia: Tales of Music and the Brain." Knopf, New York.
2250. Saele, Ø., Smáradóttir, H., and Pittman, K. (2006). Twisted story of eye migration in flatfish. *J. Morph.* **267**, 730–738.
2251. Safi, K., Seid, M.A., and Dechmann, D.K.N. (2005). Bigger is not always better: when brains get smaller. *Biol. Lett.* **1**, 283–286.
2252. Sagai, T., Hosoya, M., Mizushina, Y., Tamura, M., and Shiroishi, T. (2004). Elimination of a long-range cis-regulatory module causes complete loss of limb-specific *Shh* expression and truncation of the mouse limb. *Development* **132**, 797–803.
2253. Sagan, C. (1977). "The Dragons of Eden: Speculations on the Evolution of Human Intelligence." Random House, New York.
2254. Sagan, C. and Druyan, A. (1992). "Shadows of Forgotten Ancestors: A Search for Who We Are." Ballantine Bks. New York.
2255. Sagasti, A. (2007). Three ways to make two sides: genetic models of asymmetric nervous system development. *Neuron* **55**, 345–351.

2256. Sai, X.R. and Ladher, R.K. (2008). FGF signaling regulates cytoskeletal remodeling during epithelial morphogenesis. *Curr. Biol.* **18**, 976–981.

2257. Saijoh, Y., Oki, S., Tanaka, C., Nakamura, T., Adachi, H., Yan, Y.-T., Shen, M.M., and Hamada, H. (2005). Two Nodal-responsive enhancers control left-right asymmetric expression of *Nodal. Dev. Dynamics* **232**, 1031–1036.

2258. Saito, D., Yonei-Tamura, S., Takahashi, Y., and Tamura, K. (2006). Level-specific role of paraxial mesoderm in regulation of *Tbx5/Tbx4* expression and limb initiation. *Dev. Biol.* **292**, 79–89.

2259. Sakagami, M. and Pan, X. (2007). Functional role of the ventrolateral prefrontal cortex in decision making. *Curr. Opin. Neurobiol.* **17**, 228–233.

2260. Sakai, T., Larsen, M., and Yamada, K.M. (2003). Fibronectin requirement in branching morphogenesis. *Nature* **423**, 876–881.

2261. Sakata, N., Miyazaki, K., and Wakahara, M. (2006). Up-regulation of *P450arom* and down-regulation of *Dmrt-1* genes in the temperature-dependent sex reversal from genetic males to phenotypic females in a salamander. *Dev. Genes Evol.* **216**, 224–228.

2262. Salazar-Ciudad, I. (2007). On the origins of morphological variation, canalization, robustness, and evolvability. *Integr. Comp. Biol.* **47**, 390–400.

2263. Salazar-Ciudad, I. (2008). Making evolutionary predictions about the structure of development and morphology: beyond the neo-Darwinian and constraints paradigms. *In* "Evolving Pathways: Key Themes in Evolutionary Developmental Biology" (A. Minelli and G. Fusco, eds.). Cambridge Univ. Pr., New York, pp. 31–49.

2264. Salazar-Ciudad, I. and Jernvall, J. (2002). A gene network model accounting for development and evolution of mammalian teeth. *PNAS* **99** #12, 8116–8120.

2265. Salazar-Ciudad, I. and Jernvall, J. (2005). Graduality and innovation in the evolution of complex phenotypes: insights from development. *J. Exp. Zool. (Mol. Dev. Evol.)* **304B**, 619–631.

2266. Sami, D.A., Saunders, D., Thompson, D.A., Russell-Eggitt, I.M., Nischal, K.K., Jeffery, G., Dattani, M., Clement, R.A., Liassis, A., and Taylor, D.S. (2005). The achiasmia spectrum: congenitally reduced chiasmal decussation. *J. Ophthalmol.* **89**, 1311–1317.

2267. Samokhvalov, I.M., Samokhvalova, N.I., and Nishikawa, S.-i. (2007). Cell tracing shows the contribution of the yolk sac to adult haematopoiesis. *Nature* **446**, 1056–1061. [See also (2008) *Dev. Dynamics* **237**, 3332–3341.]

2268. Sánchez Alvarado, A. and Tsonis, P.A. (2006). Bridging the regeneration gap: genetic insights from diverse animal models. *Nature Rev. Genet.* **7**, 873–884.

2269. Sánchez, L., Gorfinkiel, N., and Guerrero, I. (2001). Sex determination genes control the development of the *Drosophila* genital disc, modulating the response to Hedgehog, Wingless and Decapentaplegic signals. *Development* **128**, 1033–1043.

2270. Sánchez, L. and Guerrero, I. (2001). The development of the *Drosophila* genital disc. *BioEssays* **23**, 698–707.

2271. Sandak, R., Mencl, W.E., Frost, S.J., and Pugh, K.R. (2004). The neurobiological basis of skilled and impaired reading: recent findings and new directions. *Sci. Studies Reading* **8**, 273–292.

2272. Sander, P.M. and Clauss, M. (2008). Sauropod gigantism. *Science* **322**, 200–201.

2273. Sanders, L.E. and Arbeitman, M.N. (2008). Doublesex establishes sexual dimorphism in the *Drosophila* central nervous system in an isoform-dependent manner by directing cell number. *Dev. Biol.* **320**, 378–390.

2274. Sanefuji, W., Ohgami, H., and Hashiya, K. (2007). Development of preference for baby faces across species in humans (*Homo sapiens*). *J. Ethology* **25**, 249–254.

2275. Sanger, T.J. and Gibson-Brown, J.J. (2004). The developmental basis of limb reduction and body elongation in squamates. *Evolution* **58**, 2103–2106.

2276. Sargis, E.J., Boyer, D.M., Bloch, J.I., and Silcox, M.T. (2007). Evolution of pedal grasping in primates. *J. Human Evol.* **53**, 103–107.

2277. Sarin, K.Y. and Artandi, S.E. (2007). Aging, graying and loss of melanocyte stem cells. *Stem Cell Rev.* **3**, 212–217.

2278. Sarnat, H.B. and Netsky, M.G. (1981). "Evolution of the Nervous System." Oxford Univ. Pr., New York.

2279. Satoh, A.K., Li, B.X., Xia, H., and Ready, D.F. (2008). Calcium-activated myosin V closes the *Drosophila* pupil. *Curr. Biol.* **18**, 951–955.

2280. Satou, Y. and Satoh, N. (2006). Gene regulatory networks for the development and evolution of the chordate heart. *Genes Dev.* **20**, 2634–2638.

2281. Sauer, E.G.F. and Sauer, E.M. (1967). Yawning and other maintenance activities in the South African ostrich. *The Auk* **84**, 571–587.

2282. Sauka-Spengler, T., Meulemans, D., Jones, M., and Bronner-Fraser, M. (2007). Ancient evolutionary origin of the neural crest gene regulatory network. *Dev. Cell* **13**, 405–420.

2283. Saunders, P.T. (1990). The epigenetic landscape and evolution. *Biol. J. Linn. Soc.* **39**, 125–134.

2284. Sawaguchi, T. and Kubota, K. (1986). A hypothesis on the primate neocortex evolution: column-multiplication hypothesis. *Internat. J. Neurosci.* **30**, 57–64.

2285. Sawyer, G.J. and Deak, V. (2007). "The Last Human: A Guide to Twenty-two Species of Extinct Humans." Yale Univ. Pr., New Haven, Conn.

2286. Sayers, K. and Lovejoy, C.O. (2008). The chimpanzee has no clothes: a critical examination of *Pan troglodytes* in models of human evolution. *Curr. Anthrop.* **49**, 87–114.

2287. Schall, J.D. (2002). The neural selection and control of saccades by the frontal eye field. *Phil. Trans. Roy. Soc. Lond. B* **357**, 1073–1082.
2288. Scheerer, P., Park, J.H., Hildebrand, P.W., Kim, Y.J., Krauss, N., Choe, H.-W., Hofmann, K.P., and Ernst, O.P. (2008). Crystal structure of opsin in its G-protein-interacting conformation. *Nature* **455**, 497–502.
2289. Schellenberg, E.G. and Peretz, I. (2008). Music, language and cognition: unresolved issues. *Trends Cogn. Sci.* **12**, 45–46.
2290. Schenker, N.M., Desgouttes, A.-M., and Semendeferi, K. (2005). Neural connectivity and cortical substrates of cognition in hominoids. *J. Human Evol.* **49**, 547–569.
2291. Scherer, G. (2002). The molecular genetic jigsaw puzzle of vertebrate sex determination and its missing pieces. *In* "The Genetics and Biology of Sex Determination," *Novartis Found. Symp. 244* (D. Chadwick and J. Goode, eds.). Wiley, New York, pp. 225–239.
2292. Scherz, P.J., McGlinn, E., Nissim, S., and Tabin, C.J. (2007). Extended exposure to Sonic hedgehog is required for patterning the posterior digits of the vertebrate limb. *Dev. Biol.* **308**, 343–354.
2293. Schierwater, B. and DeSalle, R. (2007). Can we ever identify the Urmetazoan? *Integr. Comp. Biol.* **47**, 670–676.
2294. Schillaci, M.A. (2006). Sexual selection and the evolution of brain size in primates. *PLoS ONE* **1** #1, 1–5 (e62).
2295. Schilling, T.F. (1997). Genetic analysis of craniofacial development in the vertebrate embryo. *BioEssays* **19**, 459–468.
2296. Schilling, T.F. and Kimmel, C.B. (1997). Musculoskeletal patterning in the pharyngeal segments of the zebrafish embryo. *Development* **124**, 2945–2960.
2297. Schilthuizen, M. and Davison, A. (2005). The convoluted evolution of snail chirality. *Naturwissenschaften* **92**, 504–515.
2298. Schlaggar, B.L. and McCandliss, B.D. (2007). Development of neural systems for reading. *Annu. Rev. Neurosci.* **30**, 475–503.
2299. Schlake, T., Schorpp, M., and Boehm, T. (2000). Formation of regulator/target gene relationships during evolution. *Gene* **256**, 29–34.
2300. Schlosser, G. (2002). Modularity and the units of evolution. *Theory Biosci.* **121**, 1–80.
2301. Schlosser, G. (2006). Induction and specification of cranial placodes. *Dev. Biol.* **294**, 303–351.
2302. Schlosser, G. (2008). Do vertebrate neural crest and cranial placodes have a common evolutionary origin? *BioEssays* **30**, 659–672.
2303. Schlosser, G. and Wagner, G.P., eds. (2004). "Modularity in Development and Evolution." Univ. of Chicago Pr., Chicago, Ill.
2304. Schluter, D. (2000). "The Ecology of Adaptive Radiation." Oxford Univ. Pr., New York.
2305. Schmidt-Nielsen, K. (1984). "Scaling: Why Is Animal Size So Important?" Cambridge Univ. Pr., New York.
2306. Schmolesky, M.T., Wang, Y., Creel, D.J., and Leventhal, A.G. (2000). Abnormal retinotopic organization of the dorsal lateral geniculate nucleus of the tyrosinase-negative cat. *J. Compar. Neurol.* **427**, 209–219.
2307. Schmutz, S.M. and Berryere, T.G. (2007). Genes affecting coat colour and pattern in domestic dogs: a review. *Animal Genet.* **38**, 539–549.
2308. Schober, J.M. (2007). The neurophysiology of sexual arousal. *Best Pract. Res. Clin. Endocr. Metab.* **21**, 445–461.
2309. Schoenemann, P.T. (2006). Evolution of the size and functional areas of the human brain. *Annu. Rev. Anthrop.* **35**, 379–406.
2310. Schoeninger, M.J., Bunn, H.T., Murray, S., Pickering, T., and Moore, J. (2001). Meat-eating by the fourth African ape. *In* "Meat-eating and Human Evolution" (C.B. Stanford and H.T. Bunn, eds.). Oxford Univ. Pr., New York, pp. 179–195.
2311. Schoenwolf, G.C., Bleyl, S.B., Brauer, P.R., and Francis-West, P.H. (2009). "Larsen's Human Embryology," 4th ed. Churchill Livingstone, Philadelphia.
2312. Schoenwolf, G.C. and Smith, J.L. (1990). Mechanisms of neurulation: traditional viewpoint and recent advances. *Development* **109**, 243–270.
2313. Schorpp, M., Schlake, T., Kreamalmeyer, D., Allen, P.M., and Boehm, T. (2000). Genetically separable determinants of hair keratin gene expression. *Dev. Dynamics* **218**, 537–543.
2314. Schouwey, K. and Beermann, F. (2008). The Notch pathway: hair graying and pigment cell homeostasis. *Histol. Histopathol.* **23**, 609–619.
2315. Schreiber, A.M. (2006). Asymmetric craniofacial remodeling and lateralized behavior in larval flatfish. *J. Exp. Biol.* **209**, 610–621.
2316. Schubert, M., Holland, N.D., Laudet, V., and Holland, L.Z. (2006). A retinoic acid-*Hox* hierarchy controls both anterior/posterior patterning and neuronal specification in the developing central nervous system of the cephalochordate amphioxus. *Dev. Biol.* **296**, 190–202.
2317. Schultz, A.H. (1926). Fetal growth of man and other primates. *Q. Rev. Biol.* **1**, 465–521.
2318. Schultz, A.H. (1948). The number of young at a birth and the number of nipples in primates. *Am. J. Phys. Anthrop.* **6 N.S.**, 1–23.
2319. Schultz, A.H. (1949). Ontogenetic specializations of man. *Archiv. Julius Klaus-Stiftung* **24**, 197–216.
2320. Schultz, A.H. (1949). Sex differences in the pelves of primates. *Am. J. Phys. Anthrop.* **7 (N.S.)**, 401–423.
2321. Schultz, A.H. (1950). The physical distinctions of man. *Proc. Amer. Philos. Soc.* **94**, 428–449.
2322. Schuraytz, B.C., Lindstrom, D.J., Marin, L.E., Martinez, R.R., Mittlefehldt, D.W., Sharpton, V.L., and Wentworth, S.J. (1996). Iridium metal in Chicxulub

impact melt: forensic chemistry on the K-T smoking gun. *Science* **271**, 1573–1576.

2323. Schwarz, M., Cecconi, F., Bernier, G., Andrejewski, N., Kammandel, B., Wagner, M., and Gruss, P. (2000). Spatial specification of mammalian eye territories by reciprocal transcriptional repression of Pax2 and Pax6. *Development* **127**, 4325–4334.
2324. Schweickert, A., Campione, M., Steinbeisser, H., and Blum, M. (2000). *Pitx2* isoforms: involvement of *Pitx2c* but not *Pitx2a* or *Pitx2b* in vertebrate left–right asymmetry. *Mechs. Dev.* **90**, 41–51.
2325. Schwenk, K. and Wagner, G.P. (2004). The relativism of constraints on phenotypic evolution. *In* "Phenotypic Integration. Studying the Ecology and Evolution of Complex Phenotypes." (M. Pigliucci and K. Preston, eds.). Oxford Univ. Pr., New York, pp. 390–408.
2326. Scotese, C.R. (2004). A continental drift flipbook. *J. Geology* **112**, 729–741.
2327. Searle, J.R. (2000). Consciousness. *Annu. Rev. Neurosci.* **23**, 557–578.
2328. Sears, K.E., Behringer, R.R., Rasweiler, J.J., IV, and Niswander, L.A. (2006). Development of bat flight: morphologic and molecular evolution of bat wing digits. *PNAS* **103** #17, 6581–6586.
2329. Sebastian, C., Burnett, S., and Blakemore, S.-J. (2008). Development of the self-concept during adolescence. *Trends Cogn. Sci.* **12**, 441–446.
2330. Seehausen, O., Terai, Y., Magalhaes, I.S., Carleton, K.L., Mrosso, H.D.J., Miyagi, R., van der Sluijs, I., Schneider, M.V., Maan, M.E., Tachida, H., Imai, H., and Okada, N. (2008). Speciation through sensory drive in cichlid fish. *Nature* **455**, 620–626.
2331. Seeley, R.R., Stephens, T.D., and Tate, P. (2003). "Anatomy & Physiology," 6th ed. McGraw-Hill, New York.
2332. Seifert, A.W., Harfe, B.D., and Cohn, M.J. (2008). Cell lineage analysis demonstrates an endodermal origin of the distal urethra and perineum. *Dev. Biol.* **318**, 143–152.
2333. Sekido, R. and Lovell-Badge, R. (2008). Sex determination involves synergistic action of SRY and SF1 on a specific *Sox9* enhancer. *Nature* **453**, 930–934.
2334. Sellen, D.W. (2006). Lactation, complementary feeding, and human life history. *In* "The Evolution of Human Life History" (K. Hawkes and R.R. Paine, eds.). School of Amer. Res. Pr., Santa Fe, New Mexico, pp. 155–196.
2335. Sen, S., Esen, U.I., and Sturgiss, S.N. (2002). Twin reversed arterial perfusion (TRAP) syndrome. *Int. J. Clin. Pract.* **56**, 818–819.
2336. Senju, A. and Csibra, G. (2008). Gaze following in human infants depends on communicative signals. *Curr. Biol.* **18**, 668–671.
2337. Sereno, M.I. and Tootell, R.B.H. (2005). From monkeys to humans: what do we now know about brain homologies? *Curr. Opin. Neurobiol.* **15**, 135–144.
2338. Serna, E., Gorab, E., Ruiz, M.F., Goday, C., Eirín-López, J.M., and Sánchez, L. (2004). The gene *Sex-lethal* of the Sciaridae family (Order Diptera, Suborder Nematocera) and its phylogeny in dipteran insects. *Genetics* **168**, 907–921.
2339. Sessions, S.K., Franssen, R.A., and Horner, V.L. (1999). Morphological clues from multilegged frogs: are retinoids to blame? *Science* **284**, 800–802.
2340. Sessions, S.K. and Ruth, S.B. (1990). Explanation for naturally occurring supernumerary limbs in amphibians. *J. Exp. Zool.* **254**, 38–47.
2341. Shah, M.M., Sampogna, R.V., Sakurai, H., Bush, K.T., and Nigam, S.K. (2004). Branching morphogenesis and kidney disease. *Development* **131**, 1449–1462.
2342. Shah, N.M., Pisapia, D.J., Maniatis, S., Mendelsohn, M.M., Nemes, A., and Axel, R. (2004). Visualizing sexual dimorphism in the brain. *Neuron* **43**, 313–319.
2343. Shang, J., Luo, Y., and Clayton, D.A. (1997). *Backfoot* is a novel homeobox gene expressed in the mesenchyme of developing hind limb. *Dev. Dynamics* **209**, 242–253.
2344. Shapiro, M.D., Bell, M.A., and Kingsley, D.M. (2006). Parallel genetic origins of pelvic reduction in vertebrates. *PNAS* **103** #37, 13753–13758.
2345. Shapiro, M.D., Shubin, N.H., and Downs, J.P. (2007). Limb diversity and digit reduction in reptilian evolution. *In* "Fins into Limbs: Evolution, Development, and Transformation" (B.K. Hall, ed.). Univ. Chicago Pr., Chicago, Ill., pp. 225–244.
2346. Sharpe, P.T. (1995). Homeobox genes and orofacial development. *Connect. Tissue Res.* **32**, 17–25.
2347. Shaut, C.A., Saneyoshi, C., Morgan, E.A., Knosp, W.M., Sexton, D.R., and Stadler, H.S. (2007). HOXA13 directly regulates *EphA6* and *EphA7* expression in the genital tubercle vascular endothelia. *Dev. Dynamics* **236**, 951–960.
2348. Shea, B.T. and Bailey, R.C. (1996). Allometry and adaptation of body proportions and stature in African pygmies. *Am. J. Phys. Anthrop.* **100**, 311–340.
2349. Shea, B.T. and Gomez, A.M. (1988). Tooth scaling and evolutionary dwarfism: an investigation of allometry in human pygmies. *Am. J. Phys. Anthrop.* **77**, 117–132.
2350. Shekhawat, D.S., Jangir, O.P., Prakash, A., and Pawan, S. (2001). Lens regeneration in mice under the influence of vitamin A. *J. Biosci.* **26**, 571–576.
2351. Shelby, J.A., Madewell, R., and Moczek, A.P. (2007). Juvenile hormone mediates sexual dimorphism in horned beetles. *J. Exp. Zool. (Mol. Dev. Evol.)* **308B**, 417–427.
2352. Shen, M.M. (2007). Nodal signaling: developmental roles and regulation. *Development* **134**, 1023–1034.
2353. Shepard, P. (1973). "The Tender Carnivore and the Sacred Game." Charles Scribner's Sons, New York.
2354. Sherman, P.W. (1998). The evolution of menopause. *Nature* **392**, 759–761.

2355. Sherman, P.W., Braude, S., and Jarvis, J.U.M. (1999). Litter sizes and mammary numbers of naked mole-rats: Breaking the one-half rule. *J. Mammalogy* **80**, 720–733.
2356. Shermer, M. (2004). The major unsolved problem in biology. *Sci. Am.* **290** #3, 103–105.
2357. Shermer, M. (2008). Wag the dog. *Sci. Am.* **298** #4, 44.
2358. Sherwood, C.C. (2005). Comparative anatomy of the facial motor nucleus in mammals, with an analysis of neuron numbers in primates. *Anat. Rec.* **287A**, 1067–1079.
2359. Sherwood, C.C., Hof, P.R., Holloway, R.L., Semendeferi, K., Gannon, P.J., Frahm, H.D., and Zilles, K. (2005). Evolution of the brainstem orofacial motor system in primates: a comparative study of trigeminal, facial, and hypoglossal nuclei. *J. Human Evol.* **48**, 45–84.
2360. Sherwood, C.C., Holloway, R.L., Gannon, P.J., Semendeferi, K., Erwin, J.M., Zilles, K., and Hof, P.R. (2003). Neuroanatomical basis of facial expression in monkeys, apes, and humans. *Ann. N. Y. Acad. Sci.* **1000**, 99–103.
2361. Sherwood, N.M., Adams, B.A., and Tello, J.A. (2005). Endocrinology of protochordates. *Can. J. Zool.* **83**, 225–255.
2362. Sheth, R., Bastida, M.F., and Ros, M. (2007). *Hoxd* and *Gli3* interactions modulate digit number in the amniote limb. *Dev. Biol.* **310**, 430–441.
2363. Shibazaki, Y., Shimizu, M., and Kuroda, R. (2004). Body handedness is directed by genetically determined cytoskeletal dynamics in the early embryo. *Curr. Biol.* **14**, 1462–1467.
2364. Shigetani, Y., Sugahara, F., and Kuratani, S. (2005). A new evolutionary scenario for the vertebrate jaw. *BioEssays* **27**, 331–338.
2365. Shimeld, S.M. and Holland, N.D. (2005). Amphioxus molecular biology: insights into vertebrate evolution and developmental mechanisms. *Can. J. Zool.* **83**, 90–100.
2366. Shimeld, S.M. and Holland, P.W.H. (2000). Vertebrate innovations. *PNAS* **97** #9, 4449–4452.
2367. Shimeld, S.M. and Levin, M. (2006). Evidence for the regulation of left-right asymmetry in *Ciona intestinalis* by ion flux. *Dev. Dynamics* **235**, 1543–1553.
2368. Shimeld, S.M., Purkiss, A.G., Dirks, R.P.H., Bateman, O.A., Slingsby, C., and Lubsen, N.H. (2005). Urochordate βγ-crystallin and the evolutionary origin of the vertebrate eye lens. *Curr. Biol.* **15**, 1684–1689.
2369. Shine, R. (1988). The evolution of larger body size in females: a critique of Darwin's "fecundity advantage" model. *Am. Nat.* **131**, 124–131.
2370. Shine, R. (1989). Ecological causes for the evolution of sexual dimorphism: a review of the evidence. *Q. Rev. Biol.* **64**, 419–461.
2371. Shiomi, N., Cui, X.-M., Yamamoto, T., Saito, T., and Shuler, C.F. (2006). Inhibition of Smad2 expression prevents murine palatal fusion. *Dev. Dynamics* **235**, 1785–1793.
2372. Shiota, K., Yamada, S., Komada, M., and Ishibashi, M. (2007). Embryogenesis of holoprosencephaly. *Am. J. Med. Genet. A* **143A**, 3079–3087.
2373. Shipman, P. (2008). Freed to fly again. *Am. Sci.* **96**, 20–23.
2374. Shipman, P., Walker, A., and Bichell, D. (1985). "The Human Skeleton." Harvard Univ. Pr., Cambridge, Mass.
2375. Shirangi, T.R. and McKeown, M. (2007). Sex in flies: What "body-mind" dichotomy? *Dev. Biol.* **306**, 10–19.
2376. Shiratori, H. and Hamada, H. (2006). The left–right axis in the mouse: from origin to morphology. *Development* **133**, 2095–2104.
2377. Shiratori, H., Sakuma, R., Watanabe, M., Hashiguchi, H., Mochida, K., Sakai, Y., Nishino, J., Saijoh, Y., Whitman, M., and Hamada, H. (2001). Two-step regulation of left-right asymmetric expression of *Pitx2*: initiation by Nodal signaling and maintenance by Nkx2. *Molec. Cell* **7**, 137–149.
2378. Shiratori, H., Yashiro, K., Shen, M.M., and Hamada, H. (2006). Conserved regulation and role of *Pitx2* in situs-specific morphogenesis of visceral organs. *Development* **133**, 3015–3025.
2379. Sholtis, S. and Weiss, K.M. (2005). Phenogenetics: genotypes, phenotypes, and variation. *In* "Variation: A Central Concept in Biology" (B. Hallgrímsson and B.K. Hall, eds.). Elsevier Acad. Pr., New York, pp. 499–523.
2380. Shook, D.R. and Keller, R. (2008). Morphogenic machines evolve more rapidly than the signals that pattern them: lessons from amphibians. *J. Exp. Zool. (Mol. Dev. Evol.)* **310**, 111–135.
2381. Short, R.V. (1979). Sexual selection and its component parts, somatic and genital selection, as illustrated by man and the great apes. *Adv. Study Behav.* **9**, 131–158.
2382. Shou, S., Scott, V., Reed, C., Hitzemann, R., and Stadler, H.S. (2005). Transcriptome analysis of the murine forelimb and hindlimb autopod. *Dev. Dynamics* **234**, 74–89.
2383. Shubin, N. (2008). Fish out of water. *Nat. Hist.* **117** #1, 26–31.
2384. Shubin, N. (2008). "Your Inner Fish: A Journey into the 3.5-Billion-Year History of the Human Body." Pantheon Books, New York.
2385. Sibinski, M. and Snyder, M. (2007). Obstetric brachial plexus palsy—risk factors and predictors. *Ortop. Traumatol. Rehabil.* **9**, 569–576.
2386. Siegel, J.M. (2000). Narcolepsy. *Sci. Am.* **282**#1, 76–81.
2387. Siegel, J.M. (2001). The REM sleep–memory consolidation hypothesis. *Science* **294**, 1058–1063.
2388. Siegel, J.M. (2003). Why we sleep. *Sci. Am.* **289** #5, 92–97.
2389. Siegel, J.M. (2008). Do all animals sleep? *Trends Neurosci.* **31**, 208–213.

2390. Siegel, J.M., Manger, P.R., Nienhuis, R., Fahringer, H.M., and Pettigrew, J.D. (1996). The echidna *Tachyglossus aculeatus* combines REM and non-REM aspects in a single sleep state: implications for the evolution of sleep. *J. Neurosci.* **16**, 3500–3506.

2391. Siegel, J.M., Manger, P.R., Nienhuis, R., Fahringer, H.M., and Pettigrew, J.D. (1998). Monotremes and the evolution of rapid eye movement sleep. *Phil. Trans. Roy. Soc. Lond. B* **353**, 1147–1157.

2392. Siegel, J.M., Manger, P.R., Nienhuis, R., Fahringer, H.M., Shalita, T., and Pettigrew, J.D. (1999). Sleep in the platypus. *Neuroscience* **91**, 391–400.

2393. Siegel, R.M. and Callaway, E.M. (2004). Francis Crick's legacy for neuroscience: between the α and the Ω. *PLoS Biol.* **2** #12, 2029–2032.

2394. Sikela, J.M. (2006). The jewels of our genome: the search for the genomic changes underlying the evolutionarily unique capacities of the human brain. *PLoS Genetics* **2** #5, 646–655 (e80).

2395. Sikes, J.M. and Bely, A.E. (2008). Radical modification of the A-P axis and the evolution of asexual reproduction in *Convolutriloba* acoels. *Evol. Dev.* **10**, 619–631.

2396. Sikes, N.E. (1999). Plio-Pleistocene floral context and habitat preferences of sympatric hominid species in East Africa. *In* "African Biogeography, Climate Change, and Human Evolution" (T.G. Bromage and F. Schrenk, eds.). Oxford Univ. Pr., New York, pp. 301–315.

2397. Sillitoe, R.V. and Joyner, A.L. (2007). Morphology, molecular codes, and circuitry produce the three-dimensional complexity of the cerebellum. *Annu. Rev. Cell Dev. Biol.* **23**, 549–577.

2398. Silver, S.J. and Rebay, I. (2005). Signaling circuitries in development: insights from the retinal determination gene network. *Development* **132**, 3–13.

2399. Silverman, H.B. and Dunbar, M.J. (1980). Aggressive tusk use by the narwhal (*Monodon monocerous L.*). *Nature* **284**, 57–58.

2400. Simões, P., Santos, J., Fragata, I., Mueller, L.D., Rose, M.R., and Matos, M. (2008). How repeatable is adaptive evolution? The role of geographical origin and founder effects in laboratory adaptation. *Evolution* **62**, 1817–1829.

2401. Simões-Costa, M.S., Azambuja, A.P., and Xavier-Neto, J. (2008). The search for non-chordate retinoic acid signaling: lessons from chordates. *J. Exp. Zool. (Mol. Dev. Evol.)* **310B**, 54–72.

2402. Simões-Costa, M.S., Vasconcelos, M., Sampaio, A.C., Cravo, R.M., Linhares, V.L., Hochgreb, T., Yan, C.Y.I., Davidson, B., and Xavier-Neto, J. (2005). The evolutionary origin of cardiac chambers. *Dev. Biol.* **277**, 1–15.

2403. Sinclair, A., Smith, C., Western, P., and McClive, P. (2002). A comparative analysis of vertebrate sex determination. *In* "The Genetics and Biology of Sex Determination", *Novartis Found. Symp. 244* (D. Chadwick and J. Goode, eds.). Wiley, New York, pp. 102–114.

2404. Sinclair, R. (1998). Male pattern androgenetic alopecia. *BMJ* **317**, 865–869.

2405. Sinclair, R. (2007). Female pattern hair loss, dandruff and greying of hair in twins. *J. Investig. Dermatol.* **127**, 2680.

2406. Skaer, N. and Simpson, P. (2000). Genetic analysis of bristle loss in hybrids between *Drosophila melanogaster* and *D. simulans* provides evidence for divergence of *cis*-regulatory sequences in the *achaete-scute* gene complex. *Dev. Biol.* **221**, 148–167.

2407. Slack, J.M.W. (1985). Homoeotic transformations in man: implications for the mechanism of embryonic development and for the organization of epithelia. *J. Theor. Biol.* **114**, 463–490.

2408. Slack, J.M.W. (1994). The whys and wherefores of gastrulation. *Sems. Dev. Biol.* **5**, 69–76.

2409. Slack, J.M.W. (1995). Growth factor lends a hand. *Nature* **374**, 217–218.

2410. Slack, J.M.W. (2002). Conrad Hal Waddington: the last Renaissance biologist? *Nature Rev. Genet.* **3**, 889–895.

2411. Slager, U.T., Anderson, V.M., and Handmaker, S.D. (1981). Cephalothoracopagus janiceps malformation. *Arch. Neurol.* **38**, 103–108.

2412. Slijper, E.J. (1962). "Whales." Hutchinson, London.

2413. Slingerland, E. (2008). "What Science Offers the Humanities." Cambridge Univ. Pr., New York.

2414. Smale, S.T. (2001). Core promoters: active contributors to combinatorial gene regulation. *Genes Dev.* **15**, 2503–2508.

2415. Smith, A.B. (2008). Deuterostomes in a twist: the origins of a radical new body plan. *Evol. Dev.* **10**, 493–503.

2416. Smith, H. (1991). Dental development and the evolution of life history in Hominidae. *Am. J. Phys. Anthrop.* **86**, 157–174.

2417. Smith, H.M. (1960). "Evolution of Chordate Structure: An Introduction to Comparative Anatomy." Holt, Rinehart, and Winston, New York.

2418. Smith, J.J. and Voss, S.R. (2007). Bird and mammal sex-chromosome orthologs map to the same autosomal region in a salamander (Ambystoma). *Genetics* **177**, 607–613.

2419. Smith, K.K. and Kier, W.M. (1989). Trunks, tongues, and tentacles: moving with skeletons of muscle. *Am. Sci.* **77**, 29–35.

2420. Smith, K.K. and Schneider, R.A. (1998). Have gene knockouts caused evolutionary reversals in the mammalian first arch? *BioEssays* **20**, 245–255.

2421. Smith, R.J. (1980). Rethinking allometry. *J. Theor. Biol.* **87**, 97–111.

2422. Sniegowski, P.D. and Murphy, H.A. (2006). Evolvability. *Curr. Biol.* **16**, R831–R834.

2423. Snyder, A. (2001). Paradox of the savant mind. *Nature* **413**, 251–252.

2424. Solé, R.V. and Valverde, S. (2006). Are network motifs the spandrels of cellular complexity? *Trends Ecol. Evol.* **21**, 419–422.

2425. Solounias, N. (1999). The remarkable anatomy of the giraffe's neck. *J. Zool.* **247**, 257–268.
2426. Soltis, J. (2004). The signal functions of early infant crying. *Behav. Brain Sci.* **27**, 443–490.
2427. Soukup, V., Epperlein, H.-H., Horácek, I., and Cerny, R. (2008). Dual epithelial origin of vertebrate oral teeth. *Nature* **455**, 795–798.
2428. Spector, T.D. (2007). The pleasure principle. *Nature* **445**, 822.
2429. Spéder, P., Ádám, G., and Noselli, S. (2006). Type ID unconventional myosin controls left-right asymmetry in *Drosophila. Nature* **440**, 803–807.
2430. Spéder, P., Petzoldt, A., Suzanne, M., and Noselli, S. (2007). Strategies to establish left/right asymmetry in vertebrates and invertebrates. *Curr. Opin. Gen. Dev.* **17**, 351–358.
2431. Speert, H. (1942). Supernumerary mammae, with special reference to the Rhesus monkey. *Q. Rev. Biol.* **17**, 59–68.
2432. Spemann, H. (1938). "Embryonic Development and Induction." Yale Univ. Pr., New Haven.
2433. Spencer, R. (2000). Theoretical and analytical embryology of conjoined twins. Part I. Embryogenesis. *Clin. Anat.* **13**, 36–53.
2434. Spencer, R. (2001). Parasitic conjoined twins: external, internal (fetuses in fetu and teratomas), and detached (acardiacs). *Clin. Anat.* **14**, 428–444.
2435. Spencer, R. (2003). "Conjoined Twins: Developmental Malformations and Clinical Implications." Johns Hopkins Univ. Pr., Baltimore, Md.
2436. Sperber, G.H. and Machin, G.A. (1987). Microscopic study of midline determinants in Janiceps twins. *Birth Defects Orig. Article Ser.* **23** #1, 243–275.
2437. Spieth, H.T. (1968). Evolutionary implications of the mating behavior of the species of *Antopocerus* (Drosophilidae) in Hawaii. *In* "Studies in Genetics. IV. Research Reports," Vol. 4 (Pub. #6818) (M.R. Wheeler, ed.). Univ. Texas Pr., Austin, pp. 319–333.
2438. Spieth, H.T. (1974). Courtship behavior in *Drosophila. Annu. Rev. Entomol.* **19**, 385–405.
2439. Spinka, M., Newberry, R.C., and Bekoff, M. (2001). Mammalian play: training for the unexpected. *Q. Rev. Biol.* **76**, 141–168.
2440. Spitz, L. (1996). Conjoined twins. *Brit. J. Surg.* **83**, 1028–1030.
2441. Spoon, J.M. (2001). Situs inversus totalis. *Neonatal Netw.* **20**, 59–63.
2442. Spors, H. and Sobel, N. (2007). Male behavior by knockout. *Neuron* **55**, 689–693.
2443. Spradling, A., Ganetsky, B., Hieter, P., Johnston, M., Olson, M., Orr-Weaver, T., Rossant, J., Sanchez, A., and Waterston, R. (2006). New roles for model genetic organisms in understanding and treating human disease: report from the 2006 Genetics Society of America meeting. *Genetics* **172**, 2025–2032.
2444. Sprague, G.F., Jr. (1990). Combinatorial associations of regulatory proteins and the control of cell type in yeast. *Adv. Genet.* **27**, 33–62.
2445. Springer, K., Brown, M., and Stulberg, D.L. (2003). Common hair loss disorders. *Am. Fam. Phys.* **68**, 93–102.
2446. Springer, M.S., Kirsch, J.A.W., and Case, J.A. (1997). The chronicle of marsupial evolution. *In* "Molecular Evolution and Adaptive Radiation" (T.J. Givnish and K.J. Sytsma, eds.). Cambridge Univ. Pr., New York, pp. 129–157.
2447. Spritz, R.A. and Hearing, V.J., Jr. (1994). Genetic disorders of pigmentation. *Adv. Hum. Gen.* **22**, 1–45.
2448. Spudich, J.A. (2008). Molecular motors: A surprising twist in myosin VI translocation. *Curr. Biol.* **18**, R68–R70.
2449. Spudich, J.L., Yang, C.-S., Jung, K.-H., and Spudich, E.N. (2000). Retinylidene proteins: structures and functions from archaea to humans. *Annu. Rev. Cell Dev. Biol.* **16**, 365–392.
2450. Sridharan, D., Levitin, D.J., Chafe, C.H., Berger, J., and Menon, V. (2007). Neural dynamics of event segmentation in music: converging evidence for dissociable ventral and dorsal networks. *Neuron* **55**, 521–532.
2451. Srinivasan, M. and Ruina, A. (2006). Computer optimization of a minimal biped model discovers walking and running. *Nature* **439**, 72–75. [See also (2009) *J. Human Evol.* **56**, 43–54.]
2452. Srinivasan, M.V. (1999). When one eye is better than two. *Nature* **399**, 305–307.
2453. Srivastava, D. (2006). Making or breaking the heart: from lineage determination to morphogenesis. *Cell* **126**, 1037–1048.
2454. Srivastava, S. and Srivastava, C.B.L. (1998). Occurrence of third eye (pineal eye) in a living fish (*Tachysurus sona*). *Proc. Natl. Acad. Sci. India* **68**, 241–245.
2455. Stanford, C. (1999). "The Hunting Apes: Meat Eating and the Origins of Human Behavior." Princeton Univ. Pr., Princeton, N.J.
2456. Stanford, C. (2003). "Upright: The Evolutionary Key to Becoming Human." Houghton Mifflin, New York.
2457. Stanford, C.B. (2001). A comparison of social meat-foraging by chimpanzees and human foragers. *In* "Meat-eating and Human Evolution" (C.B. Stanford and H.T. Bunn, eds.). Oxford Univ. Pr., New York, pp. 122–140.
2458. Stanford, C.B. and Allen, J.S. (1991). On strategic storytelling: current models of human behavioral evolution. *Curr. Anthrop.* **32**, 58–61.
2459. Staunton, H. (2005). Mammalian sleep. *Naturwissenschaften* **92**, 203–220.
2460. Stebbins, G.L. (1974). Adaptive shifts and evolutionary novelty: a compositionist approach. *In* "Studies in the Philosophy of Biology: Reduction and Related Problems" (F.J. Ayala and T. Dobzhansky, eds.). Univ. Calif. Pr., Berkeley, pp. 285–306.
2461. Stebbins, G.L. (1983). Mosaic evolution: an integrating principle for the modern synthesis. *Experientia* **39**, 823–834.

2462. Stein, R.C. (1973). Sound production in vertebrates: summary and prospectus. *Amer. Zool.* **13**, 1249–1255.
2463. Steklis, H.D. and Whiteman, C.H. (1989). Loss of estrus in human evolution: too many answers, too few questions. *Ethol. Sociobiol.* **10**, 417–434.
2464. Stent, G.S. (1977). Explicit and implicit semantic content of the genetic information. *In* "Foundational Problems in the Special Sciences" (R. Butts and J. Hintikka, eds.). D. Reidel, Dordrecht, The Netherlands, pp. 131–149.
2465. Stent, G.S. (1978). Genes and the embryo. *In* "Paradoxes of Progress" (G.S. Stent, ed.). Freeman, San Francisco, pp. 169–189.
2466. Stent, G.S. (2002). "Paradoxes of Free Will." Am. Philos. Soc., Philadelphia.
2467. Stern, C. (1941). The growth of testes in *Drosophila.* I. The relation between vas deferens and testis within various species. *J. Exp. Zool.* **87**, 113–158.
2468. Stern, C. (1941). The growth of testes in *Drosophila.* II. The nature of interspecific differences. *J. Exp. Zool.* **87**, 159–180.
2469. Stern, C. (1968). "Genetic Mosaics and Other Essays." Harvard Univ. Pr., Cambridge, Mass.
2470. Stern, C. (1973). "Principles of Human Genetics." 3rd ed. Freeman, San Francisco.
2471. Stern, C., Centerwall, W.R., and Sarkar, S.S. (1964). New data on the problem of Y-linkage of hairy pinnae. *Am. J. Hum. Genet.* **16**, 455–471.
2472. Stern, C.D. (1990). Two distinct mechanisms for segmentation? *Sems. Dev. Biol.* **1**, 109–116.
2473. Stern, C.D. (2006). Evolution of the mechanisms that establish the embryonic axes. *Curr. Opin. Gen. Dev.* **16**, 413–418.
2474. Stern, D. (2006). Morphing into shape. *Science* **313**, 50–51.
2475. Stern, D.L. (2000). Evolutionary developmental biology and the problem of variation. *Evolution* **54**, 1079–1091.
2476. Stern, D.L. (2000). The problem of variation. *Nature* **408**, 529–531.
2477. Stern, D.L. and Emlen, D.J. (1999). The developmental basis for allometry in insects. *Development* **126**, 1091–1101.
2478. Stern, D.L. and Orgogozo, V. (2008). The loci of evolution: how predictable is genetic evolution? *Evolution* **62**, 2155–2177.
2479. Stevens, C.A. and Qumsiyeh, M.B. (1995). Syndromal frontonasal dysostosis in a child with a complex translocation involving chromosomes 3, 7, and 11. *Am. J. Med. Genet.* **55**, 494–497.
2480. Stevens, C.F. (2005). Crick and the claustrum. *Nature* **435**, 1040–1041.
2481. Stewart, L. and Walsh, V. (2007). Music perception: sounds lost in space. *Curr. Biol.* **17**, R892–R893.
2482. Stiassny, M.L.J. (2003). Atavism. *In* "Keywords and Concepts in Evolutionary Developmental Biology" (B.K. Hall and W.M. Olson, eds.). Harvard Univ. Pr., Cambridge, Mass., pp. 10–14.
2483. Stickgold, R. (2008). Sleep: The Ebb and flow of memory consolidation. *Curr. Biol.* **18**, R423–R425.
2484. Stigler, S.M. (1995). Galton and identification by fingerprints. *Genetics* **140**, 857–860.
2485. Stiner, M.C. (2002). Carnivory, coevolution, and the geographic spread of the genus *Homo. J. Archaeol. Res.* **10**, 1–63.
2486. Stock, D.W. (2001). The genetic basis of modularity in the development and evolution of the vertebrate dentition. *Phil. Trans. Roy. Soc. Lond. B* **356**, 1633–1653.
2487. Stock, D.W. (2005). The Dlx gene complement of the leopard shark, *Triakis semifasciata,* resembles that of mammals: implications for genomic and morphological evolution of jawed vertebrates. *Genetics* **169**, 807–817.
2488. Stock, D.W., Weiss, K.M., and Zhao, Z. (1997). Patterning of the mammalian dentition in development and evolution. *BioEssays* **19**, 481–490.
2489. Stock, G.B. and Bryant, S.V. (1981). Studies of digit regeneration and their implications for theories of development and evolution of vertebrate limbs. *J. Exp. Zool.* **216**, 423–433.
2490. Stockard, C.R. (1941). "The Genetic and Endocrinic Basis for Differences in Form and Behavior." *Amer. Anat. Memoirs, No. 19,* Wistar Inst. Anat. Biol., Philadelphia.
2491. Stocum, D.L. (2004). "Tissue Restoration through Regenerative Biology and Medicine. (Advances in Anatomy, Embryology and Cell Biology)." Springer, New York, p. 176.
2492. Stoick-Cooper, C.L., Moon, R.T., and Weidinger, G. (2007). Advances in signaling in vertebrate regeneration as a prelude to regenerative medicine. *Genes Dev.* **21**, 1292–1315.
2493. Stokstad, E. (2004). Controversial fossil could shed light on early animals' blueprint. *Science* **304**, 1425.
2494. Stollewerk, A. (2008). Evolution of neurogenesis in arthropods. *In* "Evolving Pathways: Key Themes in Evolutionary Developmental Biology" (A. Minelli and G. Fusco, eds.). Cambridge Univ. Pr., New York, pp. 359–380.
2495. Stollewerk, A. and Simpson, P. (2005). Evolution of early development of the nervous system: a comparison between arthropods. *BioEssays* **27**, 874–883.
2496. Stoltzfus, A. (2006). Mutationism and the dual causation of evolutionary change. *Evol. Dev.* **8**, 304–317.
2497. Stone, G. and French, V. (2003). Evolution: Have wings come, gone and come again? Curr. Biol. **13**, R436–R438.
2498. Stone, J.R. (1995). CerioShell: a computer program designed to simulate variation in shell form. *Paleobiology* **21**, 509–519.
2499. Stone, J.R. (1996). Computer-simulated shell size and shape variation in the Caribbean land snail genus *Cerion*: a test of geometrical constraints. *Evolution* **50**, 341–347.

2500. Stone, J.R. and Hall, B.K. (2004). Latent homologues for the neural crest as an evolutionary novelty. *Evol. Dev.* **6**, 123–129.
2501. Stone, J.R. and Wray, G.A. (2001). Rapid evolution of *cis*-regulatory sequences via local point mutations. *Mol. Biol. Evol.* **18**, 1764–1770.
2502. Stone, M., Epstein, M.A., and Iskarous, K. (2004). Functional segments in tongue movements. *Clin. Linguist. Phon.* **18**, 507–521.
2503. Stopper, G.F. and Wagner, G.P. (2005). Of chicken wings and frog legs: a smorgasbord of evolutionary variation in mechanisms of tetrapod limb development. *Dev. Biol.* **288**, 21–39.
2504. Stopper, G.F. and Wagner, G.P. (2007). Inhibition of Sonic hedgehog signaling leads to posterior digit loss in *Ambystoma mexicanum*: parallels to natural digit reduction in urodeles. *Dev. Dynamics* **236**, 321–331.
2505. Storey, K.B. and Storey, J.M. (1990). Metabolic rate depression and biochemical adaptation in anaerobiosis, hibernation and estivation. *Q. Rev. Biol.* **65**, 145–174.
2506. Stothard, P. and Pilgrim, D. (2003). Sex-determination gene and pathway evolution in nematodes. *BioEssays* **25**, 221–231.
2507. Stott, R. (2003). "Darwin and the Barnacle: The Story of One Tiny Creature and History's Most Spectacular Scientific Breakthrough." Norton, New York.
2508. Stramer, B. and Martin, P. (2005). Cell biology: Master regulators of sealing and healing. *Curr. Biol.* **15**, R425–R427.
2509. Strand, A.D., Aragaki, A.K., Baquet, Z.C., Hodges, A., Cunningham, P., Holmans, P., Jones, K.R., Jones, L., Kooperberg, C., and Olson, J.M. (2007). Conservation of regional gene expression in mouse and human brain. *PLoS Genet.* **3** #4, 572–583 (e59).
2510. Strand, M.R. and Grbic, M. (1997). The development and evolution of polyembryonic insects. *Curr. Topics Dev. Biol.* **35**, 121–159.
2511. Strathmann, R.R. (1995). Peculiar constraints on life histories imposed by protective or nutritive devices for embryos. *Amer. Zool.* **35**, 426–433.
2512. Straus, W.L., Jr. (1926). The nature and inheritance of webbed toes in man. *J. Morph.* **41**, 427–439.
2513. Streit, A. (2001). Origin of the vertebrate inner ear: evolution and induction of the otic placode. *J. Anat.* **199**, 99–103.
2514. Striedter, G. (2001). Brain evolution: How constrained is it? *Behav. Brain Sci.* **24** #2, 296–297.
2515. Striedter, G.F. (2005). "Principles of Brain Evolution." Sinauer, Sunderland, Mass.
2516. Striedter, G.F. (2006). Précis of Principles of Brain Evolution. *Behav. Brain Sci.* **29**, 1–36.
2517. Struhl, K. (1999). Fundamentally different logic of gene regulation in eukaryotes and prokaryotes. *Cell* **98**, 1–4.
2518. Strutt, D. (2008). The planar polarity pathway. *Curr. Biol.* **18**, R898–R902.
2519. Stryker, M.P. (1994). Precise development from imprecise rules. *Science* **263**, 1244–1245.
2520. Sturtevant, A.H. (1923). Inheritance of direction of coiling in Limnaea. *Science* **58**, 269–270.
2521. Su, C.-Y., Luo, D.-G., Terakita, A., Shichida, Y., Liao, H.-W., Kazmi, M.A., Sakmar, T.P., and Yau, K.-W. (2006). Parietal-eye phototransduction components and their potential evolutionary implications. *Science* **311**, 1617–1621.
2522. Sugimoto, M., Tan, S.-S., and Takagi, N. (2000). X chromosome inactivation revealed by the X-linked *lacZ* transgene activity in periimplantation mouse embryos. *Int. J. Dev. Biol.* **44**, 177–182.
2523. Suh, Y., Atzmon, G., Cho, M.-O., Hwang, D., Liu, B., Leahy, D.J., Barzilai, N., and Cohen, P. (2008). Functionally significant insulin-like growth factor 1 receptor mutations in centenarians. *PNAS* **105** #9, 3438–3442.
2524. Sui, H. and Downing, K.H. (2006). Molecular architecture of axonemal microtubule doublets revealed by cryo-electron tomography. *Nature* **442**, 475–478.
2525. Sultan, S.E. and Stearns, S.C. (2005). Environmentally contingent variation: phenotypic plasiticity and norms of reaction. *In* "Variation: A Central Concept in Biology" (B. Hallgrímsson and B.K. Hall, eds.). Elsevier Acad. Pr., New York, pp. 303–332.
2526. Summerbell, D. (1976). A descriptive study of the rate of elongation and differentiation of the skeleton of the developing chick wing. *J. Embryol. Exp. Morphol.* **35**, 241–260.
2527. Summers, A. (2003). Monitor marathons. *Nat. Hist.* **112** #5, 32–33.
2528. Sun, Y.-G. (2007). A gastrin-releasing peptide receptor mediates the itch sensation in the spinal cord. *Nature* **448**, 700–703.
2529. Supp, D.M., Potter, S.S., and Brueckner, M. (2000). Molecular motors: the driving force behind mammalian left-right development. *Trends Cell Biol.* **10**, 41–45.
2530. Susman, R.L. (1983). Evolution of the human foot: evidence from Plio-Pleistocene hominids. *Foot Ankle* **3**, 365–376.
2531. Sutherland, W.J. (2005). The best solution. *Nature* **435**, 569.
2532. Suthers, R.A. (1994). Variable asymmetry and resonance in the avian vocal tract: a structural basis for individually distinct vocalizations. *J. Comp. Physiol. A* **175**, 457–466.
2533. Sutter, N.B., Bustamante, C.D., Chase, K., Gray, M.M., Zhao, K., Zhu, L., Padhukasahasram, B., Karlins, E., Davis, S., Jones, P.G., Quignon, P., Johnson, G.S., Parker, H.G., Fretwell, N., Mosher, D.S., Lawler, D.F., Satyaraj, E., Nordborg, M., Lark, K.G., Wayne, R.K., and Ostrander, E.A. (2007). A single *IGF1* allele is a major determinant of small size in dogs. *Science* **316**, 112–115.

2534. Suzuki, T., Hasso, S.M., and Fallon, J.F. (2008). Unique SMAD1/5/8 activity at the phalanx-forming region determines digit identity. *PNAS* **105** #11, 4185–4190.
2535. Suzuki, T., Takeuchi, J., Koshiba-Takeuchi, K., and Ogura, T. (2004). Tbx genes specify posterior digit identity through Shh and BMP signaling. *Dev. Cell* **6**, 43–53.
2536. Swalla, B.J. (2006). Building divergent body plans with similar genetic pathways. *Heredity* **97**, 235–243.
2537. Swami, V., Jones, J., Einon, D., and Furnham, A. (2009). Men's preferences for women's profile waist-to-hip ratio, breast size, and ethnic group in Britain and South Africa. *Br. J. Psychol.*, Epub ahead of print.
2538. Sweeney, D., Lindström, N., and Davies, J.A. (2008). Developmental plasticity and regenerative capacity in the renal ureteric bud/collecting duct system. *Development* **135**, 2505–2510.
2539. Szalay, F.S. and Dagosto, M. (1988). Evolution of hallucial grasping in the primates. *J. Human Evol.* **17**, 1–33.
2540. Szeto, D.P., Rodriguez-Esteban, C., Ryan, A.K., O'Connell, S.M., Liu, F., Kioussi, C., Gleiberman, A.S., Izpisúa-Belmonte, J.C., and Rosenfeld, M.G. (1999). Role of the Bicoid-related homeodomain factor Pitx1 in specifying hindlimb morphogenesis and pituitary development. *Genes Dev.* **13**, 484–494.
2541. Szmuk, P., Ezri, T., Evron, S., Roth, Y., and Katz, J. (2008). A brief history of tracheostomy and tracheal intubation, from the Bronze Age to the Space Age. *Intensive Care Med.* **34**, 222–228.
2542. Tabata, T. and Takei, Y. (2004). Morphogens, their identification and regulation. *Development* **131**, 703–712.
2543. Tabin, C. (1995). The initiation of the limb bud: growth factors, Hox genes, and retinoids. *Cell* **80**, 671–674.
2544. Tabin, C. and Laufer, E. (1993). Hox genes and serial homology. *Nature* **361**, 692–693.
2545. Tabin, C. and Wolpert, L. (2007). Rethinking the proximodistal axis of the vertebrate limb in the molecular era. *Genes Dev.* **21**, 1433–1442.
2546. Tabin, C.J., Carroll, S.B., and Panganiban, G. (1999). Out on a limb: parallels in vertebrate and invertebrate limb patterning and the origin of appendages. *Amer. Zool.* **39**, 650–663.
2547. Tabin, C.J. and McMahon, A.P. (2008). Grasping limb patterning. *Science* **321**, 350–352. [See also (2009) *Development* **136**, 179–190.]
2548. Tabin, C.J. and Vogan, K.J. (2003). A two-cilia model for vertebrate left-right axis specification. *Genes Dev.* **17**, 1–6.
2549. Taborsky, M. and Taborsky, B. (1993). The kiwi's parental burden. *Nat. Hist.* **102** #12, 50–57.
2550. Tadin, M., Braverman, E., Cianfarani, S., Sobrino, A.J., Levy, B., Christiano, A.M., and Warburton, D. (2001). Complex cytogenetic rearrangement of chromosome 8q in a case of Ambras syndrome. *Am. J. Med. Genet.* **102**, 100–104.
2551. Tague, R.G. (1992). Sexual dimorphism in the human bony pelvis, with a consideration of the neandertal pelvis from Kebara Cave, Israel. *Am. J. Phys. Anthrop.* **88**, 1–21.
2552. Tague, R.G. (1997). Variability of a vestigial structure: first metacarpal in *Colobus guereza* and *Ateles geoffroyi*. *Evolution* **51**, 595–605.
2553. Tague, R.G. and Lovejoy, C.O. (1986). The obstetric pelvis of A.L. 288–1 (Lucy). *J. Human Evol.* **15**, 237–255.
2554. Takada (1966). Male genitalia of some Hawaiian Drosophilidae. *In* "Studies in Genetics. III. Morgan Centennial Issue.," Vol. 3 (Pub. #6615) (M.R. Wheeler, ed.). Univ. Texas Pr., Austin, pp. 315–333.
2555. Takahashi, M., Arita, H., Hiraiwa-Hasegawa, M., and Hasegawa, T. (2008). Peahens do not prefer peacocks with more elaborate trains. *Anim. Behav.* **75**, 1209–1219.
2556. Takano-Shimizu, T. (2000). Genetic screens for factors involved in the notum bristle loss of interspecific hybrids between *Drosophila melanogaster* and D. simulans. *Genetics* **156**, 269–282.
2557. Takaoka, K., Yamamoto, M., Shiratori, H., Meno, C., Rossant, J., Saijoh, Y., and Hamada, H. (2006). The mouse embryo autonomously acquires anterior-posterior polarity at implantation. *Dev. Cell* **10**, 451–459.
2558. Takashima, S., Mkrtchyan, M., Younossi-Hartenstein, A., Merriam, J.R., and Hartenstein, V. (2008). The behaviour of *Drosophila* adult hindgut stem cells is controlled by Wnt and Hh signalling. *Nature* **454**, 651–655.
2559. Takeuchi, J.K., Koshiba-Takeuchi, K., Matsumoto, K., Vogel-Höpker, A., Naitoh-Matsuo, M., Ogura, K., Takahashi, N., Yasuda, K., and Ogura, T. (1999). Tbx5 and Tbx4 genes determine the wing/leg identity of limb buds. *Nature* **398**, 810–814.
2560. Takeuchi, J.K., Koshiba-Takeuchi, K., Suzuki, T., Kamimura, M., Ogura, K., and Ogura, T. (2003). Tbx5 and Tbx4 trigger limb initiation through activation of the Wnt/Fgf signaling cascade. *Development* **130**, 2729–2739.
2561. Takio, Y., Kuraku, S., Murakami, Y., Pasqualetti, M., Rijli, F.M., Narita, Y., Kuratani, S., and Kusakabe, R. (2007). Hox gene expression patterns in Lethenteron japonicum embryos—Insights into the evolution of the vertebrate Hox code. *Dev. Biol.* **308**, 606–620.
2562. Tamura, K., Kuraishi, R., Saito, D., Masaki, H., Ide, H., and Yonei-Tamura, S. (2001). Evolutionary aspects of positioning and identification of vertebrate limbs. *J. Anat.* **199**, 195–204.
2563. Tan, H.K., Brown, K., McGill, T., Kenna, M.A., Lund, D.P., and Healy, G.B. (2000). Airway foreign bodies (FB): a 10-year review. *Int. J. Pediatr. Otorhinolaryngol.* **56**, 91–99.

2564. Tan, U. (2006). Evidence for "Uner Tan Syndrome" as a human model for reverse evolution. *Int. J. Neurosci.* **116**, 1539–1547.
2565. Tanaka, E.M. and Gann, A.A.F. (1995). The budding role of FGF. *Curr. Biol.* **5**, 594–597.
2566. Tanaka, E.M. and Weidinger, G. (2008). Micromanaging regeneration. *Genes Dev.* **22**, 700–705.
2567. Tanaka, K., Barmina, O., and Kopp, A. (2009). Distinct developmental mechanisms underlie the evolutionary diversification of *Drosophila* sex combs. *PNAS* **106** #12, 4764–4769.
2568. Tanaka, M. and Tickle, C. (2007). The development of fins and limbs. *In* "Fins into Limbs: Evolution, Development, and Transformation" (B.K. Hall, ed.). Univ. Chicago Pr., Chicago, Ill., pp. 65–78.
2569. Tanaka, Y., Okada, Y., and Hirokawa, N. (2005). FGF-induced vesicular release of Sonic hedgehog and retinoic acid in leftward nodal flow is critical for left-right determination. *Nature* **435**, 172–177.
2570. Tang, M.K., Leung, A.K.C., Kwong, W.H., Chow, P.H., Chan, J.Y.H., Ngo-Muller, V., Li, M., and Lee, K.K.H. (2000). Bmp-4 requires the presence of the digits to initiate programmed cell death in limb interdigital tissues. *Dev. Biol.* **218**, 89–98.
2571. Taniguchi, K., Hozumi, S., Maeda, R., Ooike, M., Sasamura, T., Aigaki, T., and Matsuno, K. (2007). *D-JNK* signaling in visceral muscle cells controls the laterality of the *Drosophila* gut. *Dev. Biol.* **311**, 251–263.
2572. Tao, Y., Hartl, D.L., and Laurie, C.C. (2001). Sex-ratio segregation distortion associated with reproductive isolation in *Drosophila. PNAS* **98** #23, 13183–13188.
2573. Tapadia, M.D., Cordero, D.R., and Helms, J.A. (2005). It's all in your head: new insights into craniofacial development and deformation. *J. Anat.* **207**, 461–477.
2574. Tarchini, B. and Duboule, D. (2006). Control of *Hoxd* genes' collinearity during early limb development. *Dev. Cell* **10**, 93–103.
2575. Tattersall, I. (2002). "The Monkey in the Mirror." Harcourt, New York.
2576. Tattersall, I. (2003). Stand and deliver: Why did early hominids begin to walk on two feet? *Nat. Hist.* **112** #9, 60–64.
2577. Taub, R. (2004). Liver regeneration: from myth to mechanism. *Nature Rev. Mol. Cell Biol.* **5**, 836–847.
2578. Teaford, M.F. and Ungar, P.S. (2006). Diet and the evolution of the earliest human ancestors. *In* "The Human Evolution Source Book," 2nd ed. *Adv. Human Evol. Series,* (R.L. Ciochon and J.G. Fleagle, eds.). Pearson Prentice Hall, Upper Saddle River, N.J., pp. 189–196.
2579. Telford, M.J. (2007). A single origin of the central nervous system? *Cell* **129**, 237–239.
2580. Templeton, A.R. (1981). Mechanisms of speciation—a population genetic approach. *Annu. Rev. Ecol. Syst.* **12**, 23–48.
2581. Templeton, A.R. (2008). The reality and importance of founder speciation in evolution. *BioEssays* **30**, 470–479.
2582. Temtamy, S.A. and McKusick, V.A. (1978). "The Genetics of Hand Malformations." *Birth Defects: Original Article Series, Vol. XIV, No. 3,* Alan R. Liss, New York.
2583. Tenney, S.M. (1979). A synopsis of breathing mechanisms. *In* "Evolution of Respiratory Processes: A Comparative Approach" (S.C. Wood and C. Lenfant, eds.). Marcel Dekker, New York, pp. 51–106.
2584. Teotónio, H. and Rose, M.R. (2000). Variability in the reversibility of evolution. *Nature* **408**, 463–466.
2585. Teotónio, H. and Rose, M.R. (2001). Reversible evolution. *Evolution* **55**, 653–660.
2586. Teramitsu, I. and White, S.A. (2008). Motor learning: The FoxP2 puzzle piece. *Curr. Biol.* **18**, R335–R337.
2587. Terrace, H.S. and Metcalfe, J., eds. (2005). "The Missing Link in Cognition: Origins of Self-Reflective Consciousness." Oxford Univ. Pr., New York.
2588. Tessmar-Raible, K. (2007). The evolution of neurosecretory centers in bilaterian forebrains: insights from protostomes. *Sems. Cell Dev. Biol.* **18**, 492–501.
2589. Theissen, G. (2006). The proper place of hopeful monsters in evolutionary biology. *Theory Biosci.* **124**, 349–369.
2590. Thesleff, I., Järvinen, E., and Suomalainen, M. (2007). Affecting tooth morphology and renewal by fine-tuning the signals mediating cell and tissue interactions. *In* "Tinkering: The Microevolution of Development," *Novartis Found. Symp. 284* (G. Bock and J. Goode, eds.). Wiley, Chichester, U.K., pp. 142–157.
2591. Thewissen, J.G.M. (2007). Aquatic adaptations in the limbs of amniotes. In "Fins into Limbs: Evolution, Development, and Transformation" (B.K. Hall, ed.). Univ. Chicago Pr., Chicago, Ill., pp. 310–322.
2592. Thitamadee, S., Tuchihara, K., and Hashimoto, T. (2002). Microtubule basis for left-handed helical growth in *Arabidopsis. Nature* **417**, 193–196.
2593. Thomas, D.E. (1980). Mirror images. *Sci. Am.* **243** #6, 206–228.
2594. Thomas, R., Thieffry, D., and Kaufman, M. (1995). Dynamical behaviour of biological regulatory networks—I. Biological role of feedback loops and practical use of the concept of the loop-characteristic state. *Bull. Math. Biol.* **57**, 247–276.
2595. Thomas, R.D.K. and Reif, W.-E. (1993). The skeleton space: a finite set of organic designs. *Evolution* **47**, 341–360.
2596. Thompson, D.W. (1917). "On Growth and Form." Cambridge Univ. Pr., Cambridge.
2597. Thompson, I. (1991). Considering the evolution of vertebrate neural retina. *In* "Evolution of the Eye and Visual System" (J.R. Cronly-Dillon and R.L. Gregory, eds.). CRC Pr., Boston, pp. 136–151.

2598. Thompson, J.C. and Hardee, J.E. (2008). The first time ever I saw your face. *Trends Cogn. Sci.* **12**, 283–284.

2599. Thomson, J.A. (1932). "Riddles of Science." Liveright, New York.

2600. Thomson, K.S. (1988). "Morphogenesis and Evolution." Oxford Univ. Pr., New York.

2601. Thornhill, R. (1993). The allure of symmetry. *Nat. Hist.* **102** #9, 30–36.

2602. Thorpe, S.K.S., Holder, R.L., and Crompton, R.H. (2007). Origin of human bipedalism as an adaptation for locomotion on flexible branches. *Science* **316**, 1328–1331.

2603. Thowfeequ, S., Myatt, E.-J., and Tosh, D. (2007). Transdifferentiation in developmental biology, disease, and in therapy. *Dev. Dynamics* **236**, 3208–3217.

2604. Throckmorton, L.H. (1962). The problem of phylogeny in the genus *Drosophila*. In "Studies in Genetics. II. Research Reports on *Drosophila* Genetics, Taxonomy, and Evolution," Vol. 2 (Pub. #6205) (M.R. Wheeler, ed.). Univ. Texas Pr., Austin, pp. 207–343.

2605. Tickle, C. (2006). Making digit patterns in the vertebrate limb. *Nature Rev. Mol. Cell Biol.* **7**, 45–53.

2606. Ting, S.B., Caddy, J., Hislop, N., Wilanowski, T., Auden, A., Zhao, L.-l., Ellis, S., Kaur, P., Uchida, Y., Holleran, W.M., Elias, P.M., Cunningham, J.M., and Jane, S.M. (2005). A homolog of *Drosophila grainy head* is essential for epidermal integrity in mice. *Science* **308**, 411–413.

2607. Titze, I.R. (2008). The human instrument. *Sci. Am.* **298** #1, 94–101.

2608. Tobias, M.L., Marin, M.L., and Kelley, D.B. (1991). Development of functional sex differences in the larynx of *Xenopus laevis*. *Dev. Biol.* **147**, 251–259.

2609. Tobias, M.L., Marin, M.L., and Kelley, D.B. (1991). Temporal constraints on androgen directed laryngeal masculinization in *Xenopus laevis*. *Dev. Biol.* **147**, 260–270.

2610. Tobias, P.V. (1995). The brain of the first hominids. *In* "Origins of the Human Brain" (J.-P. Changeux and J. Chavaillon, eds.). Clarendon Pr., Oxford, pp. 61–83.

2611. Tocheri, M.W., Orr, C.M., Larson, S.G., Sutikna, T., Jatmiko, Saptomo, E.W., Due, R.A., Djubiantono, T., Morwood, M.J., and Jungers, W.L. (2007). The primitive wrist of *Homo floresiensis* and its implications for hominin evolution. *Science* **317**, 1743–1745.

2612. Tokunaga, C. (1982). Curt Stern, 1902–1981, in memoriam. *Jpn. J. Genet.* **57**, 459–466.

2613. Tomasello, M., Hare, B., Lehmann, H., and Call, J. (2007). Reliance on head versus eyes in the gaze following of great apes and human infants: the cooperative eye hypothesis. *J. Human Evol.* **52**, 314–320.

2614. Tomita, Y. (1993). Tyrosinase gene mutations causing oculocutaneous albinisms. *J. Invest. Dermatol.* **100**, 186S–190S.

2615. Tooby, J. and DeVore, I. (1987). The reconstruction of hominid behavioral evolution through strategic modeling. *In* "The Evolution of Human Behavior: Primate Models" (W.G. Kinzey, ed.). SUNY Pr., Albany, N.Y., pp. 183–237.

2616. Torday, J.S., Rehan, V.K., Hicks, J.W., Wang, T., Maina, J., Weibel, E.R., Hsia, C.C.W., Sommer, R.J., and Perry, S.F. (2007). Deconvoluting lung evolution: from phenotypes to gene regulatory networks. *Integr. Comp. Biol.* **47**, 601–609.

2617. Torgersen, J. (1947). Transposition of viscera, bronchiectasis and nasal polyps. *Acta Radiol.* **28**, 17–24.

2618. Tosini, G. (1997). The pineal complex of reptiles: physiological and behavioral roles. *Ethol. Ecol. Evol.* **9**, 313–333.

2619. Touboul, M., Kleine, T., Bourdon, B., Palme, H., and Wieler, R. (2007). Late formation and prolonged differentiation of the Moon inferred from W isotopes in lunar metals. *Nature* **450**, 1206–1209.

2620. Tour, E., Hittinger, C.T., and McGinnis, W. (2005). Evolutionarily conserved domains required for activation and repression functions of the *Drosophila* Hox protein Ultrabithorax. *Development* **132**, 5271–5281.

2621. Towers, M., Mahood, R., Yin, Y., and Tickle, C. (2008). Integration of growth and specification in chick wing digit-patterning. *Nature* **452**, 882–886.

2622. Traut, W., Niimi, T., Ikeo, K., and Sahara, K. (2006). Phylogeny of the sex-determining gene Sex-lethal in insects. *Genome* **49**, 254–262.

2623. Travis, J. (2007). A close look at Urbisexuality. *Science* **316**, 390–391.

2624. Treffert, D.A. and Christensen, D.D. (2005). Inside the mind of a savant. *Sci. Am.* **293** #6, 108–113.

2625. Treffert, D.A. and Wallace, G.L. (2002). Islands of genius. *Sci. Am.* **286** #6, 76–85.

2626. Tregenza, T., Wedell, N., and Chapman, T. (2006). Sexual conflict: a new paradigm? *Phil. Trans. Roy. Soc. B* **361**, 229–234.

2627. Treisman, J. and Lang, R. (2002). Development and evolution of the eye: Fondation des Treilles, September, 2001. *Mechs. Dev.* **112**, 3–8.

2628. Treisman, J.E. (2004). How to make an eye. *Development* **131**, 3823–3827.

2629. Trelstad, R.L. (1982). The bilaterally asymmetric architecture of the submammalian corneal stroma resembles a cholesteric liquid crystal. *Dev. Biol.* **92**, 133–134.

2630. True, J.R. (2008). Combing evolution. *Evol. Dev.* **10**, 400–402.

2631. True, J.R. and Carroll, S.B. (2002). Gene co-option in physiological and morphological evolution. *Annu. Rev. Cell Dev. Biol.* **18**, 53–80.

2632. True, J.R. and Haag, E.S. (2001). Developmental system drift and flexibility in evolutionary trajectories. *Evol. Dev.* **3**, 109–119. [See also (2008) *Nature Rev. Genet.* **9**, 965–974.]

2633. Trueb, L. (1973). Bones, frogs, and evolution. *In* "Evolutionary Biology of the Anurans: Contemporary Research on Major Problems" (J.L. Vial, ed.). Univ. Missouri Pr., Columbia, pp. 65–132.
2634. Trujillo-Cenóz, O. (1985). The eye: development, structure and neural connections. In "Comprehensive Insect Physiology, Biochemistry, and Pharmacology," Vol. 6 (G.A. Kerkut and L.I. Gilbert, eds.). Pergamon Pr., Oxford, pp. 171–223.
2635. Trut, L.N. (1999). Early canid domestication: the farm-fox experiment. *Am. Sci.* **87**, 160–169. [See also (2009) *BioEssays* **31**, 349–360.]
2636. Tsien, J.Z. (2007). The memory code. *Sci. Am.* **297** #1, 52–59.
2637. Tsong, A.E., Miller, M.G., Raisner, R.M., and Johnson, A.D. (2003). Evolution of a combinatorial transcriptional circuit: a case study in yeasts. *Cell* **115**, 389–399.
2638. Tsong, A.E., Tuch, B.B., Li, H., and Johnson, A.D. (2006). Evolution of alternative transcriptional circuits with identical logic. *Nature* **443**, 415–420.
2639. Tsonis, P.A., Vergara, M.N., Spence, J.R., Madhavan, M., Kramer, E.L., Call, M.K., Santiago, W.G., Vallance, J.E., Robbins, D.J., and Rio-Tsonis, K.D. (2004). A novel role of the hedgehog pathway in lens regeneration. *Dev. Biol.* **267**, 450–461.
2640. Tsuda, M., Sasaoka, Y., Kiso, M., Abe, K., Haraguchi, S., Kobayashi, S., and Saga, Y. (2003). Conserved role of nanos proteins in germ cell development. *Science* **301**, 1239–1241.
2641. Tu, D.C., Batten, M.L., Palczewski, K., and Van Gelder, R.N. (2004). Nonvisual perception in the chick iris. *Science* **306**, 129–131.
2642. Tuch, B.B., Li, H., and Johnson, A.D. (2008). Evolution of eukaryotic transcription circuits. *Science* **319**, 1797–1799.
2643. Tucker, A. and Sharpe, P. (2004). The cutting-edge of mammalian development; how the embryo makes teeth. *Nature Rev. Genet.* **5**, 499–508.
2644. Tucker, A.S., Matthews, K.L., and Sharpe, P.T. (1998). Transformation of tooth type induced by inhibition of BMP signaling. *Science* **282**, 1136–1138.
2645. Tufro-McReddie, A., Norwood, V.F., Aylor, K.W., Botkin, S.J., Carey, R.M., and Gomez, R.A. (1997). Oxygen regulates vascular endothelial growth factor-mediated vasculogenesis and tubulogenesis. *Dev. Biol.* **183**, 139–149.
2646. Turecki, G. (2005). Dissecting the suicide phenotype: the role of impulsive-aggressive behaviours. *J. Psychiatry Neurosci.* **30**, 398–408.
2647. Turke, P.W. (2008). Williams's theory of the evolution of senescence: still useful at fifty. *Q. Rev. Biol.* **83**, 243–256.
2648. Turkeltaub, P.E., Gareau, L., Flowers, D.L., Zeffiro, T.A., and Eden, G.F. (2003). Development of neural mechanisms for reading. *Nature Neurosci.* **6**, 767–773.
2649. Turner, C.W. (1939). "The Comparative Anatomy of the Mammary Glands." Univ. Coop. Store, Columbia, Mo.
2650. Turner, C.W. (1939). The mammary glands. *In* "Sex and Internal Secretions: A Survey of Recent Research" (E. Allen, ed.). Williams & Wilkins: Baltimore, Md, pp. 740–803.
2651. Turner, J.S. (2007). "The Tinkerer's Accomplice: How Design Emerges from Life Itself." Harvard Univ. Pr., Cambridge, Mass.
2652. Turnpenny, P.D., Alman, B., Cornier, A.S., Giampietro, P.F., Offiah, A., Tassy, O., Pourquié, O., Kusumi, K., and Dunwoodie, S. (2007). Abnormal vertebral segmentation and the Notch signaling pathway in man. *Dev. Dynamics* **236**, 1456–1474.
2653. Turpin, R., Lejeune, J., Lafourcade, J., Chigot, P.-L., and Salmon, C. (1961). Présomption de monozygotisme en dépit d'un dimorphisme sexuel: sujet masculin XY et sujet neutre Haplo X. *Comptes Rendus l'Acad. Sci. Série III.* **252**, 2945–2946.
2654. Tuteja, G. and Kaestner, K.H. (2007). Forkhead Transcription Factors I. *Cell* **130**, 1160. [See also (2009) *Nature Rev. Genet.* **10**, 233–240.]
2655. Tuteja, G. and Kaestner, K.H. (2007). Forkhead Transcription Factors II. *Cell* **131**, 192.
2656. Twigg, S.R.F., Kan, R., Babbs, C., Bochukova, E.G., Robertson, S.P., Wall, S.A., Morriss-Kay, G.M., and Wilkie, A.O.M. (2004). Mutations of ephrin-B1 (*EFNB1*), a marker of tissue boundary formation, cause craniofrontonasal syndrome. *PNAS* **101**#23, 8652–8657.
2657. Tyson, N.D. (2005). The perimeter of ignorance. *Nat. Hist.* **114** #9, 28–34.
2658. Ueshima, R. and Asami, T. (2003). Single-gene speciation by left-right reversal. *Nature* **425**, 679.
2659. Uller, T., Pen, I., Wapstra, E., Beukeboom, L.W., and Komdeur, J. (2007). The evolution of sex ratios and sex-determining systems. *Trends Ecol. Evol.* **22**, 292–297.
2660. Ungar, P.S., Grine, F.E., and Teaford, M.F. (2006). Diet in early *Homo*: a review of the evidence and a new model of adaptive versatility. *Annu. Rev. Anthropol.* **35**, 209–228.
2661. Unwin, D.M., Frey, E., Martill, D.M., Clarke, J.B., and Riess, J. (1996). On the nature of the pteroid in pterosaurs. *Proc. Roy. Soc. Lond. B* **263**, 45–52.
2662. Upchurch, P. (2008). Gondwanan break-up: legacies of a lost world? *Trends Ecol. Evol.* **23**, 229–236.
2663. Valentine, J.W. (2004). "On the Origin of Phyla." Univ. Chicago Pr., Chicago, Ill.
2664. Valentine, J.W. (2006). Ancestors and urbilateria. *Evol. Dev.* **8**, 391–393.
2665. Valenzano, M., Paoletti, R., Rossi, A., Farinini, D., Garlaschi, G., and Fulcheri, E. (1999). Sirenomelia. Pathological features, antenatal ultrasonographic clues, and a review of current embryological theories. *Hum. Reprod. Update* **5**, 82–86.
2666. Vallortigara, G., Cozzutti, C., Tommasi, L., and Rogers, L.J. (2001). How birds use their eyes:

opposite left-right specialization for the lateral and frontal visual hemifield in the domestic chick. *Curr. Biol.* **11**, 29–33.

2667. Vallortigara, G., Snyder, A., Kaplan, G., Bateson, P., Clayton, N.S., and Rogers, L.J. (2008). Are animals autistic savants? *PLoS Biol.* **6** #2, 208–214 (e42).

2668. van den Akker, E., Forlani, S., Chawengsaksophak, K., de Graaff, W., Beck, F., Meyer, B.I., and Deschamps, J. (2002). *Cdx1* and *Cdx2* have overlapping functions in anteroposterior patterning and posterior axis elongation. *Development* **129**, 2181–2193.

2669. van Deusen, H.M. (1966). The seventh Archbold expedition. *BioScience* **16**, 449–455.

2670. Van Essen, D.C. (2005). Surface-based comparisons of macaque and human cortical organization. *In* "From Monkey Brain to Human Brain" (S. Dehaene, J.-R. Duhamel, M.D. Hauser, and G. Rizzolatti, eds.). MIT Pr., Cambridge, Mass., pp. 3–19.

2671. Van Gelder, R.N. (2008). Non-visual photoreception: sensing light without sight. *Curr. Biol.* **18**, R38–R39.

2672. van Holde, K. and Zlatanova, J. (1996). Chromatin architectural proteins and transcription factors: a structural connection. *BioEssays* **18**, 697–700.

2673. van Tienhoven, A. (1983). "Reproductive Physiology of Vertebrates," 2nd ed. Cornell Univ. Pr., Ithaca, N.Y.

2674. Vanhaeren, M. (2005). Speaking with beads: the evolutionary significance of personal ornaments. In "From Tools to Symbols: From Early Hominids to Modern Humans" (F. d'Errico and L. Backwell, eds.). Witwatersrand Univ. Pr., Johannesburg, South Africa, pp. 525–553.

2675. Vargas, A.O. and Aboitiz, F. (2005). How ancient is the adult swimming capacity in the lineage leading to euchordates? *Evol. Dev.* **7**, 171–174.

2676. Vargas, A.O. and Fallon, J.F. (2005). The digits of the wing of birds are 1, 2, and 3. A review. *J. Exp. Zool. (Mol. Dev. Evol.)* **304B**, 206–219.

2677. Varki, A., Geschwind, D.H., and Eichler, E.E. (2008). Explaining human uniqueness: genome interactions with environment, behaviour and culture. *Nature Rev. Genet.* **9**, 749–763.

2678. Várkonyi, P.L., Meszéna, G., and Domokos, G. (2006). Emergence of asymmetry in evolution. *Theor. Pop. Biol.* **70**, 63–75.

2679. Vasey, N. and Walker, A. (2001). Neonate body size and hominid carnivory. *In* "Meat-eating and Human Evolution" (C.B. Stanford and H.T. Bunn, eds.). Oxford Univ. Pr., New York, pp. 332–349.

2680. Veitia, R.A., Bottani, S., and Birchler, J.A. (2008). Cellular reactions to gene dosage imbalance: genomic, transcriptomic and proteomic effects. *Trends Genet.* **24**, 390–397.

2681. Vélez, M.M. and Clandinin, T.R. (2008). Neural circuitry: seeing the parts that make the picture. *Curr. Biol.* **18**, R378–R380.

2682. Verheyden, J.M. and Sun, X. (2008). An Fgf/*Gremlin* inhibitory feedback loop triggers termination of limb bud outgrowth. *Nature* **454**, 638–641.

2683. Verhulst, J. (1996). Atavisms in *Homo sapiens*: a Bolkian heterodoxy revisited. *Acta Biotheor.* **44**, 59–73.

2684. Vermeij, G.J. (1975). Evolution and distribution of left-handed and planispiral coiling in snails. *Nature* **254**, 419–420.

2685. Vernikos, J. (1996). Human physiology in space. *BioEssays* **18**, 1029–1037.

2686. Verra, F., Escudier, E., Lebargy, F., Bernaudin, J.F., De Crémoux, H., and Bignon, J. (1995). Ciliary abnormalities in bronchial epithelium of smokers, ex-smokers, and nonsmokers. *Am. J. Respir. Crit. Care Med.* **151**, 630–634.

2687. Vickaryous, M.K. and Hall, B.K. (2006). Human cell type diversity, evolution, development, and classification with special reference to cells derived from the neural crest. *Biol. Rev.* **81**, 425–455. [See also (2008) *Nature Rev. Genet.* **9**, 868–882.]

2688. Viggiano, D., Pirolo, L., Cappabianca, S., and Passiatore, C. (2002). Testing the model of optic chiasm formation in human beings. *Brain Res. Bull.* **59**, 111–115.

2689. Villee, C.A. (1942). The phenomenon of homeosis. *Am. Nat.* **76**, 494–506.

2690. Vincent, S., Perkins, L.A., and Perrimon, N. (2001). Doublesex surprises. *Cell* **106**, 399–402.

2691. Vincent, S.D., Norris, D.P., Le Good, J.A., Constam, D.B., and Robertson, E.J. (2004). Asymmetric *nodal* expression in the mouse is governed by the combinatorial activities of two distinct regulatory elements. *Mechs. Dev.* **121**, 1403–1415.

2692. Vincent, T.L. and Brown, J.S. (2005). "Evolutionary Game Theory, Natural Selection, and Darwinian Dynamics." Cambridge Univ. Pr., New York.

2693. Vinckier, F., Dehaene, S., Jobert, A., Dubus, J.P., Sigman, M., and Cohen, L. (2007). Hierarchical coding of letter strings in the ventral stream: dissecting the inner organization of the visual word-form system. *Neuron* **55**, 143–156.

2694. Vinicius, L. (2005). Human encephalization and developmental timing. *J. Human Evol.* **49**, 762–776.

2695. Virtanen, H.E., Cortes, D., Rajpert-De Meyts, E., Ritzén, E.M., Nordenskjöld, A., Skakkebaek, N.E., and Toppari, J. (2007). Development and descent of the testis in relation to cryptorchidism. *Acta Paediatrica* **96**, 622–627.

2696. Vitruvius (1960). "The Ten Books on Architecture," Morgan, M. H., translator. Dover, New York.

2697. Vogel, G. (2005). The unexpected brains behind blood vessel growth. *Science* **307**, 665–667.

2698. Vogel, S. (1988). "Life's Devices: The Physical World of Animals and Plants." Princeton Univ. Pr., Princeton, N.J.

2699. Vogel, S. (1998). "Cats' Paws and Catapults: Mechanical Worlds of Nature and People." W.W. Norton, New York.

2700. Vogel, S. (2003). "Comparative Biomechanics: Life's Physical World." Princeton Univ. Pr., Princeton, N.J.
2701. Vokes, S.A., Ji, H., Wong, W.H., and McMahon, A.P. (2008). A genome-scale analysis of the cis-regulatory circuitry underlying sonic hedgehog-mediated patterning of the mammalian limb. *Genes Dev.* **22**, 2651–2663.
2702. Vollrath, F. (1998). Dwarf males. *Trends Ecol. Evol.* **13**, 159–163.
2703. Volman, S.F. and Konishi, M. (1990). Comparative physiology of sound localization in four species of owls. *Brain Behav. Evol.* **36**, 196–215.
2704. von Gall, C., Stehle, J.H., and Weaver, D.R. (2002). Mammalian melatonin receptors: molecular biology and signal transduction. *Cell Tissue Res.* **309**, 151–162.
2705. Vonk, F.J., Admiraal, J.F., Jackson, K., Reshef, R., de Bakker, M.A.G., Vanderschoot, K., van den Berge, I., van Atten, M., Burgerhout, E., Beck, A., Mirtschin, P.J., Kochva, E., Witte, F., Fry, B.G., Woods, A.E., and Richardson, M.K. (2008). Evolutionary origin and development of snake fangs. *Nature* **454**, 630–633.
2706. Vonk, F.J. and Richardson, M.K. (2008). Serpent clocks tick faster. *Nature* **454**, 282–283.
2707. Vorbach, C., Capecchi, M.R., and Penninger, J.M. (2006). Evolution of the mammary gland from the innate immune system? *BioEssays* **28**, 606–616.
2708. Vulliemoz, S., Raineteau, O., and Jabaudon, D. (2005). Reaching beyond the midline: why are human brains cross wired? *Lancet Neurol.* **4**, 87–99.
2709. Wächtershäuser, G. (1987). Light and life: on the nutritional origins of sensory perception. *In* "Evolutionary Epistemology, Rationality, and the Sociology of Knowledge" (G. Radnitzky and W.W. Bartley, III, eds.). Open Court, La Salle, Ill., pp. 121–138.
2710. Wada, H. (2001). Origin and evolution of the neural crest: a hypothetical reconstruction of its evolutionary history. *Develop. Growth Differ.* **43**, 509–520.
2711. Waddington, C.H. (1940). Genes as evocators in development. *Growth* **1 (Suppl.)**, 37–44.
2712. Waddington, C.H. (1940). "Organizers and Genes." Cambridge Univ. Pr., Cambridge.
2713. Waddington, C.H. (1957). "The Strategy of the Genes: A Discussion of Some Aspects of Theoretical Biology." George Allen & Unwin, London.
2714. Waddington, C.H. (1960). Experiments on canalizing selection. *Genet. Res. Camb.* **1**, 140–150.
2715. Wager, T.D., Davidson, M.L., Hughes, B.L., Lindquist, M.A., and Ochsner, K.N. (2008). Prefrontal-subcortical pathways mediating successful emotion regulation. *Neuron* **59**, 1037–1050.
2716. Wagner, A. (2005). "Robustness and Evolvability in Living Systems." Princeton Univ. Pr., Princeton, N.J.
2717. Wagner, G.P. (1989). The biological homology concept. *Annu. Rev. Ecol. Syst.* **20**, 51–69.
2718. Wagner, G.P. (2005). The developmental evolution of avian digit homology: an update. *Theory Biosci.* **124**, 165–183.
2719. Wagner, G.P. (2007). The developmental genetics of homology. *Nature Rev. Genet.* **8**, 473–479.
2720. Wagner, G.P. (2008). Pleiotropic scaling of gene effects and the "cost of complexity." *Nature* **452**, 470–472.
2721. Wagner, G.P., Chiu, C.-H., and Laubichler, M. (2000). Developmental evolution as a mechanistic science: the inference from developmental mechanisms to evolutionary processes. *Amer. Zool.* **40**, 819–831.
2722. Wagner, G.P. and Gauthier, J.A. (1999). 1,2,3 = 2,3,4: A solution to the problem of the homology of the digits in the avian hand. *Proc. Natl. Acad. Sci. USA* **96**, 5111–5116.
2723. Wagner, G.P. and Larsson, H.C.E. (2007). Fins and limbs in the study of evolutionary novelties. *In* "Fins into Limbs: Evolution, Development, and Transformation" (B.K. Hall, ed.). Univ. Chicago Pr., Chicago, Ill., pp. 49–61. (See also (2008)). *Nature* **456**, 636–638.]
2724. Wagner, G.P. and Laubichler, M.D. (2004). Rupert Riedl and the re-synthesis of evolutionary and developmental biology: body plans and evolvability. *J. Exp. Zool. (Mol. Dev. Evol.)* **302B**, 92–102.
2725. Wagner, G.P. and Lynch, V.J. (2005). Molecular evolution of evolutionary novelties: the vagina and uterus of therian mammals. *J. Exp. Zool. (Mol. Dev. Evol.)* **304B**, 580–592.
2726. Wagner, G.P. and Lynch, V.J. (2008). The gene regulatory logic of transcription factor evolution. *Trends Ecol. Evol.* **23**, 377–385.
2727. Wagner, G.P., Pavlicev, M., and Cheverud, J.M. (2007). The road to modularity. *Nature Rev. Genet.* **8**, 921–931.
2728. Wagner, P.J. (1995). Testing evolutionary constraint hypotheses with early Paleozoic gastropods. *Paleobiology* **21**, 248–272.
2729. Wagner, P.J. (1996). Contrasting the underlying patterns of active trends in morphologic evolution. *Evolution* **50**, 990–1007.
2730. Wahlsten, D. (1999). Single-gene influences on brain and behavior. *Annu. Rev. Psychol.* **50**, 599–624.
2731. Wake, M.H., ed. (1979). "Human's Comparative Vertebrate Anatomy," 3rd ed. Univ. Chicago Pr., Chicago, Ill.
2732. Wald, G. (1973). Origin of death. In "The End of Life: A Discussion at the Nobel Conference (Gustavus Adolphus College, St. Peter, Minnesota, 1972)" (J.D. Roslansky, ed.). North Holland, Amsterdam, pp. 1–20.
2733. Wallace, B. (1985). Reflections on the still-"hopeful monster." *Quart. Rev. Biol.* **60**, 31–42.
2734. Wallen, K. and Lloyd, E.A. (2008). Inappropriate comparisons and the weakness of cryptic choice: a reply to Vincent J. Lynch and D. J. Hosken. *Evol. Dev.* **10**, 398–399.

2735. Wallman, J. and Winawer, J. (2004). Homeostasis of eye growth and the question of myopia. *Neuron* **43**, 447–468.
2736. Walls, G.L. (1942). "The Vertebrate Eye." Cranbrook Pr., Bloomfield Hills, Mich.
2737. Walter, C. (2006). "Thumbs, Toes, and Tears . . . and Other Traits That Make Us Human." Walker, New York.
2738. Walter, C. (2008). Affairs of the lips. *Sci. Am. Mind* **19** #1, 24–29.
2739. Walter, M.F., Black, B.C., Afshar, G., Kermabon, A.-Y., Wright, T.R.F., and Biessmann, H. (1991). Temporal and spatial expression of the *yellow* gene in correlation with cuticle formation and DOPA decarboxylase activity in *Drosophila* development. *Dev. Biol.* **147**, 32–45.
2740. Walther, C. and Gruss, P. (1991). Pax-6, a murine paired box gene, is expressed in the developing CNS. *Development* **113**, 1435–1449.
2741. Wandelt, J. and Nagy, L.M. (2004). Left-right asymmetry: more than one way to coil a shell. *Curr. Biol.* **14**, R654–R656.
2742. Wang, C., Rüther, U., and Wang, B. (2007). The Shh-independent activator function of the full-length Gli3 protein and its role in vertebrate limb digit patterning. *Dev. Biol.* **305**, 460–469.
2743. Wang, H.-Y., Chien, H.-C., Osada, N., Hashimoto, K., Sugano, S., Gojobori, T., Chou, C.-K., Tsai, S.-F., Wu, C.-I., and Shen, C.-K.J. (2007). Rate of evolution in brain-expressed genes in humans and other primates. *PLoS Biol.* **5** #2, 335–342 (e13).
2744. Wang, J. and Laurie, G.W. (2004). Organogenesis of the exocrine gland. *Dev. Biol.* **273**, 1–22.
2745. Wang, Q., Li, W., Liu, X.S., Carroll, J.S., Jänne, O.A., Keeton, E.K., Chinnaiyan, A.M., Pienta, K.J., and Brown, M. (2007). A hierarchical network of transcription factors governs androgen receptor-dependent prostate cancer growth. *Molec. Cell* **27**, 380–392.
2746. Wang, W.C.H. and Shashikant, C.S. (2007). Evidence for positive and negative regulation of the mouse Cdx2 gene. *J. Exp. Zool. (Mol. Dev. Evol.)* **308B**, 308–321.
2747. Wang, Y. and Nathans, J. (2007). Tissue/planar cell polarity in vertebrates: new insights and new questions. *Development* **134**, 647–658.
2748. Warburton, D. (2008). Order in the lung. *Nature* **453**, 733–734.
2749. Ward, C. (2003). The evolution of human origins. *Amer. Anthrop.* **105**, 77–88.
2750. Ward, P.D. (1991). "On Methuselah's Trail: Living Fossils and the Great Extinctions." Freeman, New York.
2751. Ward, S.J. (1998). Numbers of teats and pre- and post-natal litter sizes in small diprotodont marsupials. *J. Mammalogy* **79**, 999–1008.
2752. Ware, S., Reiskind, J., Blackburn, D.C., and Shipman, P. (2008). Leaping lizards [correspondence]. *Am. Sci.* **96**, 179.
2753. Washburn, S.L. (1960). Tools and human evolution. *Sci. Am.* **203** #3, 62–75.
2754. Wasiak, S. and Lohnes, D. (1999). Retinoic acid affects left-right patterning. *Dev. Biol.* **215**, 332–342.
2755. Watanabe, T. and Costantini, F. (2004). Real-time analysis of ureteric bud branching morphogenesis in vitro. *Dev. Biol.* **271**, 98–108.
2756. Waters, P.D., Wallis, M.C., and Graves, J.A.M. (2007). Mammalian sex—Origin and evolution of the Y chromosome and *SRY*. *Sems. Cell Dev. Biol.* **18**, 389–400.
2757. Watson, C.J. and Khaled, W.T. (2008). Mammary development in the embryo and adult: a journey of morphogenesis and commitment. *Development* **135**, 995–1003.
2758. Wawersik, M., Milutinovich, A., Casper, A.L., Matunis, E., Williams, B., and Van Doren, M. (2005). Somatic control of germline sexual development is mediated by the JAK/STAT pathway. *Nature* **436**, 563–567.
2759. Wawersik, S. and Maas, R.L. (2000). Vertebrate eye development as modeled in *Drosophila*. *Hum. Mol. Genet.* **9**, 917–925.
2760. Waxman, D.J. and Celenza, J.L. (2003). Sexual dimorphism of hepatic gene expression: novel biological role of KRAB zinc finger repressors revealed. *Genes Dev.* **17**, 2607–2613.
2761. Weale, R.A. (1966). Why does the human retina possess a fovea? *Nature* **212**, 255–256.
2762. Weatherbee, S.D. and Carroll, S.B. (1999). Selector genes and limb identity in arthropods and vertebrates. *Cell* **97**, 283–286.
2763. Weatherbee, S.D., Halder, G., Kim, J., Hudson, A., and Carroll, S. (1998). Ultrabithorax regulates genes at several levels of the wing-patterning hierarchy to shape the development of the *Drosophila* haltere. *Genes Dev.* **12**, 1474–1482.
2764. Weatherbee, S.D., Nijhout, H.F., Grunert, L.W., Halder, G., Galant, R., Selegue, J., and Carroll, S. (1999). *Ultrabithorax* function in butterfly wings and the evolution of insect wing patterns. *Curr. Biol.* **9**, 109–115.
2765. Webb, G.J.W. and Cooper-Preston, H. (1989). Effects of incubation temperature on crocodiles and the evolution of reptilian oviparity. *Amer. Zool.* **29**, 953–971.
2766. Webb, P.W. (1988). Simple physical principles and vertebrate aquatic locomotion. *Amer. Zool.* **28**, 709–725.
2767. Wegner, M. (1999). From head to toes: the multiple facets of Sox proteins. *Nucl. Acids Res.* **27**, 1409–1420.
2768. Weil, A. (2003). Teeth as tools. *Nature* **422**, 128.
2769. Weinberger, N.M. (2004). Music and the brain. *Sci. Am.* **291** #5, 88–95.
2770. Weiner, J. (1994). "The Beak of the Finch: A Story of Evolution in Our Time." Knopf, New York.
2771. Weiner, L., Han, R., Scicchitano, B.M., Li, J., Hasegawa, K., Grossi, M., Lee, D., and Brissette, J.L.

(2007). Dedicated epithelial recipient cells determine pigmentation patterns. *Cell* **130**, 932–942.
2772. Weinstein, B.M. (2005). Vessels and nerves: marching to the same tune. *Cell* **120**, 299–302.
2773. Weintraub, H. (1993). The MyoD family and myogenesis: redundancy, networks, and thresholds. *Cell* **75**, 1241–1244.
2774. Weiss, J.R., Moysich, K.B., and Swede, H. (2005). Epidemiology of male breast cancer. *Cancer Epidemiol. Biomarkers Prev.* **14**, 20–26.
2775. Weiss, K.M. (1990). Duplication with variation: metameric logic in evolution from genes to morphology. *Yrbk. Phys. Anthrop.* **33**, 1–23.
2776. Weiss, K.M. (1993). A tooth, a toe, and a vertebra: the genetic dimensions of complex morphological traits. *Evol. Anthrop.* **2**#4, 121–134.
2777. Weiss, K.M. (2005). The phenogenetic logic of life. *Nature Rev. Genet.* **6**, 36–45.
2778. Weiss, K.M. (2008). Tilting at Quixotic Trait Loci (QTL): an evolutionary perspective on genetic causation. *Genetics* **179**, 1741–1756.
2779. Weiss, K.M. and Buchanan, A.V. (2004). "Genetics and the Logic of Evolution." J. Wiley & Sons, Hoboken, N.J.
2780. Weiss, K.M., Stock, D.W., and Zhao, Z. (1998). Dynamic interactions and the evolutionary genetics of dental patterning. *Crit. Rev. Oral Biol. Med.* **9**, 369–398.
2781. Weiss, P. (1955). Beauty and the beast: life and the rule of order. *Sci. Monthly* **81**, 286–299.
2782. Weissengruber, G.E., Forstenpointner, G., Peters, G., Kübber-Heiss, A., and Fitch, W.T. (2002). Hyoid apparatus and pharynx in the lion (*Panthera leo*), jaguar (*Panthera onca*), tiger (*Panthera tigris*), cheetah (*Acinonyx jubatus*), and domestic cat (*Felis silvestris f. catus*). *J. Anat. (Lond.)* **201**, 195–209.
2783. Welchman, A.E., Deubelius, A., Conrad, V., Bülthoff, H.H., and Kourtzi, Z. (2005). 3D shape perception from combined depth cues in human visual cortex. *Nature Neurosci.* **8**, 820–827.
2784. Wellik, D.M. (2007). *Hox* patterning of the vertebrate axial skeleton. *Dev. Dynamics* **236**, 2454–2463.
2785. Wellik, D.M. and Capecchi, M.R. (2003). *Hox10* and *Hox11* genes are required to globally pattern the mammalian skeleton. *Science* **301**, 363–367.
2786. Welte, M.A. (2004). Bidirectional transport along microtubules. *Curr. Biol.* **14**, R525–R537.
2787. Wendelin, D.S., Pope, D.N., and Mallory, S.B. (2003). Hypertrichosis. *J. Am. Acad. Dermatol.* **48**, 161–179.
2788. Werblin, F. and Roska, B. (2007). The movies in our eyes. *Sci. Am.* **296** #4, 73–79.
2789. West-Eberhard, M.J. (1989). Phenotypic plasticity and the origins of diversity. *Annu. Rev. Ecol. Syst.* **20**, 249–278.
2790. West-Eberhard, M.J. (2003). "Developmental Plasticity and Evolution." Oxford Univ. Pr., New York.
2791. Weston, E.M., Friday, A.E., and Liò, P. (2007). Biometric evidence that sexual selection has shaped the hominin face. *PLoS ONE* **8**, e710.
2792. Wheeler, G.L., Miranda-Saavedra, D., and Barton, G.J. (2008). Genome analysis of the unicellular green alga *Chlamydomonas reinhardtii* indicates an ancient evolutionary origin for key pattern recognition and cell-signaling protein families. *Genetics* **179**, 193–197.
2793. Wheeler, M.R. and Takada, H. (1971). Male genitalia of some representative genera of American Drosophilidae. In "Studies in Genetics," Vol. 6 (Pub. #7103) (M.R. Wheeler, ed.). Univ. Texas Pr., Austin, pp. 225–240.
2794. Wheeler, P.E. (1984). The evolution of bipedality and loss of functional body hair in hominids. *J. Human Evol.* **13**, 91–98.
2795. Whitcome, K.K., Shapiro, L.J., and Lieberman, D.E. (2007). Fetal load and the evolution of lumbar lordosis in bipedal hominins. *Nature* **450**, 1075–1078.
2796. White, T.D., Asfaw, B., DeGusta, D., Gilbert, H., Richards, G.D., Suwa, G., and Howell, F.C. (2003). Pleistocene *Homo sapiens* from Middle Awash, Ethopia. *Nature* **423**, 742–747.
2797. Whitehead, H. and Mann, J. (2000). Female reproductive strategies of cetaceans: life histories and calf care. *In* "Cetacean Societies Field Studies of Dolphins and Whales" (J. Mann, R.C. Connor, P.L. Tyack, and H. Whitehead, eds.). Univ. Chicago Pr., Chicago, pp. 219–246.
2798. Whitfield, J. (1999). Heard but not seen. *Nature* **399**, 24.
2799. Whiting, M.F., Bradler, S., and Maxwell, T. (2003). Loss and recovery of wings in stick insects. *Nature* **421**, 264–267.
2800. Wicht, H. and Lacalli, T.C. (2005). The nervous system of amphioxus: structure, development, and evolutionary significance. *Can. J. Zool.* **83**, 122–150.
2801. Widelitz, R.B., Veltmaat, J.M., Mayer, J.A., Foley, J., and Chuong, C.-M. (2007). Mammary glands and feathers: comparing two skin appendages which help define novel classes during vertebrate evolution. *Sems. Cell Dev. Biol.* **18**, 255–266.
2802. Wiedersheim, R. (1895). "The Structure of Man: An Index to His Past History." MacMillan, New York.
2803. Wiens, J.A. (1989). "The Ecology of Bird Communities. Vol. 1. Foundations and Patterns." Cambridge Univ. Pr., New York.
2804. Wiens, J.J. (2001). Widespread loss of sexually selected traits: how the peacock lost its spots. *Trends Ecol. Evol.* **16**, 517–523.
2805. Wiens, J.J. and Hoverman, J.T. (2008). Digit reduction, body size, and paedomorphosis in salamanders. *Evol. Dev.* **10**, 449–463.
2806. Wilder, H.H. (1923). "The History of the Human Body." Holt, New York.
2807. Wildman, D.E., Chen, C., Erez, O., Grossman, L.I., Goodman, M., and Romero, R. (2006). Evolution of the mammalian placenta revealed by phylogenetic analysis. *PNAS* **103** #9, 3203–3208.

2808. Wilhelm, D. (2007). *R-spondin1*—discovery of the long-missing, mammalian female-determining gene? *BioEssays* **29**, 314–318.
2809. Wilhelm, D., Palmer, S., and Koopman, P. (2007). Sex determination and gonadal development in mammals. *Physiol. Rev.* **87**, 1–28.
2810. Wilhelm, H. (2008). The pupil. *Curr. Opin. Neurol.* **21**, 36–42.
2811. Wilkie, A.O.M. and Morriss-Kay, G.M. (2001). Genetics of craniofacial development and malformation. *Nature Rev. Genet.* **2**, 458–468.
2812. Wilkie, A.O.M., Patey, S.J., Kan, S.-H., Van den Ouweland, A.M.W., and Hamel, B.C.J. (2002). FGFs, their receptors, and human limb malformations. *Am. J. Med. Genet.* **112**, 266–278.
2813. Wilkins, A.S. (1997). Canalization: a molecular genetic perspective. *BioEssays* **19**, 257–262.
2814. Wilkins, A.S. (2002). "The Evolution of Developmental Pathways." Sinauer, Sunderland, Mass.
2815. Wilkins, A.S. (2005). Recasting developmental evolution in terms of genetic pathway and network evolution... and the implications for comparative biology. *Brain Res. Bull.* **66**, 495–509.
2816. Wilkins, A.S. (2007). Between "design" and "bricolage": Genetic networks, levels of selection, and adaptive evolution. *PNAS* **104 Suppl. 1**, 8590–8596.
2817. Wilkins, A.S. (2007). Genetic networks as transmitting and amplifying devices for natural genetic tinkering. *In* "Tinkering: The Microevolution of Development," *Novartis Found. Symp. 284* (G. Bock and J. Goode, eds.). Wiley, Chichester, U.K., pp. 71–89.
2818. Wilkinson, G.S. and Johns, P.M. (2005). Sexual selection and the evolution of mating systems in flies. In "The Evolutionary Biology of Flies" (D.K. Yeates and B.M. Wiegmann, eds.). Columbia Univ. Pr. New York, pp. 312–339.
2819. Wilkinson, M.T. (2007). Sailing the skies: the improbable aeronautical success of the pterosaurs. *J. Exp. Biol.* **210**, 1663–1671.
2820. Wilkinson, M.T., Unwin, D.M., and Ellington, C.P. (2006). High lift function of the pteroid bone and forewing of pterosaurs. *Proc. Roy. Soc. Lond. B* **273**, 119–126.
2821. Willey, A. (1911). "Convergence in Evolution." John Murray, London.
2822. Williams, G.C. (1957). Pleiotropy, natural selection, and the evolution of senescence. *Evolution* **11**, 398–411.
2823. Williams, G.C. (1975). "Sex and Evolution." Princeton Univ. Pr., Princeton, N.J.
2824. Williams, G.C. (1992). "Natural Selection: Domains, Levels, and Challenges." *Oxford Series in Ecology and Evolution*, Vol. **4**. Oxford Univ. Pr., New York.
2825. Williams, M.E., Lehoczky, J.A., and Innis, J.W. (2006). A group 13 homeodomain is neither necessary nor sufficient for posterior prevalence in the mouse limb. *Dev. Biol.* **297**, 493–507.
2826. Williams, N. (2007). Sounds big to me. *Curr. Biol.* **17**, R532.
2827. Williams, P.D. and Day, T. (2003). Antagonistic pleiotropy, mortality source interactions, and the evolutionary theory of senescence. *Evolution* **57**, 1478–1488.
2828. Williams, R.J. (1971). "You Are Extraordinary." Pyramid Books, New York, N.Y.
2829. Williams, R.W. and Herrup, K. (1988). The control of neuron number. *Annu. Rev. Neurosci.* **11**, 423–453.
2830. Williams, T.M., Selegue, J.E., Werner, T., Gompel, N., Kopp, A., and Carroll, S.B. (2008). The regulation and evolution of a genetic switch controlling sexually dimorphic traits in Drosophila. *Cell* **134**, 610–623.
2831. Wilson, G.N. (1988). Heterochrony and human malformation. *Am. J. Med. Genet.* **29**, 311–321.
2832. Wilson, S.I., Shafer, B., Lee, K.J., and Dodd, J. (2008). A molecular program for contralateral trajectory: Rig-1 control by LIM homeodomain transcription factors. *Neuron* **59**, 413–424.
2833. Wistow, G. (1993). Lens crystallins: gene recruitment and evolutionary dynamism. *Trends Biochem. Sci.* **18**, 301–306.
2834. Witschi, E. (1948). Migration of the germ cells of human embryos from the yolk sac to the primitive gonadal folds. *Carnegie Inst. Wash. Publ. 575 (Contributions to Embryology, No. 209)*, 68–80.
2835. Wittkopp, P.J. (2005). Genomic sources of regulatory variation in *cis* and in trans. *Cell. Mol. Life Sci.* **62**, 1779–1783.
2836. Wittkopp, P.J. (2006). Evolution of cis-regulatory sequence and function in Diptera. *Heredity* **97**, 139–147.
2837. Wittkopp, P.J., Carroll, S.B., and Kopp, A. (2003). Evolution in black and white: genetic control of pigment patterns in *Drosophila*. *Trends Genet.* **19**, 495–504.
2838. Wittkopp, P.J., Haerum, B.K., and Clark, A.G. (2008). Regulatory changes underlying expression differences within and between *Drosophila* species. *Nature Genet.* **40**, 346–350.
2839. Wittkopp, P.J., Vaccaro, K., and Carroll, S.B. (2002). Evolution of *yellow* gene regulation and pigmentation in *Drosophila*. *Curr. Biol.* **12**, 1547–1556.
2840. Wittman, A.B. and Wall, L.L. (2007). The evolutionary origins of obstructed labor: bipedalism, encephalization, and the human obstetric dilemma. *Obstet. Gynecol. Surv.* **62**, 739–748.
2841. Wöhr, M. and Schwarting, R.K. (2007). Ultrasonic communication in rats: can playback of 50-kHz calls induce approach behavior? *PLoS ONE* **2** #12, e1365.
2842. Wolf, M. (2007). "Proust and the Squid: The Story and Science of the Reading Brain." New York, HarperCollins.
2843. Wolpert, L. (1969). Positional information and the spatial pattern of cellular differentiation. *J. Theor. Biol.* **25**, 1–47.

2844. Wolpert, L., Jessell, T., Lawrence, P., Meyerowitz, E., Robertson, E., and Smith, J. (2007). "Principles of Development," 3rd ed. Oxford Univ. Pr., New York.
2845. Wolpoff, M.H. (1969). The effects of mutations under conditions of reduced selection. *Soc. Biol.* **16**, 11–23.
2846. Wong, B.B.M. and Rosenthal, G.G. (2006). Female disdain for swords in the swordtail fish. *Am. Nat.* **167**, 136–140.
2847. Wood, B. (2002). Hominid revelations from Chad. *Nature* **418**, 133–135.
2848. Wood, W.B. (2005). The left-right polarity puzzle: determining embryonic handedness. *PLoS Biol.* **3** #8, 1348–1351 (e292).
2849. Wrangham, R.W. (2005). The Delta Hypothesis: Homonoid ecology and hominin origins. *In* "Interpreting the Past: Essays on Human, Primate, and Mammal Evolution in Honor of David Pilbeam" (D.E. Lieberman, R.J. Smith, and J. Kelley, eds.). Brill Acad. Pub., Boston, Mass., pp. 231–242.
2850. Wray, G.A. (1998). Promoter logic. *Science* **279**, 1871–1872.
2851. Wray, G.A. (2003). Transcriptional regulation and the evolution of development. *Int. J. Dev. Biol.* **47**, 675–684.
2852. Wray, G.A. (2006). Spot on (and off). *Nature* **440**, 1001–1002.
2853. Wray, G.A. (2007). The evolutionary significance of *cis*-regulatory mutations. *Nature Rev. Genet.* **8**, 206–216.
2854. Wray, G.A. and Abouheif, E. (1998). When is homology not homology? *Curr. Opin. Gen. Dev.* **8**, 675–680.
2855. Wray, G.A., Hahn, M.W., Abouheif, E., Balhoff, J.P., Pizer, M., Rockman, M.V., and Romano, L.A. (2003). The evolution of transcriptional regulation in eukaryotes. *Mol. Biol. Evol.* **20**, 1377–1419.
2856. Wright, S. (1929). Fisher's theory of dominance. *Am. Nat.* **63**, 274–279.
2857. Wright, S. (1932). The roles of mutation, inbreeding, crossbreeding and selection in evolution. *Proc. 6th Internat. Congr. Genet.* **1**, 356–366.
2858. Wright, S. (1986). The material basis of evolution. *In* "Evolution: Selected Papers (of Sewall Wright)" (W.B. Provine, ed.). Univ. Chicago Pr., Chicago, Ill., pp. 389–394.
2859. Wright, S.J. (2005). "A Photographic Atlas of Developmental Biology." Morton, Englewood, Colo.
2860. Wrischnik, L.A., Timmer, J.R., Megna, L.A., and Cline, T.W. (2003). Recruitment of the proneural gene *scute* to the *Drosophila* sex-determination pathway. *Genetics* **165**, 2007–2027.
2861. Wroe, S. and Milne, N. (2007). Convergence and remarkably consistent constraint in the evolution of carnivore skull shape. *Evolution* **61**, 1251–1260.
2862. Wu, P., Hou, L., Plikus, M., Hughes, M., Scehnet, J., Suksaweang, S., Widelitz, R.B., Jiang, T.-X., and Chuong, C.-M. (2004). *Evo-Devo* of amniote integuments and appendages. *Int. J. Dev. Biol.* **48**, 249–270.
2863. Wutz, A. and Gribnau, J. (2007). X inactivation Xplained. *Curr. Opin. Gen. Dev.* **17**, 387–393.
2864. Wynn, T. and Coolidge, F.L. (2006). The effect of enhanced working memory on language. *J. Human Evol.* **50**, 230–231.
2865. Wynn, T. and Coolidge, F.L. (2008). A Stone-Age meeting of minds. *Am. Sci.* **96**, 44–51.
2866. Wynne, C.D.L. (2004). "Do Animals Think?" Princeton Univ. Pr., Princeton, N.J.
2867. Xu, P.-X., Zhang, X., Heaney, S., Yoon, A., Michelson, A.M., and Maas, R.L. (1999). Regulation of *Pax6* expression is conserved between mice and flies. *Development* **126**, 383–395.
2868. Yajima, I., Endo, K., Sato, S., Toyoda, R., Wada, H., Shibahara, S., Numakunai, T., Ikeo, K., Gojobori, T., Goding, C.R., and Yamamoto, H. (2003). Cloning and functional analysis of ascidian Mitf in vivo: insights into the origin of vertebrate pigment cells. *Mechs. Dev.* **120**, 1489–1504.
2869. Yalcinkaya, T.M., Siiteri, P.K., Vigne, J.-L., Licht, P., Pavgi, S., Frank, L.G., and Glickman, S.E. (1993). A mechanism for virilization of female spotted hyenas in utero. *Science* **260**, 1929–1931.
2870. Yamada, G., Suzuki, K., Haraguchi, R., Miyagawa, S., Satoh, Y., Kamimura, M., Nakagata, N., Kataoka, H., Kuroiwa, A., and Chen, Y. (2006). Molecular genetic cascades for external genitalia formation: an emerging organogenesis program. *Dev. Dynamics* **235**, 1738–1752.
2871. Yamaguchi, T. (1977). Studies on the handedness of the fiddler crab, Uca lactea. *Biol. Bull.* **152**, 424–436.
2872. Yamamoto, D. (2007). The neural and genetic substrates of sexual behavior in *Drosophila*. *Adv. Genet.* **59**, 39–66.
2873. Yang, A.S. (2001). Modularity, evolvability, and adaptive radiations: a comparison of the hemi- and holometabolous insects. *Evol. Dev.* **3**, 59–72.
2874. Yang, X., Schadt, E.E., Wang, S., Wang, H., Arnold, A.P., Ingram-Drake, L., Drake, T.A., and Lusis, A.J. (2006). Tissue-specific expression and regulation of sexually dimorphic genes in mice. *Genome Res.* **16**, 995–1004.
2875. Yang, Z., Ding, K., Pan, L., Deng, M., and Gan, L. (2003). Math5 determines the competence state of retinal ganglion cell progenitors. *Dev. Biol.* **264**, 240–254.
2876. Yashiro, K., Shiratori, H., and Hamada, H. (2007). Haemodynamics determined by a genetic programme govern asymmetric development of the aortic arch. *Nature* **450**, 285–288.
2877. Ye, X., Hama, K., Contos, J.J., Anliker, B., Inoue, A., Skinner, M.K., Suzuki, H., Amano, T., Kennedy, G., Arai, H., Aoki, J., and Chun, J. (2005). LPA3-mediated lysophosphatidic acid signalling in embryo implantation and spacing. *Nature* **435**, 104–108.

2878. Yekta, S., Tabin, C.J., and Bartel, D.P. (2008). MicroRNAs in the Hox network: an apparent link to posterior prevalence. *Nature Rev. Genet.* **9**, 789–796.
2879. Yeo, R.A. and Gangestad, S.W. (1994). Developmental origins of variation in human hand preference. *In* "Developmental Instability: Its Origins and Evolutionary Implications" (T.A. Markow, ed.). Kluwer, London, pp. 283–298.
2880. Yi, W. and Zarkower, D. (1999). Similarity of DNA binding and transcriptional regulation by *Caenorhabditis elegans* MAB-3 and *Drosophila melanogaster* DSX suggests conservation of sex-determining mechanisms. *Development* **126**, 873–881.
2881. Yin, V.P. and Poss, K.D. (2008). New regulators of vertebrate appendage regeneration. *Curr. Opin. Gen. Dev.* **18**, 381–386.
2882. Yokoyama, T., Copeland, N.G., Jenkins, N.A., Montgomery, C.A., Elder, F.F.B., and Overbeek, P.A. (1993). Reversal of left-right asymmetry: a situs inversus mutation. *Science* **260**, 679–682.
2883. Yoshida, K. and Saiga, H. (2008). Left-right asymmetric expression of Pitx is regulated by the asymmetric nodal signaling through an intronic enhancer in *Ciona intestinalis. Dev. Genes Evol.* **218**, 353–360.
2884. Yost, H.J. (1998). Left-right development in Xenopus and zebrafish. *Sems. Cell Dev. Biol.* **9**, 61–66.
2885. Yost, H.J. (2003). Left-right asymmetry: nodal cilia make and catch a wave. *Curr. Biol.* **13**, R808–R809.
2886. Young, D. and Robertson, B. (2001). Genomics: Leprosy—a degenerative disease of the genome. *Curr. Biol.* **11**, R381–R383.
2887. Young, N.M. and Hallgrímsson, B. (2005). Serial homology and the evolution of mammalian limb covariation structure. *Evolution* **59**, 2691–2704.
2888. Youngsteadt, E. (2008). Simple sleepers. *Science* **321**, 334–337.
2889. Yu, D. and Small, S. (2008). Precise registration of gene expression boundaries by a repressive morphogen in *Drosophila. Curr. Biol.* **18**, 868–876.
2890. Yu, J.-K., Satou, Y., Holland, N.D., Shin-I, T., Kohara, Y., Satoh, N., Bronner-Fraser, M., and Holland, L.Z. (2007). Axial patterning in cephalochordates and the evolution of the organizer. *Nature* **445**, 613–617.
2891. Yu, J.Y. and Dickson, B.J. (2006). Sexual behaviour: Do a few dead neurons make the difference? *Curr. Biol.* **16**, R23–R25.
2892. Yu, J.Y. and Dickson, B.J. (2008). Hidden female talent. *Nature* **453**, 41–42.
2893. Yu, Z., Bhandari, A., Mannik, J., Pham, T., Xu, X., and Andersen, B. (2008). Grainyhead-like factor Get1/Grhl3 regulates formation of the epidermal leading edge during eyelid closure. *Dev. Biol.* **319**, 56–67.
2894. Yurgelun-Todd, D. (2007). Emotional and cognitive changes during adolescence. *Curr. Opin. Neurobiol.* **17**, 251–257.
2895. Zacchetti, G., Duboule, D., and Zakany, J. (2007). Hox gene function in vertebrate gut morphogenesis: the case of the caecum. *Development* **134**, 3967–3973.
2896. Zacchigna, S., Ruiz de Almodovar, C., and Carmeliet, P. (2008). Similarities between angiogenesis and neural development: what small animal models can tell us. *Curr. Topics Dev. Biol.* **80**, 1–55.
2897. Zahradnicek, O., Horacek, I., and Tucker, A.S. (2008). Viperous fangs: Development and evolution of the venom canal. *Mechs. Dev.* **125**, 786–796.
2898. Zallen, J.A. (2007). Planar polarity and tissue morphogenesis. *Cell* **129**, 1051–1063.
2899. Zarkower, D. (2002). Invertebrates may not be so different after all. In "The Genetics and Biology of Sex Determination," *Novartis Found. Symp.* 244 (D. Chadwick and J. Goode, eds.). Wiley, New York, pp. 115–135.
2900. Zatorre, R. (1997). Sound work. *Sci. Am.* **277** #3, 97–98.
2901. Zeidler, M.P. and Perrimon, N. (2000). Sex determination: Co-opted signals determine gender. *Curr. Biol.* **10**, R682–R684.
2902. Zeil, J., Hemmi, J.M., and Backwell, P.R.Y. (2006). Fiddler crabs. *Curr. Biol.* **16**, R40–R41.
2903. Zelditch, M.L. (2005). Developmental regulation of variability. *In* "Variation: A Central Concept in Biology" (B. Hallgrímsson and B.K. Hall, eds.). Elsevier Acad. Pr., New York, pp. 249–276.
2904. Zeller, R. and Deschamps, J. (2002). First come, first served. *Nature* **420**, 138–139.
2905. Zhang, F., Zhou, Z., Xu, X., Wang, X., and Sullivan, C. (2008). A bizarre Jurassic maniraptoran from China with elongate ribbon-like feathers. *Nature* **455**, 1105–1108.
2906. Zhang, G.J. and Cohn, M.J. (2008). Genome duplication and the origin of the vertebrate skeleton. *Curr. Opin. Gen. Dev.* **18**, 387–393.
2907. Zhang, J. (2003). Evolution of the human *ASPM* gene, a major determinant of brain size. *Genetics* **165**, 2063–2070.
2908. Zhang, S.-D. and Odenwald, W.F. (1995). Misexpression of the white (w) gene triggers male-male courtship in *Drosophila. Proc. Natl. Acad. Sci.*USA **92**, 5525–5529.
2909. Zhao, C., Deng, W., and Gage, F.H. (2008). Mechanisms and functional implications of adult neurogenesis. *Cell* **132**, 645–660.
2910. Zhu, J., Nakamura, E., Nguyen, M.-T., Bao, X., Akiyajma, H., and Mackem, S. (2008). Uncoupling Sonic Hedgehog control of pattern and expansion of the developing limb bud. *Dev. Cell* **14**, 624–632.
2911. Zhu, L., Wilken, J., Phillips, N.B., Narendra, U., Chan, G., Stratton, S.M., Kent, S.B., and Weiss, M.A. (2000). Sexual dimorphism in diverse metazoans is regulated by a novel class of intertwined zinc fingers. *Genes Dev.* **14**, 1750–1764.
2912. Zhurov, V., Terzin, T., and Grbic, M. (2004). Early blastomere determines embryo proliferation and

caste fate in a polyembryonic wasp. *Nature* **432**, 764–769.

2913. Zimmer, C. (2000). Calling a bluff. *Nat. Hist.* **109** #2, 20–22.
2914. Zimmer, C. (2000). "Parasite Rex: Inside the Bizarre World of Nature's Most Dangerous Creatures." Simon & Schuster, New York.
2915. Zimmer, C. (2004). Faster than a hyena? Running may make humans special. *Science* **306**, 1283.
2916. Zimmer, C. (2005). The neurobiology of the self. *Sci. Am.* **293** #5, 92–101.
2917. Zimmer, C. (2007). Evolved for cancer? *Sci. Am.* **296** #1, 68–75.
2918. Zimmerman, J.E., Naidoo, N., Raizen, D.M., and Pack, A.I. (2008). Conservation of sleep: insights from non-mammalian model systems. *Trends Neurosci.* **31**, 371–376.
2919. Zinzen, R.P., Cande, J., Ronshaugen, M., Papatsenko, D., and Levine, M. (2006). Evolution of the ventral midline in insect embryos. *Dev. Cell* **11**, 895–902.
2920. Zou, Y. and Lyuksyutova, A.I. (2007). Morphogens as conserved axon guidance cues. *Curr. Opin. Neurobiol.* **17**, 22–28.
2921. Zuber, M.E., Gestri, G., Viczian, A.S., Barsacchi, G., and Harris, W.A. (2003). Specification of the vertebrate eye by a network of eye field transcription factors. *Development* **130**, 5155–5167.
2922. Zufall, R.A. and Rausher, M.D. (2004). Genetic changes associated with floral adaptation restrict future evolutionary potential. *Nature* **428**, 847–850.
2923. Zuker, C.S. (1994). On the evolution of eyes: would you like it simple or compound? *Science* **265**, 742–743.
2924. Zuzarte-Luis, V. and Hurle, J.M. (2005). Programmed cell death in the embryonic vertebrate limb. *Sems. Cell Dev. Biol.* **16**, 261–269.

Index